Microbial Exopolysaccharides

Current Research and Developments

Edited by:

Özlem Ateş Duru

Nişantaşı Üniversitesi, Istanbul, Turkey

Caister Academic Press

www.caister.com

Caister Academic Press
Norfolk, UK

www.caister.com

British Library Cataloguing-in-Publication Data
A catalogue record for this book is available from the British Library

ISBN: 978-1-912530-26-7 (book)
ISBN: 978-1-912530-27-4 (ebook)

Contents

Current books of interest

Exploring the Capacity of Halophilic Microorgansisms to Synthesize Exopolysaccharides with Interesting Properties

1

Margarita Kambourova

Institute of Microbiology, Bulgarian Academy of Sciences, Sofia, Bulgaria
Corresponding author: margikam@microbio.bas.bg

DOI: https://doi.org/10.21775/9781912530267.01

Abstract

Although microbial exopolysacchaides (EPSs) have been intensively studied during the last two decades, the knowledge on these compounds is still scarce due to the phylogenetic diversity of microbial producers and the complexity of their molecules. Halophilic producers have been described in both bacterial and archaeal domains. Most of them require sugar and organic nitrogen for high yield synthesis. Halophiles synthesize molecules containing different monosaccharides, aminosugars, sulphates, uronic acids. Investigations into the halophilic EPS properties reveal their biotechnological potential in the field of the bio-medical applications, food products as emulsifiers and gelling agents, bioremediation for removing of toxic compounds, heavy metal removal, and in the petroleum industry for an enhanced oil recovery. The current review aims to increase the basic knowledge on EPS production by halophiles focusing on the recent investigations on halophilic producers of EPSs, culture conditions for attaining high yield, chemical composition and structure of their molecules, and the potential applications based on their physico-chemical properties.

Introduction

Although the polysaccharide market is dominated by plant polysaccharides, the amount of microbial products increases annually. Among the microbial polysaccharides widely distributed in the polymer market such as xanthan, dextran, gellan, curdlan etc, none is produced by extremophilic microorganisms. Extremophiles have gained a lot of attention from the research community as the accumulated knowledge on the properties reveal the advantages of their exploration and novelty of their metabolic products. Like other extremophiles halophiles provide the possibility of new natural products and although only some of them currently have reached the biotechnological market scientists believe they are potential resource for many novel processes. Despite the significantly high phylogenetic diversity among halophilic microorganisms, suggesting a variety of metabolic pathways, and hence the possibility for isolation of new products, they are a biotechnologically highly neglected group (Oren, 2010). Most studies on EPS production, properties and potential for biotechnological exploration by halophiles have been published in the last few years (Biswas and Paul, 2017). Their novel structural and rheological properties are different from more traditional polymers and suggest new areas for applications in the food industry, pharmacy and cosmetics, remediation, petrol industry as thickeners, stabilizers, emulsifiers, coagulants, and suspending agents. The processes with exploration of halophilic microorganisms combine relatively short time course of fermentations and high polymer yield of mesophilic microorganisms with typicallly for extremophilic microorganisms a lack of pathogenicity. An additional advantage in the utilization of halophilic organisms is the reduced risk of microbial contamination due to the increased salt concentration. Such a high concentration suggests a possibility for exploration in open and continuous fermentation process which minimizes the cost of production (Yin et al., 2015). Synthesized by halophiles EPS are characterized by a unique stability under the extreme conditions of the most industrial processes (high temperature, salinity, extreme pH).

Halophilic microorganisms have adapted to live in saline environments like marine salterns, salt mines, natural saline lakes, coastal lagoons, saline soils. Much of our knowledge about halophilic microorganisms is based on the microbiological studies of salterns that harbor specific microbial communities in different geographic areas including Tunisia, Israel, Australia, Mexico, India, Spain, and Bulgaria (Garabito et

Figure 1. Burgas salterns, Bulgaria. The red (pink) color of saltern's water is determined by the crystallization of the magnesium salts in high concentration, which are red in color, as well as the domination of the single-cell algae *Dunaliella salina*, that contains a large amount of beta-carotenes.

al., 1998; Burns et al., 2004; Sørensen et al., 2005; Baati et al., 2008; Manikandan et al., 2009; Trigui et al., 2011; Dillon et al., 2013; Boyadzhieva et al., 2018) (Figure 1).

The required salt concentration for an optimal microbial growth determines their affiliation to different halophilic groups, extreme halophiles (optimal growth above 15% NaCl), moderate halophiles (optimal growth at 3–15% NaCl) and slight halophiles (optimal growth at 1–3% NaCl) as well as halotolerant microorganisms growing in the absence of salt and in the same time tolerant to relatively high salt concentrations (Ventosa et al., 2015). Halophilic and halotolerant microorganisms are often found together in the phylogenetic tree.

Not only high salt concentration can limit biodiversity in hypersaline environments but also other environmental factors like low oxygen concentration, high or low temperature, sometimes alkaline pH, and presence of heavy metals (Ventosa, 2006).

Another factor that can affect biodiversity is pressure, low nutrient content, solar radiation and other toxic components. Study of the halophilic microorganisms contributes to our overall understanding of the physiology of these unusual organisms that have evolved to function optimally at high saline concentrations by cellular, structural, and molecular adaptations to hostile environment resulted in novel molecules biosynthesis, from simple osmolytes and lipids to complex secondary metabolites (Wilson and Breamble, 2009). They have developed various strategies in order to survive the high salt concentration causing an osmotic stress like Na+ /H+ antiporter systems for exclusion of sodium from the cytoplasm and accumulation of small electrically neutral organic solutes ("compatible solutes") to prevent loss of water. Among other protective mechanisms EPS biosynthesis is often observed adaptation strategy to halophilic environments.

This review summarizes the limited information on EPS from halophiles, their structures and properties that could contribute their future applications.

Microbial producers of exopolysaccharides

Microorganisms adapted to live at high salt concentrations are found in both domains: Archaea and Bacteria. Recent investigations have changed the common opinion that the square archaeon *Haloquadratum waslbyi*, the bacteroidete *Salinibacter ruber* and nanohaloarchaea predominate in hypersaline environments (Ventosa et al., 2015). Some eukaryotic organisms like the brine shrimp *Artemia* and the green algae *Dunaliella* are also able to live at high salt concentrations up to 1.50–2.0 %. Representatives of taxonomically emerging groups with very different tolerance and requirements for the amount of salt belong to halophilic bacteria and they could be found in different families and even phyla. Moderately or extremely halophilic species belong to eight among the validly described 28 phyla in the domain Bacteria (http://www.bacterio.cict.fr/ classifphyla.html), namely *Proteobacteria, Firmicutes, Actinobacteria, Spirochaetes, Bacteroidetes, Thermotogae, Cyanobacteria*, and *Tenericutes* (Ventosa et al., 2012). Most of the EPS producers so far reported are representatives of the family *Halomonadaceae* and *Alteromonadaceae*. Significant part of the described EPS producers belongs to the genera *Halomonas* (Quesada et al., 1993; Arias et al., 2003; Mata et al. 2006; Poli et al., 2007; Llamas et al., 2012; Poli et al., 2013), *Alteromonas* (Cambon-Bonavita et al., 2002), *Chromohalobacter* (Radchenkova et al., 2018), *Idiomarina*

(Martínez-Cánovas et al., 2004), *Saccharophagus* (González-García et al., 2015), *Pseudoalteromonas* (Roca et al., 2016). *Vibrio harveyi* strain VB23 has been described as another Gamma-proteobacterial producer of EPS with interesting structures (Bramhachari and Dubey, 2006). The marine bacterium *Zoogloea* sp. KCCM10036 belonging to *Beta-Proteobacteria* produces several EPSs (Lim et al., 2007). Some representatives of *Alphaproteobacteria* like *Palleronia* (Martínez-Checa et al., 2005) and *Salipiger* (Llamas et al., 2010), and of *Firmicutes* like *Bacillus* (Osama et al., 2015) have been also described as good EPS producers. *Zunongwangia profunda* SM-A87, *Flavobacteriaceae* is among the best producers with 8.9 g/l EPS (Liu et al., 2011).

Except for the methanogens, all halophilic archaea are extremely halophilic and belong to the archeal phylum Euryarchaeota. They have developed specialized adaptation strategies to high salt concentration like cell ability to float by production of gas vesicles, activity of halozymes at high salt concentration, and catching light by rhodopsins synthesis. Most of the halophilic archaeal EPS producers belong to the family *Halobacteriaceae*, genera *Haloferax*, *Haloarcula*, *Halobacterium*, *Halococcus*, and *Halopiger* (Anton et al., 1988; Parolis et al., 1996; Paramonov et al.,1998; Nicolaus et al., 1999; Hezayen et al., 2010). *Haloferax mediterranei* has been the first reported and still the best archaeal EPS producer with 3.5 g/l (Anton et al., 1988). *Natrialba aegyptiaca* has synthesized an extracellular polymer in which carbohydrate content is quite low (12.5 % w/w) (Hezayen et al., 2001).

Fermentation processes for EPS production by halophiles and EPS recovery

EPS production processes are characterized by an extreme rheology of the fermentation, product concentration, and sometimes conformational changes of the product. Such negative influence on the exopolysaccharide yields could be avoided by an adjustment of an appropriate equipment design and operation at specific culture conditions like carbon and nitrogen sources, salt concentration and mineral requirements, temperature and pH optima. The highest EPS yield for a halophilic microorganism has been reported for *Alteromonas macleodii* strain - 23.4 g/l exopolysaccharide (Mehta et al., 2014) (Table 1).

Exopolysaccharide producing halophilic organisms have been isolated mainly on a complex media or rarely on a synthetic media with high carbon/nitrogen ratio.

According to the common opinion the choice of a carbon source strongly affects the yield, the total amount of the polysaccharide, and features of the produced molecule. For example, marine bacterium *Saccharophagus degradans* synthesizes at least twice the value of EPS in glucose than in other sugars (González-García et al., 2015). However, a lack of an influence of the used carbon source has been observed for *Halomonas ventosae* and *Halomonas anticariensis* (Mata et al., 2006). Halophilic EPS producers use a wide variety of carbon sources, with sugars as a preferable carbon source including sucrose, glucose, lactose, maltose, but also mannitol, sorbitol, methanol. Glucose is one of the most common carbon sources that favors highest yield (Table 1). Among the microorganisms that prefer lactose as a carbon source are *Chromohalobacter canadensis* 28, *Rhodothermus marinus, Alteromonas macleodii* and *Zunongwangia profunda* SM-A87 (Liu et al., 2011; Mehta et al., 2014; Sardari et al., 2017; Radchenkova et al., 2018). As the presence of sugars in the composition of the culture medium results in an enhanced product cost much hope is oriented towards the exploration of industrial and agro wastes like cane molasses, milk whey, rice bran. EPS production by *Halomonas* sp. AAD6 has reached up to 12.4 g/l after cultivation in sugar beet molasses (Küçükaşik et al., 2011). The type of sugar could change the molecular mass, conformation, and monosaccharide ratios in the EPS molecule, and thus finally its functional properties (Sutherland, 1994). An influence of the C-source on EPS composition has been reported for *Halomonas alkaliantarctica* strain CRSS that produces a mannan in a complex medium and a fructo-glucan in a minimal medium with sodium acetate as a sole carbon source (Poli et al., 2007).

Table 1. Examples of good halophilic producers that synthesize higher than 1g/l EPS.

Max EPS production (g/l)	**Microorganisms**	**Carbon/ nitrogen sources**	**Salt concentration (%)**	**Chemical composition of EPS (%)**	**Reference**
23.4	*Alteromonas macleodii* PA2	lactose/ peptone	2.0	No data	Mehta et al., 2014
21.30	*Pantoea* strain BM39	glucose/ $NaNO_3$+ peptone	4.0	arabinose, glucose galactose, gulcuronic acid	Silvi et al., 2013

9.8	*Pseudoalteromonas* sp. AM	glucose/meat extract	3.0	glucose	Al-Nahas et al., 2011
9.23	*Hahella chejuensis*	sucrose/ tryptone	No data	No data	Ko et al., 2000
8.90	*Zunongwangia profunda* SM-A87	lactose/ peptone	No data	No data	Liu et al., 2011
4.12	*Saccharophagus degradans*	glucose/ NH_4Cl	2.4	glucose, mannose, galactose	González-García et al., 2015
3.89	*Halomonas stenophila*	glucose/ peptone	No data	glucose 24, glucuronic acid 7.5, mannose 5.5, fucose 4.5, galactose 1.2, rhamnose 1	Amjres et al., 2015
3.8	*Halomonas maura* S-30	glucose/ peptone	2.5	mannose 34.8, galactose 14, glucose 29.3, glucuronic acid 21.9	Arias et al., 2003
3.5	*Haloferax mediterranei*	glucose/ yeast extract	25.0	Mannose, glucose, galactose, amino sugars, uronic acids, sulfate	Anton et al., 1988
2.8	*Vulcaniella (Halomonas) eurihalina* Al-12	glucose/ peptone	7.5	glucose, galactose, mannose and rhamnose	Quesada et al., 1993
2.56	*Halomonas xianhensis* SUR308	glucose/ peptone	10	No data	Biswas et al., 2015
1.85	*Halomonas* sp. AAD6	sucrose/ peptone+ $(NH_4)_2SO_4$	No data	fructose	Poli et al., 2009
1.7	*Halomonas almeriensis* MST	glucose/ peptone	7.5	mannose 72, glucose 27.5, rhamnose 0.5	Llamas et al., 2012
1.5	*Idiomarina ramblicola* R-22T	glucose/ peptone	7.5	EPS 1: mannose 68.2, glucose 25, rhamnose 6.8 EPS2: mannose 53.6, glucose 18.9, galacturonic acid 25-29	Mata et al., 2008
1.2	*Salipiger mucosus* A3T	glucose/ peptone +yeast extract	7.5	glucose 19.7, mannose 34, galactose 32.9, fucose 13.4	Llamas et al., 2010
1.0	*Alteromonas hispanica* F32T	galactose/ peptone	7.5	mannose 62.75, glucose 18.15, xylose 12.25, rhamnose 6.85	Mata et al., 2008

The nitrogen sources currently being used for exopolysaccharide production could be organic like peptone, tryptone, urea or yeast extract or inorganic, like ammonium sulfate, sodium nitrate etc. According to the common opinion the EPS yield is higher in a medium with organic nitrogen than with inorganic nitrogen sources (Biswas and Paul, 2017). Generally, the increase in the C : N ratio promotes the exopolysaccharide production and the depletion of nitrogen might be a signal for EPS synthesis (Vergas-Garcia et al., 2001). Even at high value of C : N ratio additional increase from 50 to 100 results in 3.5 fold higher EPS production by *Saccharophagus degradans* (González-García et al., 2015). Reports for best EPS production in low C : N ratio are also available. The highest EPS yield by *Hahella chejuensis* has been reported at sucrose/tryptone ratio of 2:1 and sharply decreases at higher than 3:1 (Ko et al., 2000).

It is commonly accepted that EPS synthesis is an adaptive mechanism for surviving in a hostile environment and stressful conditions like temperature, salinity and pressure increase EPS yield. Such influence has been proved in *Colwellia psychrerythraea* strain 34H (Marx et al., 2009), while the production of heteropolysaccharide synthesized by the euryarchaeal producer *Haloferax mediterranei* seems to be independent of most environmental factors and is mostly linked to the total biomass (Anton et al., 1988). *H. ventosae* and *H. anticariensis* strains excrete highest EPS quantities when cultivated under optimal growth conditions, which ensure highest speed of cell division and correspondingly the biomass accumulation (Mata et al., 2006). Optimal salt concentration for EPS synthesis varies from 0.5 % w/v (Osama et al., 2015) to 7.5 % w/v (Quesada et al., 1993; Mata et al., 2006, 2008; Llamas et al., 2010, 2012). Unusually high salt concentration has provided the highest polymer concentration by *Haloferax mediterranei* (25% NaCl) and *Chromohalobacter canadensis* 28 (15% NaCl) (Anton et al., 1988; Radchenkova et al., 2018). Most of the halophilic microorganisms grow at mesophilic temperatures and an increase or decrease of the incubation temperature with several degree results in greater EPS production. The cryoprotective role of EPS in low temperatures has been confirmed in the case of the strain *Pseudoalteromonas* sp. CAM025 when lowering of the temperature with 10-18°C resulted in 30 fold increase of EPS yield (Mancuso et al., 2005). As most of saline environments are characterized by a neutral pH, numerous halophilic microorganisms produce exopolysaccharides in neutral pH media. High stability to pH and inorganic salts has been reported for the EPS produced by *Volcaniella eurihalina* F2-7 and *Chromohalobacter canadensis* 28 (Calvo et al. 1995; Radchenkova et al., 2018). As the exopolysaccharide secretion in the culture broth leads to the lack of homogeneity in mixing, mass transfer, oxygen and heat conductivity in the bioreactor the proper choice of aeration and agitation rates including the impeller type could affect the rheology and correspondingly exopolysaccharide production and quality. The high aeration rate has enhanced EPS production and viscosity of the culture broth of marine bacterium *Hahella chejuensis* (Lee et al., 2001).

EPS synthesis could be growth associated or observed mainly in the late exponential and stationary phase when the medium is depleted with one or more nutrients and seems to be a a feature of a particular genera. Identification of higher quantities of EPS in late exponential and early stationary phase is explained by some authors by the close contact with the cell in earlier exponential phase and further diffusion throughout the medium as an amorphous slime in the late growth stage (Mata et al., 2006). EPS production by *Alteromonas macleodii* subsp. *fijiensis* begins at the end of the exponential phase and continues throughout the stationary phase (Poli et al., 2010). The marine bacterium *Pseudomonas* sp. strain S9 produces EPS during both exponential phase and stationary phase during energy and nutrient starvation (Wrangstadh et al., 1990). The synthesis by *Pantoea* strain BM39 is growth associated and persisted in the exponential phase (Silvi et al., 2013). Although continuous cultivation processes from halophiles still are not described, maximum EPS levels at continuous cultures are expected at low dilution rates when longer residence time of microorganisms in the bioreactor promotes the utilization of excess carbon source (Kumar et al., 2007).

The steps in the process of EPS recovery depend of the yield, the presence of undesired molecules in the culture broth and the level of purity required for biomedical applications. The comparatively low EPS concentration in the fermentation broth and the presence of contaminating solid (e.g. cells) or soluble medium components in the culture liquid determine the need in further EPS recovering that could significantly influence production cost and commercial value of the exopolysaccharide. EPS is not tightly bound with the cell in the case of slime synthesis and could be isolated from the cells by centrifugation; in the case of capsular EPS a preliminary alkaline or temperature treatment is necessary. Despite of the common application of low molecular alcohol or acetone precipitation as a further step in EPS recovery it has some disadvantages as a high solvent consummation in the case of low EPS yield and co-precipitation of proteins and nucleic acids available in the culture supernatant. Alternative solution is the ultrafiltration that easily separate high molecular weight EPS from lower molecular weight proteins. As the salt concentration in the precipitate is significant in the case of halophilic producers, desalting treatment is necessary, usually by a dialysis. Special requirements to EPS assigned for medical applications determine the need of high degree of purity usually reached by chromatographic methods. The precipitated EPS are finally recovered by centrifugation, dissolved with distilled water and dried.

Chemical composition and structure of EPS from halophiles

According to their chemical composition EPSs are divided into two main groups, homopolysaccharides and heteropolysaccharides. The reports on halophilic homopolysaccharide producers are quite limited (Poli et al., 2009). Halophiles synthesize predominantly heteropolysaccharides containing three, four or even more different monosaccharides as D-xylose, D-arabinose, D-ribose; D-glucose, D-rhamnose, D-galactose, and D-mannose. Additionally they could contain amino sugars (D-glucosamine and D-galactosamine) and uronic acids (D-glucuronic acids and D-galacturonic acids). The 1,4-β- or 1,3-β- linkages between monosaccharides are most commonly found in the rigid backbone of EPSs and the 1,2-α- or 1,6-α-linkages in the flexible ones (Sutherland, 1994). As can be seen from the Table 1, most of the polymers contain glucose, mannose, and galactose. The presence of uronic acids is found in most acidic polymers like those from *Halomonas eurihalina* and *Halobacterium* sp. SM5 (Bejar et al., 1998; Lungmann et al., 2007). Many EPSs possess acyl, sulfate or phosphate groups along with the monosaccharide residues. High sulfate content and a significant amount of uronic acids are typical for exopolysaccharides synthesized by *Halomonas* (Biswas and Paul, 2014). EPS from *Bacillus marinus* is composed by glucose and glucouronic acid in a molar ratio of 3:1and contains sulfate group (20.2%) (Osama et al., 2015). The EPS from *Rhodococcus marinus* is highly sulfated and contains arabinose and xylose (Sardari et al., 2017).

Inclusion of non-sugar components into a complex carbohydrate molecule contributes the unique diversity of the tertiary structure and functional properties of the carbohydrate molecules. The protein component could add to the foaming and emulsifying properties of an exopolysaccharide produced by a marine bacterium *Enterobacter cloaceae* (Iyer et al., 2005). Polyglutamic acid fraction (72%) of the exopolymer from *Chromohalobacter canadensis* determines unusually high swelling water retention capacity of the compound (Radchenkova et al., 2018). Similarly, the halophilic archaea *Natrialba aegyptiaca* has synthesized an extracellular polymer composed predominantly of glutamic acid (85 % w/w) and carbohydrates (12.5 % w/ w) (Hezayen et al., 2001). The exopolymer from *Halobacterium* sp. SM5is water-soluble, acidic, with 70.40% protein, 7.30% sulfate, 5.02 % total sugar, 3.15% neutral sugar and2.05% uronic acid (Lungmann et al., 2007). Atypical exopolymer synthesized by *Shewanella colwelliana* contains Ca, S, P and Si (40%–45%), carbohydrates (15%–35%, mainly glucose), lipids (10%), and proteins (<5%) (Abu et al., 1991). FR2 synthesized by *Rhodococcus erythropolis* PR4 is composed by galactose, glucose, mannose, pyruvic acid, glucuronic acid, stearic and palmitic acids (Urai et al., 2007).

The polysaccharides are essentially very long, thin molecular chains with molecular mass of the order at 0.5–2.0 x 10^3 kDa, most often 1.10^2 -3.10^2 (Poli et al., 2010). Low Mw of 20 and 46 kDa has been reported for EPSs from *Halomonas ventosae* and *Halomonas anticariensis* correspondingly, 77 kDa from *Halobacterium* sp. SM5, 80 kDa from *Rhodothermus marinus* (Mata et al., 2006; Lungmann et al., 2007; Sardari et al., 2017).

Despite of the restricted technical tools for revealing of EPS structure some authors suggest that the chemical diversity of the complex carbohydrates is far greater than proteins and nucleic acids (Merritt et al., 2013). Polysaccharide biosynthesis is more complicated than template based synthesis of nucleic acids and proteins and is regulated by a number of parameters such as sugar precursors, number of sugar residues, the possibility for branching and the expression level of enzymes (Delbarre-Ladrat et al., 2014). The enzymes involved in the exopolysaccharide synthesis are active in different locations. Hexokinase, phosphoglucomutase, and diphosphate-glucose pyrophosphorylase are intracellular; glycosyl transferases could be found in the cell periplasmic membrane; enzymes involved in the polymerization of the macromolecules are extracellular (Kumar et al., 2007).

Applications

EPSs from halophilic producers express many valuable properties like pseudoplastic, thixotropic and shear thinning rheology, high viscosity or gelling ability that determine their applicability in various fields ranging from medical to environmental applications. Some of them are enlisted below:

Medical applications

The established fibrinolytic, antitumor, antiviral and immunological activity of halophilic EPSs determine their perspective as substituents of chemical drugs. EPS from *H. stenophila* blocks the growth of human T-lymphocyte tumors while that from *H. eurihalina* enhances the unspecific proliferation of human lymphocytes in a presence of the anti-CD3 mononuclear antibody (Ruiz-Ruiz et al., 2011; Llamas et al., 2012). The sulphate and phosphate groups often determine biomedical capacity of halophilic EPS and highly sulfated EPS's could be used in pharmacy (Arias et al., 2003). Over-sulphation of EPSs from *Alteromonas infernus* results in activation of their bioactivity (Jouault et al., 2001). The sulphated EPS from *H. eurihalina* H2-7 enhances the unspecific proliferation of human lymphocytes in the presence of the mononuclear antibody (Pérez-Fernández et al., 2000). Fibrinolytic activity of EPS from *Pseudoalteromonas* sp. AM is comparable to that of a fibrinolytic drug (Al-Nahas et al., 2011). The EPS from *Bacillus marinus* expresses antiviral activity against MDCK and H1N1 virus as well as antitumor activity against human breast adenocarcinoma cell line (MCF-7) and human alveolar basal epithelial cell line (Osama et al., 2015). The established high compatibility of halophilic exopolymers with cell lines could favor heavily their exploration (Lungmann et al., 2007). EPS derivatives of *Alteromonas infernus* could be applied in lipemia and arteriosclerosis treatment (Guezennec et al., 1998). EPS from *Vibrio diabolicus* is found to be strong bone healing material (Zanchetta et al., 2003). The performed investigations suggest some correlation between polymer structure and molecular weight and its antitumor activity. Antitumor activity of the high molecular weight glucans is higher than those of low molecular weight and the presence of ß-(1-3) linkages in the main polymer chain and ß-(1-6) in branch points determines more effective action than ß-(1-6) linkages in the main chain.

Emulsifiers and gelling agents

Many of the halophilic EPSs form viscous solutions and their extremophilic origin makes determines solution stability over a wide pH range and temperatures. In general those of the species of *Halomonas* are known as possessing valuable rheological properties (Biswas and Paul, 2017). EPS produced by *H. eurihalina* has been identified as a viscosity and gelling agent at acidic pH appropriate for food industries (Arias et al., 2003). EPSs produced by the marine bacterium *Zoogloea* sp. KCCM10036 have shown an excellent flocculating activity (Lim et al., 2007). The viscous exopolymer synthesized by the halophilic luminescent *Vibrio harveyi* strain VB23 possesses a good emulsifying activity (Bramhachari et al., 2006). Good emulsifying efficiency is demonstrated by an EPSs produced by *H. maura, H. almeriensis, H. xianhensis* and *Salipiger mucosus* (Bouchotroch et al., 2000; Biswas et al., 2015). Emulsifying properties of the acidic EPS from *Enterobacter cloacae* is comparable to other commercial gums (Iyer et al., 2005). The EPS from *Pseudoalteromonas* sp. AM formes very stable emulsion of sunflower oil-water in very low concentration (0.01% w/v) (Al-Nahas et al., 2011). Very stable emulsions (100%) are reported for the exopolymer from *Chromohalobacter canadensis* 28 (Radchenkova et al., 2018). A branched EPS that has gelling properties is produced by *Pseudoalteromonas* strain HYD721 (Rougeaux et al., 1999). Very often halophilic polymers contain substantial protein fraction that enhance emulsifying capacity of EPS and could be used in food products as replacers of artificial emulsifier (Mata et al., 2006).

Bioremediation

Halophilic EPSs could be employed in environmental biotechnology for salted soil and marine water bioremediation and decontamination processes. EPS synthesized by *Planococcus maitriensis* Anita I has shown an oil spreading ability similar to those of Triton X100 and Tween 80, that makes it suitable for bioremediation, enhanced oil recovery and cosmetic applications (Kumar et al., 2007). EPS from *Halomonas* spp. TG39 is capable of removing methylene blue (Gutierrez et al., 2009), while that from *H. eurihalina* H-28 removes polycyclic aromatic hydrocarbons (Martinez-Checa et al., 2007). An approach for resolving the problem with restricted water content in hypersaline soils is enrichment with biofilm forming halotolerant bacteria. The excreted biofilm containing mainly EPS enhances soil aggregation, which in turn improves water stability and decreases Na-uptake by plants (Costa et al., 2018).

Heavy metal removal

The presence of electrically charged groups like uronic acids and sulfates in the EPS molecule favors attachment of toxic metal ions in environment friendly processes for removing toxic metals from polluted environments and wastewater. Negatively surface charged EPS from a marine isolate of the species *Enterobacter cloacae* contains 9.23% uronic acids and expresses high metal complexing capacity when it grows on surfaces coated with toxic compounds like cuprous oxide (Iyer et al., 2005). Similarly, EPSs from *Pseudoalteromonas* strain SM9913 and *Alteromonas* strain 1644 are characterized by a high degree of acetylation and show flocculation behavior and a bio-sorption capacity for binding heavy metals (Bozzi et al., 1996; Li et al., 2008). Native mauran from *H. maura* has a high affinity to Pb whilst the deacetylated EPS binds efficiently both, Pb and Ag (Arias et al., 2003). EPS produced by *H. almeriensis* catches copper, lead and cobalt and this from *H. ventosae* strains is effective in copper uptake (Mata et al., 2006; Llamas et al., 2012).

Enhanced oil recovery

Stability of the halophilic EPSs to high salt concentrations determines the perspective in the exploration of their adhesive properties in oil recovery and in the petroleum industry. The emulsifying activity of *Halomonas* EPSs is suggested to have a role as an emulsifying agent in oil recovery from oil deposits with high salinity (Bouchotroch et al., 2000; Mata et al., 2006; Poli et al., 2007). EPS produced by *Halomonas* sp. TG39 was observed to increase the solubilization of aromatic hydrocarbons and enhance their biodegradation from oil-contaminated surface waters (Gutierrez et al., 2013). The enrichment of EPS producers and formation of oil aggregates by the synthesized amphiphilic EPS was observed during the oil spill.

Acknowledgement

The author is grateful to the National Fund for Scientific Research, Bulgaria (Grant No B02/26) for supporting the investigations on halophilic microorganisms and EPS by them.

References

Abu, G.O., Weiner, R.M., Rice, J., and Colwell, R.R. (1991). Properties of an extracellular adhesive polymer from the marine bacterium, *Shewanella colwelliana*. Biofouling 3, 69–84. https://doi.org/10.1080/08927019109378163

Al-Nahas, M.O., Darwish, M.M., Ali, A.E., and Amin, M.A. (2011). Characterization of an exopolysaccharide-producing marine bacterium, isolate *Pseudoalteromonas* sp. AM. Afr. J. Microbiol. Res. 5, 3823–3831. https://doi.org/10.5897/AJMR11.757

Amjres, H., Béjar, V., Quesada, E., Carranza, D., Abrini, J., Sinquin, C., Ratiskol, J., Colliec-Jouault, S., and Llamas, I. (2015). Characterization of haloglycan, an exopolysaccharide produced by *Halomonas stenophila* HK30. Int. J. Biol. Macromol. 72, 117-124. https://doi.org/10.1016/j.ijbiomac.2014.07.052

Anton, J., Meseguer, I., Rodriguez-Valera, F. (1988). Production of an extracellular polysaccharide by *Haloferax mediterranei*. Appl. Environ. Microbiol. 54, 2381–2386.

Arias, S., del Moral, A., Ferrer, M.R., Tallon, R., Quesada, E., and Béjar, V. (2003). Mauran, an exopolysaccharide produced by the halophilic bacterium *Halomonas maura*, with a novel composition and interesting properties for biotechnology. Extremophiles 7, 319-326. https://doi.org/10.1007/s00792-003-0325-8

Baati, H., Guermazi, S., Amdouni, R., Gharsallah, N., Sghir, A., and Ammar, E. (2008). Prokaryotic diversity of a Tunisian multipond solar saltern. Extremophiles 12, 505–518. https://doi.org/10.1007/s00792-008-0154-x.

Béjar, V., Llamas, I., Calvo, C., and Quesada, E. (1998). Characterization of exopolysaccharides produced by 19 halophilic strains of the species *Halomonas eurihalina*. J. Biotechnol. 61, 135-141. https://doi.org/10.1016/S0168-1656(98)00024-8

Biswas, J., Mandal, S., and Paul, A.K. (2015). Production, partial purification and some bio-physicochemical properties of EPS produced by *Halomonas xianhensis* SUR308 isolated from a saltern environment. JBAPN 5, 108-119. https://doi.org/10.1080/22311866.2015.1038852

Biswas, J., and Paul, A.K. (2014). Production of extracellular polymeric substances by halophilic bacteria of solar salterns. Chin. J. Biol. Article ID 205731. http://dx.doi.org/10.1155/2014/205731

Biswas, J., and Paul, A.K. (2017). Diversity and production of extracellular polysaccharide by halophilic microorganisms. Biodiversity Int. J. 1, ID 00006. https://doi.org/10.15406/bij.2017.01.00006

Bouchotroch, S., Quesada, E., Izquierdo, I., Rodriguez, M., and Bejar, V. (2000). Bacterial exopolysaccharides produced by new discovered bacteria belonging to the genus *Halomonas* isolated from hypersaline habitats in Morocco. J. Ind. Microbiol. Biotech. 24, 374-378. https://doi.org/10.1038/sj.jim.7000002

Boyadzhieva, I., Tomova, I., Radchenkova, N., Kambourova, M., Poli, A., and Vasileva-Tonkova, E. (2018). Diversity of heterotrophic halophilic bacteria isolated from coastal solar salterns, Bulgaria and their ability to synthesize bioactive molecules with biotechnological impact. Microbiology (Moscow) 87, 519–528. https://doi.org/10.1134/S0026261718040033

Bozzi, L., Milas, M., and Rinaudo, M. (1996). Characterization and solution properties of a new exopolysaccharide excreted by the bacterium *Alteromonas* sp. strain 1644. Int. J. Biol. Macromol. 18, 9–17. https://doi.org/10.1016/0141-8130(95)01038-6

Bramhachari, P.V., and Dubey, S.K. (2006). Production and characterization of exopolysaccharide produced by *Vibrio harveyi* strain VB23. Lett. Appl. Microbiol. 43, 571–577. https://doi.org/10.1111/j.1472-765X.2006.01967.x

Burns, D.G., Camakaris, H.M., Janssen, P.H., and Dyall-Smith, M.L. (2004). Combined use of cultivation- dependent and cultivation-independent methods indicates that members of most haloarchaeal groups in an Australian crystallizer pond are cultivable. Appl. Environ. Microbiol. 70, 5258–5265. https://doi.org/10.1128/AEM.70.9.5258-5265.2004

Calvo, C., Ferrer, M.R., Martinez-Checa, F., Bejar, V., and Quesada, E. (1995). Some rheological properties of the extracellular polysaccharide produced by *Volcaniella eurihalina* F2-7. Appl. Biochem. Biotechnol. 55, 45–54. https://doi.org/10.1007/BF02788747

Cambon-Bonavita, M.A., Raguenes, G., Jean, J., Vincent, P., and Guezennec, J. (2002). A novel polymer produced by a bacterium isolated from a deep-sea hydrothermal vent polychaete annelid. J. Appl. Microbiol. 93, 310-315. https://doi.org/10.1046/j.1365-2672.2002.01689.x

Costa, O.Y., Raaijmakers, J.M., and Kuramae, E.E. (2018). Microbial extracellular polymeric substances: ecological function and impact on soil aggregation. Front. Microbiol. 9, Article 1636. https://doi.org/10.3389/fmicb.2018.01636

Delbarre-Ladrat, C., Sinquin, C., Lebellenger, L., Zykwinska, A., and Colliec-Jouault, S. (2014). Exopolysaccharides produced by marine bacteria and their applications as glycosaminoglycan-like molecules. Front. Chem. 2, 85. https://doi.org/10.3389/fchem.2014.00085

Dillon, J.G., Carlin, M., Gutierrez, A., Nguyen, V., and McLain, N. (2013). Patterns of microbial diversity along a salinity gradient in the GuerreroNegro solar saltern, Baja CA Sur,Mexico. Front. Microbiol. 4, article 399, 13 pages. https://doi.org/10.3389/fmicb.2013.00399

Garabito, M.J., Marquez, M.C., and Ventosa, A. (1998). Halotolerant Bacillus diversity in hypersaline environments. Can. J. Microbiol. 44, 95- 102. https://doi.org/10.1139/w97-125

González-García, Y., Heredia, A., Meza-Contreras, J. C., Escalante, F. M., Camacho-Ruiz, R. M., and Córdova, J. (2015). Biosynthesis of extracellular polymeric substances by the marine bacterium *Saccharophagus degradans* under different nutritional conditions. Int. J. Polym. Sci. Article ID 526819. http://dx.doi.org/10.1155/2015/526819

Guezennec, J., Pignet, P., Lijour, Y., Gentric, E., Ratiskol, J., and Colliec-Jouault, S. (1998). Sulfation and depolymerization of a bacterial exopolysaccharide of hydrothermal origin. Carbohydr. Polym. 37, 19-24. https://doi.org/10.1016/S0144-8617(98)00006-X

Gutierrez, T., Berry, D., Yang, T., Mishamandani, S., McKay, L., Teske, A., and Aitken, M.D.

(2013). Role of bacterial exopolysaccharides (EPS) in the fate of the oil released during the deepwater horizon oil spill. PLoS One 8, e67717. https://doi.org/10.1371/journal.pone.0067717

Gutierrez, T., Morris, G., and Green, D.H. (2009). Yield and physicochemical properties of EPS from *Halomonas* spp. strain TG39 identifies a role for protein and anionic residues (sulfate and phosphate) in emulsification of n-hexadecane. Biotechnol. Bioeng. 103, 207-216. https://doi.org/10.1002/bit.22218

Hezayen, F.F., Gutierrez, M.C., Steinbuchel, A., Tindall, B.J., and Rehm, B.H.A. (2010). *Halopiger aswanensis* spp. nov., a polymer producing and extremely halophilic archaeon isolated from hypersaline soil. Int. J. Syst. Evol. Microbiol. 60, 633-637. https://doi.org/10.1099/ijs.0.013078-0

Hezayen, F.F., Rehm, B.H., Tindall, B.J., and Steinbüchel, A. (2001). Transfer of *Natrialba asiatica* B1T to *Natrialba taiwanensis* sp. nov. and description of *Natrialba aegyptiaca* sp. nov., a novel extremely halophilic, aerobic, non-pigmented member of the Archaea from Egypt that produces extracellular poly (glutamic acid). Int. J. Syst. Evol. Microbiol. 51, 1133-1142. https://doi.org/10.1099/00207713-51-3-1133

Iyer, A., Mody, K., and Jha, B. (2005). Characterization of an exopolysaccharide produced by a marine, *Enterobacter cloacae*. Indian J. Exp. Biol. 43, 467-471.

Jouault, S.C., Chevolot, L., Helley, D., Ratiskol, J., Bros, A., Sinquin, C., Roger, O., and Fischer, A.M. (2001). Characterization, chemical modifications and *in vitro* anticoagulant properties of an exopolysaccharide produced by *Alteromonas infernus*. Biochim. Biophys. Acta 1528, 141–151. https://doi.org/10.1016/S0304-4165(01)00185-4

Ko, S.H., Lee, H.S., Park, S.H., and Lee, H.K. (2000). Optimal conditions for the production of exopolysaccharide by marine microorganism *Hahella chejuensis*. Biotechnol. Bioprocess Eng. 5, 181–185. https://doi.org/10.1007/BF02936591

Küçükaşik, F., Kazak, H., Güney, D., Finore, I., Poli, A., Yenigün, O., Nicolaus, B., and Öner, E.T. (2011). Molasses as fermentation substrate for levan production by *Halomonas* sp. Appl. Microbiol. Biotechnol. 89, 1729-1740. https://doi.org/10.1007/s00253-010-3055-8

Kumar, A.S., Mody, K., and Jha, B. (2007). Bacterial exopolysaccharides: A perception. J. Basic Microbiol. 47, 103–117. https://doi.org/10.1002/jobm.200610203

Lee, H.S., Park, H.S., Lee, J.H., and Lee, H.K. (2001). Effect of aeration rates on production of extracellular polysaccharide, EPS-R, by marine bacterium *Hahella chejuensis*. Biotechnol. Bioprocess Eng. 6, 359–362. https://doi.org/10.1007/BF02933006

Li, W.W., Zhou, W.Z., Zhang, Y.Z., Wang, J., and Zhu, X.B. (2008). Flocculation behavior and mechanism of an exopolysaccharide from the deep-sea psychrophilic bacterium *Pseudoalteromonas* sp. SM9913. Bioresour. Technol. 99, 6893–6899. https://doi.org/10.1016/j.biortech.2008.01.050

Lim, D.J., Kim, J.D., Kim, M.Y., Yoo, S.H., and Kong, J.Y. (2007). Physicochemical properties of the exopolysaccharides produced by marine bacterium *Zoogloea* sp. KCCM10036. J. Microbiol. Biotechnol. 17, 979–984.

Liu, S.B., Qiao, L.P., He, H.L., Zhang, Q., Chen, X.L., Zhou, W.Z., Zhou, B.C., and Zhang, Y.Z. (2011). Optimization of fermentation conditions and rheological properties of exopolysaccharide produced by deep-sea bacterium *Zunongwangia profunda* SM-A87. PLoS One 6, e26825. https://doi.org/10.1371/journal.pone.0026825

Llamas, I., Amjres, H., Mata, J.A., Quesada, E., and Bejar, V. (2012). The potential biotechnological applications of the exopolysaccharide produced by the halophilic bacterium *Halomonas almeriensis*. Molecules 17, 7103-7120. https://doi.org/10.3390/molecules17067103

Llamas, I., Mata, J.A., Tallon, R., Bressollier, P., Urdaci, M.C., Quesada, E., and Béjar, V. (2010). Characterization of the exopolysaccharide produced by *Salipiger mucosus* A3T, a halophilic species belonging to the Alphaproteobacteria, isolated on the spanish mediterranean seaboard. Mar. Drugs 8, 2240–2251. https://doi.org/10.3390/md8082240

Lungmann, P., Choorit, W., and Prasertsan, P. (2007). Physio-chemical and biological properties of partially purified exopolymers from newly isolated halophilic bacterial strain SM5. Sci. Technol. 29, 1571-1582.

Mancuso Nichols, C., Bowman, J.P., and Guezennec, J. (2005). Effects of incubation temperature on growth and production of exopolysaccharide by an Antarctic sea ice bacterium grown in batch culture. Appl. Environ. Microbiol. 71, 3519–3523. https://doi.org/10.1128/AEM.71.7.3519-3523.2005

Manikandan, M., Kannan, V., and Pǎsić, L. (2009). Diversity of microorganisms in solar salterns of Tamil Nadu, India. World J. Microbiol. Biotechnol. 25, 1007–1017. https://doi.org/10.1007/s11274-009-9980-y

Martínez-Cánovas, M.J., Béjar, V., Martínez-Checa, F., Páez, R., and Quesada, E. (2004). *Idiomarina fontislapidosi* sp. nov. and *Idiomarina ramblicola* sp. nov. isolated from
inland hypersaline habitats in Spain. Int. J. Syst. Evol. Microbiol. 54, 1793–1797. https://doi.org/10.1099/ijs.0.63172-0

Martínez-Checa, F., Quesada, E., Martínez-Cánovas, M.J., Llamas, I., and Béjar, V. (2005). *Palleronia marisminoris* gen. nov., sp. nov., a moderately halophilic, exopolysaccharide
producing bacterium belonging to the 'Alphaproteobacteria', isolated from a saline soil. Int.
J. Syst. Evol. Microbiol. 55, 2525–2530. https://doi.org/10.1099/ijs.0.63906-0

Martinez-Checa, F., Toledo, F.L., El Mabrouki, K., Quesada, E., and Calvo, C. (2007). Characteristics of bioemulsifier V2-7 synthesized in culture media added of hydrocarbons: Chemical composition, emulsifying activity and rheological properties. Bioresour. Technol. 98, 3130– 3135. https://doi.org/10.1016/j.biortech.2006.10.026

Marx, J.G., Carpenter, S.D., and Deming, J.W. (2009). Production of cryoprotectant extracellular polysaccharide substances (EPS) by the marine psychrophilic bacterium *Colwellia psychrerythraea* strain 34H under extreme conditions. Can. J. Microbiol. 55, 63–72. https://doi.org/10.1139/W08-130

Mata, J.A., Béjar, V., Llamas, I., Arias, S., Bressollier, P., Tallon, R., Urdaci, M.C., and Quesada, E. (2006). Exopolysaccharides produced by the recently described halophilic bacteria *Halomonas ventosae* and *Halomonas anticariensis*. Res. Microbiol. 157, 827-835. https://doi.org/10.1016/j.resmic.2006.06.004

Mata, J.A., Bejar, V., Bressollier, P., Tallon, R., Urdaci, M.C., Quesada, E., Llamas, I. (2008). Characterization of exopolysaccharides produced by three moderately halophilic bacteria belonging to the family *Alteromonadaceae*. J. Appl. Microbiol. 105, 521-528. https://doi.org/10.1111/j.1365-2672.2008.03789.x

Mehta, A., Sidhu, C., Pinnaka, A.K., and Roy Choudhury, A. (2014). Extracellular polysaccharide production by a novel osmotolerant marine strain of *Alteromonas macleodii* and its application towards biomineralization of silver. PLoS ONE 9, e98798. https://doi.org/10.1371/journal.pone.0098798

Merritt, J.H., Ollis, A.A., Fisher, A.C., and Delisa, M.P. (2013). Glycans-by-design: engineering bacteria for the biosynthesis of complex glycans and glycoconjugates. Biotechnol. Bioeng. 110, 1550–1564. https://doi.org/10.1002/bit.24885

Nicolaus, B., Lama, L., Esposito, E., Manca, M.C., Improta, R., Bellitti, M.R., Duckworth, A.W., Grant, W.D., and Gambacorta, A. (1999). *Haloarcula* spp. able to biosynthesize exo- and endopolymers. J. Ind. Microbiol. Biotechnol. 23, 489-496. https://doi.org/10.1038/sj.jim.2900738

Oren, A. (2010). Industrial and environmental applications of halophilic microorganisms. Environ. Technol. 31, 825-834. https://doi.org/10.1080/09593330903370026

Osama, H. El Sayed, Kader, A.E., El Sayed, M., Salem, H.M., Mahmoud, M.G., Asker, M.S., and Saher, M.S. (2015). Isolation, characterization and biological activities of exopolysaccharide produced by *Bacillus marinus*. Pharm. Chem. 7, 200-208.

Paramonov, N.A., Parolis, L.A.S., Parolis, H., Boan, I.F., Anton, J., and Rodríguez-Valera F. (1998). The structure of the exocellular polysaccharide produced by the

Archaeon *Haloferax gibbonsii* (ATCC 33959). Carbohydr. Res. 309, 89-94. https://doi.org/10.1016/S0008-6215(98)00102-5

Parolis, H., Parolis, L.A.S., Boan, I.F., Rodriguez-Valera, F., Widmalm, G., Manca, M.C., Jansson, P.-E., and Sutherland, I.W. (1996). The structure of the exopolysaccharide produced by the halophilic Archaeon *Haloferax mediterranei* strain R4 (ATCC 33500). Carbohydr. Res. 295, 147-156. https://doi.org/10.1016/S0008-6215(96)90134-2

Pérez-Fernández, M.E., Quesada, E., Gálvez, J., and Ruíz, C. (2000). Effect of exopolysaccharide V2-7 isolated from *Halomonas eurihalina* on the proliferation *in vitro* of human peripheral blood lymphocites. Immunopharmacol. Immunotoxicol. 22, 131–141.
https://doi.org/10.3109/08923970009016411

Poli, A., Esposito, E., Orlando, P., Lama, L., Giordano, A., de Appolonia, F., Nicolaus, B., and Gambacorta, A. (2007). *Halomonas alkaliantarctica* sp. nov., isolated from saline lake Cape Russell in Antarctica, an alkalophilic moderately halophilic, exopolysaccharide-producing bacterium. Syst. Appl. Microbiol. 30, 31–38. https://doi.org/10.1016/j.syapm.2006.03.003

Poli, A., Kazak, H., Gürleyendağ, B., Tommonaro, G., Pieretti, G., Öner, E.T., and Nicolaus, B. (2009). High level synthesis of levan by a novel *Halomonas* species growing on defined media. Carbohydr. Polym. 78, 651-657. https://doi.org/10.1016/j.carbpol.2009.05.031

Poli, A., Anzelmo, G., and Nicolaus, B. (2010). Bacterial exopolysaccharides from extreme marine habitats: production, characterization and biological activities. Mar. Drugs 8, 1779–1802. https://doi.org/10.3390/md8061779

Poli, A., Nicolaus, B., Denizci, A.A., Yavuzturk, B., and Kazan, D. (2013). *Halomonas smyrnensis* sp. nov., a moderately halophilic, exopolysaccharide-producing bacterium. Int. J. Syst. Evol. Microbiol. 63, 10–18. https://doi.org/10.1099/ijs.0.037036-0

Quesada, E., Bejar, V., and Calvo, C. (1993). Exopolysaccharide production by *Volcaniella eurihalina*. Experimentia 49, 1037-1041. https://doi.org/10.1007/BF01929910

Radchenkova, N., Boyadzhieva, I., Atanasova, N., Poli, A., Finore, I., Di Donato, P., Nicolaus, B., Panchev, I., Kuncheva, M., and Kambourova, M. (2018). Extracellular polymer substance synthesized by a halophilic bacterium *Chromohalobacter canadensis* 28. Appl. Microbiol. Biotechnol. 102, 4937-4949. https://doi.org/10.1007/s00253-018-8901-0

Roca, C., Lehmann, M., Torres, C.A., Baptista, S., Gaudêncio, S.P., Freitas, F., and Reis, M.A. (2016). Exopolysaccharide production by a marine *Pseudoalteromonas* sp. strain isolated from Madeira Archipelago ocean sediments. New Biotechnology 33, 460-466. https://doi.org/10.1016/j.nbt.2016.02.005

Rougeaux, H., Guezennec, J., Carlson, R.W., Kervarec, N., Pichon, R., and Talaga, P. (1999). Structural determination of the exopolysaccharide of *Pseudoalteromonas* strain HYD 721 isolated from a deep-sea hydrothermal vent. Carbohydr. Res. 315, 273–285.
https://doi.org/10.1016/S0008-6215(99)00019-1

Ruiz-Ruiz, C., Srivastava, G.K., Carranza, D., Mata, J.A., Llamas I, Santamaría, M., Quesada, E., and Molina, I.J. (2011). An exopolysaccharide produced by the novel halophilic bacterium *Halomonas stenophila* strain B100 selectively induces apoptosis in human T leukaemia cells. Appl. Microbiol. Biotechnol. 89, 345-355. https://doi.org/10.1007/s00253-010-2886-7

Sardari, R.R., Kulcinskaja, E., Ron, E.Y., Björnsdóttir, S., Friðjónsson, Ó.H., Hreggviðsson, G.Ó., and Karlsson, E.N. (2017). Evaluation of the production of exopolysaccharides by two strains of the thermophilic bacterium *Rhodothermus marinus*. Carbohydr. Polym. 156, 1-8. https://doi.org/10.1016/j.carbpol.2016.08.062

Silvi, S., Barghini, P., Aquilanti, A., Juarez-Jimenez, B., and Fenice, M. (2013). Physiologic and metabolic characterization of a new marine isolate (BM39) of *Pantoea* sp. producing high levels of exopolysaccharide. Microb. Cell Fact. 12, 10. https://doi.org/10.1186/1475-2859-12-10.

Sørensen, B., Canfield, D.E., Teske, A.P., and Oren, A. (2005). Community composition of a hypersaline endoevaporitic microbial mat. Appl. Environ. Microbiol. 71, 7352–7365. https://doi.org/10.1128/AEM.71.11.7352-7365.2005

Sutherland, I.W. (1994). Structure-function relationship in microbial exopolysaccharides. Biotechnol. Adv. 12, 393–448. https://doi.org/10.1016/0734-9750(94)90018-3

Trigui, H., Masmoudi, S., Brochier-Armanet, C., Barani, A., Grégori, G., Denis, M., Dukan, S., and Maalej, S. (2011). Characterization of heterotrophic prokaryote subgroups in the Sfax coastal solar salterns by combining flow cytometry cell sorting and phylogenetic analysis. Extremophiles 15, 347–358. https://doi.org/10.1007/s00792-011-0364-5

Urai, M., Yoshizaki, H., Anzai, H., Ogihara, J., Iwabuchi, N., Harayama, S., Sunairi, M., and Nakajima, M. (2007). Structural analysis of mucoidan, an acidic extracellular polysaccharide produced by a pristane-assimilating marine bacterium, *Rhodococcus erythropolis* PR4. Carbohydr. Resour. 342, 927–932. https://doi.org/10.1016/j.carres.2007.02.002

Ventosa, A., Carmen-Marquez, M., Sanchez-Porro, C., and de la Haba, R.R. (2012). Taxonomy of halophilic Archea and Bacteria. In Advances in Understanding the Biology of Halophilic Microorganisms, R.H. Vreeland, ed. (Springer Science +Business Media Dordrecht, Netherlands), pp. 59-80. https://doi.org/10.1007/978-94-007-5539-0_3

Ventosa, A. (2006). Unusual micro-organisms from unusual habitats: hypersaline environments. In Prokaryotic diversity: Mechanisms and Significance, N.A. Logan, H.M. Lappin-Scott, P.C.F. Oyston, eds. (Cambridge University Press, Cambridge), pp. 223-253. https://doi.org/10.1017/CBO9780511754913.015

Ventosa, A., de la Haba, R.R., Sánchez-Porro, C., and Papke, R.T. (2015). Microbial diversity of hypersaline environments: a metagenomic approach. Curr. Opin. Microbiol. 25, 80-87. https://doi.org/10.1016/j.mib.2015.05.002

Vergas-Garcia, M.C., Lopez, M.J., Elorrieta, Suarez F., and Moreno, J. (2001). Influence of nutritional and environmental factors on polysaccharide production by *Azotobacter vinelandii* cultured on 4-hydroxybenzoic acid. J. Ind. Microbiol. Biotechnol. 27, 5–10. https://doi.org/10.1038/sj/jim/7000152

Wilson, Z.E., and Brimble, M.A. (2009). Molecules derived from the extremes of life. Nat. Prod. Rep. 26, 44-71. https://doi.org/10.1039/B800164M

Wrangstadh, M., Szewzyk, U., Ostling, J., and Kjelleberg, S. (1990). Starvation-specific formation of a peripheral exopolysaccharide by a marine *Pseudomonas* sp., strain S9. Appl. Environ. Microbiol. 56, 2065–2072.

Yin, J., Chen, J.C., Wu, Q., and Chen, G.Q. (2015). Halophiles, coming stars for industrial biotechnology. Biotechnol. Adv. 33, 1433-1442. https://doi.org/10.1016/j.biotechadv.2014.10.008

Zanchetta, P., Lagarde, N., and Guezennec, J. (2003). A new bone-healing material: a hyaluronic acid-like bacterial exopolysaccharide. Calcif. Tissue Int. 72, 74-79. https://doi.org/10.1007/s00223-001-2091-x

Sources, Biosynthesis, Properties, Structures and Applications of Halophilic Exopolysaccharides

2

Annarita Poli[1*], Paola Di Donato[1,2], Ilaria Finore[1], Luigi Leone[1], Barbara Nicolaus[1]

[1]Institute of Biomolecular Chemistry, Consiglio Nazionale delle Ricerche, Via Campi Flegrei, 34, 80078 Pozzuoli, NA, Italy
[2]Department of Science and Technology, University of Naples "Parthenope", Centro Direzionale, Isola C4, 80143 Naples, Italy
*Corresponding author: apoli@icb.cnr.it

DOI: https://doi.org/10.21775/9781912530267.02

Abstract
Saline and hypersaline environments such as natural saline seas, lakes, coastal lagoons, marine salterns, deserts, saline soils, etc. are considered extreme habitats for microbial life. They harbor halophiles as well as halotolerant microorganisms growing in the absence of salt and in the same time tolerant to relatively high salt concentrations. Halophilic microorganisms suggest many advantages for biotechnological exploration like relatively short time course of fermentations, lack of pathogenicity and the production of stable biomolecules, such as exopolysaccharides (EPS), under the extreme conditions of the most industrial processes. The aim of this review is to present a summary of the status of the research about the structures and the biosynthesis of exopolysaccharides by halophilic microorganisms, including medium-released and biofilm embedded polysaccharides. Moreover, relationships between the structure and the function of the exopolysaccharides and their possible biotechnological and industrial applications, are discussed.

Introduction

The term "halophilic" derives from greek άλς= salt + φιλέω= to love and indicates all those organisms able to colonize salty environments. However, the first studies concerning halophilic microorganisms date back to earlier 20th century and relied on salted foods, such as salted fish that become red because of the microbial contamination (Poli et al., 2017). Salt-loving organisms have, then, been detected in both natural and artificial environments; some examples are the following: seawater, deep sea brines, salt lakes, warm brines, hypersaline soils, salted food products, salt mine and all places wherein the salt concentration tends to the saturation. The best known salty lakes are the Owens Lake in California, the Great Lake in Utah and the Dead Sea; fermentation salty food sources of halophilic microorganisms are soy sauce, Chinese fermented beans, salted cod, salted anchovies, sauerkraut, etc.

Nowadays, there are many definitions to explain what is a halophile; the most accepted description of halophiles identifies them as microorganisms that grow optimally at Na^+ concentrations greater than 0.2 M (Oren et al., 2006). According to the optimal salt concentration for growth, these organisms are recognized into three grouping: extreme halophiles that grow in an environment with 3.4–5.1 M (20% to 30%, w/v) NaCl; moderate halophiles that grow in an environment with 0.85–3.4 M (3% to 25%, w/v) NaCl; and slight halophiles that grow in an environment with 0.2–0.85 M (1% to 5%, w/v) NaCl (DasSarma et al., 2012). Halophiles are surprising organisms, highly specialized, able to proliferate in environments characterized by the presence of salt. From the taxonomy point of view, halophiles fall back to *Archaea, Bacteria* and *Eucarya* domains. During last decades, the understanding of biodiversity in salt-saturated environments underwent a substantial upgrade, thanks to the improving of culture methods, chemotaxonomic and molecular biology techniques. Several halophilic microorganisms are being exploited in biotechnological fields. There are promising studies on halophilic bacteria and archaea, as they have ability for producing biochemicals, which possess a significant potential use in industrial and environmental technology (Oren, 2010). Furthermore, halophilic bacteria have the ability to produce biopolymers that are used in industrial and medical fields. Moreover, a number of moderate halophiles are known to change the accumulated concentrations of the osmoprotectants such as ectoine, hydroxyectoine and proline in order to protect their cytoplasm during stressing exposure. These molecules are desired for a variety of uses in biotechnology, for example in the enzyme protection against different factors such as freeze-thawing, freeze-drying and heating and they can be used for applications in medicine, cosmetology, dermatology and nutrition. Another application fields of halophilic bacteria are related to the biodegradation of dyes, polyaromatic hydrocarbons (PAHs), organic pollutants and the desalinization of wastewater (Poli et al., 2017).

An objective of this section is to provide basic information on the chemical structure of prokaryotic EPS, mechanisms for their synthesis and genetic approach for improving the yield and modification of EPS and functional properties and applications of commercially important EPS.

Table 1. Examples of halophilic microorganisms isolated from salty habitats.

Microorganism	**Site and Temperature of Isolation**	**References**
Halomonas rifensis	saltern in the Brikcha Rural Commune village in Chefchaouen in the Rif Mountains, northern Morocco	Amjres et al., 2011
H. stenophila	soils from the San Pedro de Pinatar and Sabinar saline-wetlands in Murcia (south-east, Spain)	Amjres et al., 2015
H. maura	salt soil bordering a solar saltern in Asilah, Morocco	Arias et al., 2003
H. eurihalina	Saline soil in Alicante (Southern Spain)	Béjar et al., 1998
H. nitroreducens strain WB1	hydrothermal vent	Chikkanna et al., 2018
H. smyrnensis AAD6^T	Solar saltern, Turkey	Poli et al., 2013
H. cerina	different saline soils in Spain	Gonzalez-Domenech et al., 2008
H. almeriensis M8^T	soil sample from Cabo de Gata, Almería, Spain	Llamas et al., 2012
H. ventosae	saline soil in Jaen, south-eastern Spain	Martínez-Cánovas et al., 2004a
H. anticariensis strains FP35^T	saline soils at Fuente de Piedra, an endorheic wetland in the province of Málaga in southern Spain	Martínez-Cánovas et al., 2004c
H. alkaliantarctica strain CRSS	Cape Russell in Antarctica (Ross Sea, 74 52.35 S 163 53.03 E)	Poli et al., 2007
H. alkaliphila	salt pool located in Montefredane in Campania Region (South of Italy)	Romano et al., 2006a
Bacillus licheniformis strain B3-15	shallow marine hot springs at Vulcano island (Italy)	Maugeri et al., 2002
Colwellia psychrerythraea 34H	enriched Arctic marine sediments	Deming et al., 1988
Flavobacterium uliginosus MP-55	homogenates of sea weed collected in Japan	Umezawa et al.,1983

Hahella chejuensis strain 96 CJ10356	marine sediments from Cheju Island (Republic of Korea)	Lee et al., 2001
Polaribacter sp. SM1127	brown alga Laminaria Arctic and Antarctic regions	Dong et al., 2012
Salipiger nanhaiensis strain ZH114^T	deep water of the South China Sea	Dai et al., 2015
S. mucescens A3^T	saline soil bordering a saltern on the Mediterranean coast at Calblanque (Murcia, south-eastern Spain)	Martınez-Canovas et al., 2004b
Haloarcula japonica strains T5–T7	Tunisian marine saltern	Nicolaus et al., 1999
H. hispanica	Solar saltern in Spain	Juez et al., 1986
Alteromonas hispanica F-32^T	hypersaline water sample collected from Fuente de Piedra (southern Spain)	Martınez-Checa et al., 2005
Idiomarina fontislapidosi F23^T, *I. ramblicola* strain R22^T	hypersaline habitats in Spain	Martınez-Canovas et al., 2004d
Bacillus strain B3-15	water of a shallow, marine hot spring at Vulcano Island (Eolian Islands, Italy)	Maugeri et al., 2002
Aphanothece halophytica GR02	Guangrao saltwork in the Shandong province of People's Republic of China	Morris et al., 2000
Haloferax mediterranei	Mediterranean Sea	Anton et al., 1988
Pantoea sp. strain BM39	sediments collected in the Tyrrhenian Sea	Silvi et al., 2013
Haloterrigena turkmenica	sulfate saline soil in Turkmenistan	Ventosa et al., 1999
Aphanocapsa halophytia strain MN-11	salt lake	Sudo et al., 1995
Amorphus orientalis strain YIM D10^T	salt mine sediment	Wang et al., 2010

Exopolysaccharide producing halophilic microorganisms

Many halophilic microorganisms show the ability to synthesize exopolysaccharides (Table 1). Most probably, these biomolecules protect the microbial cell against the environmental stress factors, by regulating the osmotic pressure in the presence of salts. Indeed, the polymers, surrounding the microorganism, attenuate the physical stress associated to the salinity by protecting membrane integrity (DasSarma and DasSarma, 2001; Poli et al., 2010; Oren, 2013).

Archaea

In the frame of the *Archaea* domain, halophilic microorganisms producing EPS belong to *Haloarcula*, *Halococcus*, *Haloferax* and *Natronococcus* genera (Nicolaus et al., 2010). *Haloarcula japonica* strains T5, isolated from the Tunisian marine saltern, produces an exopolysaccharide, when grown in a minimal medium supplemented with glucose with a yield of 370 mg/L. The polymer showed a following monosaccharide composition: Man, Gal and GlcA (Nicolaus et al., 1999). Also *Haloarcula hispanica* resulted EPS producer; the polymer in this case did not presented uronic acids, but only Man, Gal with traces of Glc and sulphate groups (Yang et al., 2017). *Haloferax mediterranei* strain R4, was isolated from saltern area in Spain; it is well-known for its ability to secrete an acidic exopolymer (Parolis et al., 1996). The extreme halophilic archaea *Haloterrigena turkmenica* DSM-5511 have been studied for its capability to produce an exopolymer, showing a higher yield of 20.68 mg per 100 ml when grown in the presence of glucose. The highest EPS production was observed in the stationary phase; the polymer was composed of two main fractions of 801.7 and 206.0 kDa and contained Glu, Gal, GlcN, GalN and GlcA (Squillaci et al., 2016).

Bacteria

Also in the *Bacteria* domain there are many halophilic organisms, good producers of EPS. For example, the moderately halophilic *Halomonas* spp. show this ability.

H. maura is the most famous specie eps-producing, isolated from a saline soil bordering a solar saltern in Asilah, Morocco (Bouchotroch et al., 2001). The exopolymer, named mauran is an anionic, sulfated heteropolysaccharide with a high uronic-acid content with interesting functional properties and its suitability for use in biotechnology (Arias et al., 2003). Amjres et al. (2015) and (2011), described the EPS production in *H. stenophila* strain HK30 and *H. rifensis* strain $HK31^T$, respectively, both isolated from a soil sample taken from a solar saltern in the Brikcha Rural Commune village in Chefchaouen in the Rif Mountains, northern Morocco. The exopolymer secreted by *H. stenophila* was named haloglycan and the composition was 44 ± 0.1 carbohydrates, 2.35 ± 0.1 acetyls, 2 ± 0.01 sulphates, 1.93 ± 0.02 sodium, 1.2 ± 0.2 uronic acids, 1 ± 0.01 phosphorus and 1 ± 0.01 calcium (% w/v). Béjar et al. (1998) compared the EPS yields released from 19 strains of *H. eurihalina,* isolated from saline soils, in two diverse media obtaining as highest yield of 1.6 g/L by strain H212. *H. smyrnensis* $AAD6^T$ isolated from soil samples taken from Çamaltı Saltern area in Turkey, was found to produce high levels of levan by using sucrose or sugar beet molasses and starch molasses as cheap and alternative carbon sources (Finore et al., 2014; Kucukasik et al., 2011; Poli et al., 2013). *H.*

alkaliantarctica strain CRSS, isolated from salt sediments near the salt lake in Cape Russell in Antarctica (Figure 1), produces different EPS, according to the different growth conditions, by means of a mannan and a xylo-mannan on complex media and fructo-glucan on minimal medium (Poli et al., 2007; Poli et al., 2004). The halophilic *H. nitroreducens* strain WB1, isolated from a hydrothermal vent, produces an anionic exopolysaccharide during the log and early stationary phases, when grown on malt extract/yeast extract medium. The composition of EPS consists of Glu, Man and Gal (Chikkanna et al., 2018). *H. cerina* SP4^T, isolated from saline soil in Spain, is a moderately halophilic microorganism EPS producing (González-Domenech et al., 2008). *H. almeriensis* strain M8^T also secretes a polymer constituted by two different fractions with a different molecular weights: 6.3×10^6 and of 1.5×10^4 Da. The monosaccharide composition of the high-molecular-weight fraction was Man, Glu and Rha; while the low-molecular-weight fraction contained Man and Glu (Llamas et al., 2012). *H. anticariensis* FP35^T and isolated from soil samples taken from Fuente de Piedra, a saline wetland in the province of Màlaga in southern Spain. The EPS yield was of 29.65 mg for 100 mL of broth with a molecular weight of 50 KDa, mainly composed by Glu, Man and Gal (Mata et al., 2006). *H.*

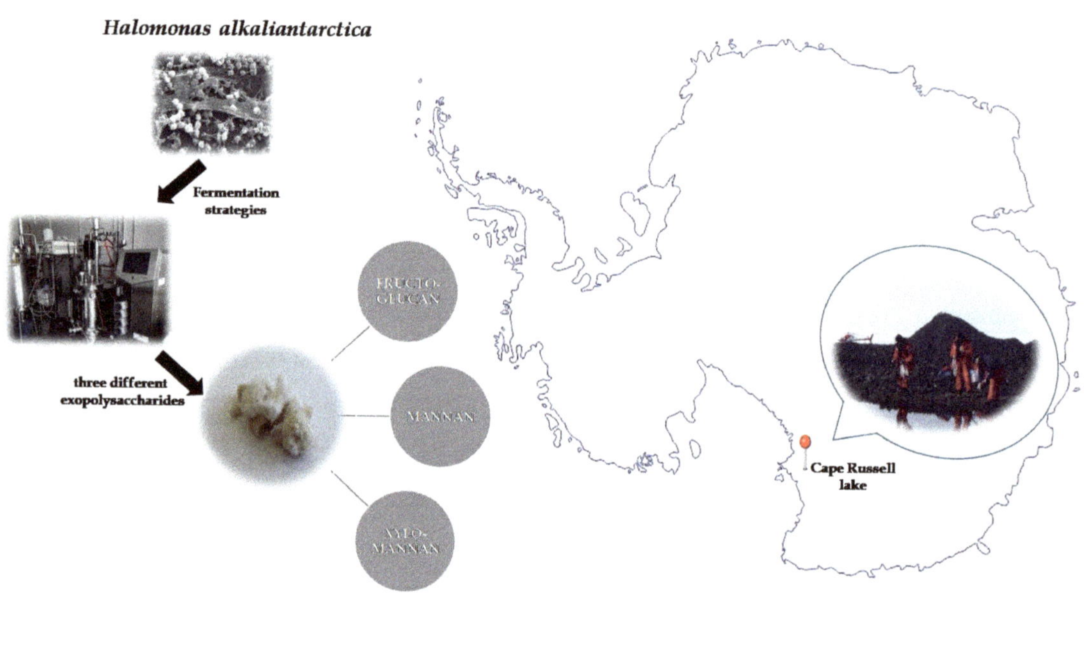

Figure 1. *Halomonas alkaliantarctica* strain CRSS, isolated from salt sediments near the salt lake in Cape Russell in Antarctica, produces different EPS, according to the different growth conditions, by means of a mannan and a xylo-mannan when grow on complex media, and a fructo-glucan polymer when grow on minimal medium (Poli et al., 2007; Poli et al., 2004).

ventosae AI12^T, isolated from a saline soil in Jaén, south-eastern Spain, produced an EPS associated with the cell surface and also in the medium surrounding the bacterium (Martínez-Cánovas et al., 2004a). In more detail, during the exponential growth phase, the EPS was present as thin capsule on the cell surface but it was progressively released into the medium as the cells reached the stationary phase, with a yield of 28.35 mg for 100 mL of broth. The molecular mass of the polymers was about 50 kDa and they were composed of Glu, Man and Gal (Mata et al., 2006). *H. alkaliphila* strain 18bAG^T is a halotolerant alkaliphilic bacterium isolated from a salt pool in Campania (Italy) able to synthesize an extracellular material containing 66.2% of carbohydrate. The production resulted more abundant (32.55 mg 100 ml^{-1}) when collected in the stationary growth phase in a medium containing 20% maltose and 1% yeast extract (Romano et al., 2006a).

Oceanobacillus oncorhynchi subsp. *incaldanensis* strain 20AG^T is an alkalitolerant halophile microorganism isolated from an algal mat collected from a sulfurous spring in Campania (Italy); during the stationary growth phase, it excreted 9 mg L^{-1} of a polysaccharide when grown in minimal medium containing trehalose (10 gL^{-1}) (Romano et al., 2006b). *Alteromonas hispanica* F32^T isolated from hypersaline habitat in Spain is a moderate halophilic bacterium able to produce an exopolysaccharide with a molecular weight of 1.9 X 10^7 Da composed of Glc, Man, Rha and Xyl as monosaccharide and sulphate and phosphate groups (Mata et al., 2008). *Bacillus licheniformis* strain B3-15 is a halophilic and thermotolerant bacterium, isolated from a sample of shallow marine hot springs located in Vulcano island (Italy). The exopolysaccharide, produced with a yield of 165 mg L^{-1}, was a tetrasaccharide repeating unit essentially constituted by sugars having a manno-pyranosidic configuration (Maugeri et al., 2002). The halophilic *Colwellia psychrerythraea* 34H was isolated from Arctic marine sediments; it synthesizes both extracellular and capsular polymers as cryoprotectant when grown at 4°C (Marx et al., 2009). The polysaccharides are composed by amino sugars and uronic acids with amino acidic decorations. When the bacterium is incubated at 8°C it produces another polymer with different monosaccharide composition not involved into anti-freezing activity (Casillo et al., 2017). *Salipiger nanhaiensis* ZH114^T is a moderate halophilic, facultatively anaerobic, isolated from deep water of the South China Sea and it has been resulted an EPS producer (Dai et al., 2015). *Salipiger mucosus* A3^T, isolated on the Mediterranean seaboard, produced an heteropolysaccharide, mainly during its exponential and stationary growth phases with a molecular mass of 250 kDa and composed by Glu, Man, Gal and Fuc (Llamas et al., 2010). *Salipiger mucescens* A3^T was isolated from a saline soil bordering a saltern on the Mediterranean coast at Calblanque (Murcia, south-eastern Spain); it produced an EPS associated with the cell surface (Martínez-Cánovas et al., 2004b). *Pantoea* sp. strain BM39, isolated from sediments collected in the Tyrrhenian Sea, is a microorganism able to secrete very high levels of EPS, ca. 21 g L^{-1} in 24 h on glucose medium, during the exponential growth. The exopolymer resulted to be essentially composed by glucose and showed a molecular mass of ca. 830 kDa (Silvi et al., 2013). *Chromohalobacter canadensis* 28, isolated from Pomorie salterns is a moderate halophilic bacterium producing an extracellular polymer substance, that is composed by a exopolysaccharide fraction (14.3% w/w) and protein fraction (72% w/w); the EPS is mainly composed by GluN, Glu, Rha and Xyl (Radchenkova et al., 2018).

The halophilic cyanobacterium *Aphanothece halophytica* GR02 isolated from the Guangrao saltwork in the Shandong province of People's Republic of China secreted an EPS composed by two main fractions, of which the major fraction consisted of Glu, Fuc, Man, Ara and GlcA; while, the minor fraction consisted of Rha, Man, Fuc, Glu, Gal and GluA (Li et al., 2001). The halophilic cyanobacterium *Aphanocapsa halophytia* MN-11 was isolated from a hypersaline environment and resulted to be able to produce large quantities of exopolysaccharide; the amount of nitrogen and phosphorus affected the EPS production by reaching maximum 40 mg L^{-1} (Sudo et al., 1995).

Exopolysaccharide biosynthesis

Exopolysaccharides are either homo- or heteropolymers which are frequently decorated by non-carbohydrate substituents, such as acetyl, pyruvyl, or succinyl groups conferring anionic properties to the polymers. Bacteria exopolysaccharides are synthesized via different biosynthesis pathways and the genes responsible for the synthesis are often clustered within the genome. The knowledge related to EPS biosynthetic processes and the genetic regulation, are essential to produce tailor-made biopolymers. Polysaccharide synthesis is a complex process that proceeds via assembly in the cells of sugar nucleotide precursors in which different enzymes and regulatory molecules are involved. It begins with the entry of sugars in cell, which are catabolized by periplasmic oxidation or intracellular phosphorylation; sugars that do not take part in the central metabolic pathways, act as a raw material for EPS manufacture (Freitas et al., 2011). The energy rich precursor monosaccharides are in the form of nucleotide diphosphate/monophosphate sugars (NDP/ NMP-sugar). This is a crucial step of biosynthesis: sugars often are in the form of sugar-1P and rarely in form of sugar-2P or sugar-6P, and serve as activated primary residues (Madhuri and Prabhakar, 2014). Some intermediates like Fructose-6P or Glucose-1P, in majority of cases lead to synthesis of uridine di-phosphate-N-acetyl glucosamine (UDP-GlcNAc), uridine di-phosphate-N-acetyl galactosamine (UDP- GalNAc) and dTDP- rhamnose, precursor molecules for EPS synthesis. In subsequent step, phospho-glucomutase enzyme catalyses the conversion of sugar-6P to sugar-1P and UDP-glucose-pyrophosphorylase and dTDP-glucose-pyrophosphorylase catalyse the conversion of sugar-1P to UDP-glucose and dTDP-glucose. After the synthesis of the nucleotide sugar precursors, the next steps are the synthesis of oligosaccharide repeat units, their assembly in order to form the polysaccharide and the product export. Nucleotide diphosphates (NDPs) or nucleotide monophosphates (NMPs) serve as activated donors for the glycosyltransferase-catalyzed transfer of the sugar to a lipid carrier or a carbohydrate. Lipid carriers are long-chain phosphate esters that are located in the cytoplasmic membrane and facilitate the assembly of the repeating unit in a specific order and also its transport through the cell membrane. Polysaccharides are assembled and exported by one of these known distinctive types of mechanisms: the synthase- depend (Whitney and Howell, 2013), ATP-Binding Cassette (ABC) transporter-(Willis and Whitfield, 2013), and Wzx/Wzy dependent (Islam and Lam, 2014) pathways and the extracellular synthesis by use of a single sucrase protein. In the first three biosynthesis pathways, the precursor molecules, which are necessary for the stepwise elongation of the polymer strands, are realized inside the cell, while for the extracellular production, the polymer strand is elongated by direct addition of monosaccharides obtained by cleavage of di- or trisaccharides.

In the Wzx/Wzy dependent pathway, the repeating units, linked to an undecaprenol diphosphate anchor (C55) at level of the inner membrane, are assembled by different glycosyltransferases (GTs) and translocated across the cytoplasmic membrane by a Wzx protein (flippase protein). Wzy polymerase and the polysaccharide co-polymerase protein (PCP) are then responsible for the polymerization process. The polysaccharide export protein families (OPX) and the PCP carry the transport across the membranes. EPS assembled by the Wzx/Wzy pathway are heteropolymers (e.g., xanthan) and possess different sugar pattern, up to four or five types of sugars in their chemical structure. All strains using this pathway carry the genes for the flippase (Wzx) and the polymerase (Wzy) within their extracellular polysaccharide operons. This pathway is responsible for the biosynthesis of diutan, gellan, welan, xanthan and colanic acid also known as the M antigen (an EPS with no commercial application, but is of high interest due to pathogenicity in *Enterobacteria* studies).

In the ABC transporter-dependent pathway, the EPS chain is assembled on a lipid carrier situated in the inner leaflet of the inner membrane before the carriage across the inner membrane by an ABC transporter. The EPS is then exported through the periplasm and across the outer membrane by the OPX and the PCP protein families. In the synthase-dependent pathway, complete polymer chain is polymerised and secreted across the inner membrane by an inner membrane synthase protein. The activity of the polysaccharide synthase is post-translationally regulated by an inner-membrane c-di-GMP receptor. The EPS is then exported across the outer membrane by TRP-containing protein and an integral outer-membrane beta-barrel.

In the synthase-dependent pathway complete polymer strands are secreted across the membranes and the cell wall: this pathway is separate from a flippase for the translocation of the repeat units. Both polymerization and translocation processes are performed by a single synthase protein. In the case of alginate and cellulose, for example, the synthase protein is a single subunit of an envelope-spanning multiprotein complex (Whitney and Howell, 2013). This pathway is often utilized for the synthesis of homopolymers: this is the case of curdlan biosynthesis for example, in which only β-(1-3)-linked glucose is present or in the case of bacterial cellulose, consisting only of β-(1-4)-linked glucose units. In addition to curdlan, also alginate, cellulose and hyaluronic acid are examples of EPS in which the synthase dependent mechanism is responsible for their synthesis.

In the extracellular synthesis pathway, the biosynthesis of polymers such as dextran, levan and mutan, occurs via GTs, which are secreted and covalently linked to the cell surface. In this case the responsible enzymes involved in the process, transfer the activated precursor monosaccharides from substrate to growing polysaccharide that assembles in a final structure by the formation of linkage pattern and degree of branching.

Regulation of microbial polysaccharide production

Since microbial EPS production is usually not confined to just one type of EPS as product but rather a mixture of different polymers, each being synthesized by a certain gene cluster, the availability of the precursors encoded by these genes has a high impact on the yield and structure of the EPS excreted by the cell.

For a systems-based approach to the EPS biosynthetic pathways, microbial genome sequence is considered as a starting point, and from this point of view, Next Generation Sequencing (NGS) technologies play a significant role by enabling high-throughput genomic data at very high speed with a relatively low cost. NGS technologies triggered a rapid increase in whole genome sequencing (WGS) projects providing valuable data that could be used for comparative and functional genomics as well as other systems biology applications. Such approaches either aim to elucidate biosynthesis mechanisms, to improve the microbial productivity via strain improvement strategies or to modify physicochemical and/or rheological properties of the biopolymer by changing its composition, length or degree of branching. For example, genome data of *Xanthomonas campestris* revealed that xanthan biosynthesis was encoded by a single gene cluster of 12 kb (Vorhölter et al., 2008). Similarly, whole genome sequence of the opportunistic human pathogen *Pseudomonas aeruginosa* PA01 revealed that 24 genes were involved which were clustered in two separate operons (Hay et al., 2010). *P. aeruginosa* has the genetic capacity to produce three different extracellular polysaccharides, namely Pel, Psl and alginate, that are synthesized via an isoprenoid lipid carrier dependent mechanism (for Psl) and a lipid carrier independent mechanism (for alginate and likely for Pel) (Franklin et al., 2011). In *P. aeruginosa*, levels of polyP, linear polymers of tens or hundreds of orthophosphate residues linked by high-energy phosphoanhydride bonds, were found to be co-regulated with AlgR, which is a regulatory protein for the synthesis of alginate. Bis-(3',5')-cyclic-dimeric-guanosine monophosphate (c-di-GMP) is known to act as an allosteric regulator of cellulose and alginate biosynthesis, and as both an allosteric and transcriptional regulator of Pel EPS. Analysis of the curdlan producing *Agrobacterium* sp. ATCC 31749 genome pointed to the presence of numerous proteins associated with the synthesis and degradation of c-di-GMP and that an unidentified sigma factor was involved in the regulation of curdlan biosynthesis (Ruffing and Chen, 2012). In these examples, these gene clusters are usually located on the chromosome whereas in most lactic acid bacteria, these genes are found to be confined to plasmids. When the whole genome sequence of the EPS producer microorganism is not available, sequence data of a taxonomically close species could also be used for systems-based studies. Such an approach has been used for the improvement of levan production by the halophilic strain *Halomonas smyrnensis* AAD6^T, where first, the available whole genome sequence of a taxonomically close microorganism, *Chromohalobacter salexigens* DSM 3043, was used to construct a comprehensive genome-scale metabolic model (Ateş et al., 2011) and then this model was recruited and adopted to the producer strain via integration of the available biochemical, physiological and phenotypic features of *H. smyrnensis* AAD6^T. With metabolic system analysis of this generic metabolic model, significant improvement in levan yields was obtained (Ateş et al., 2013).

Engineering strategies can be subdivided into different categories. One is an increased volumetric productivity: these studies were mostly aiming at increasing of sugar nucleotide precursors to enhance the carbon flux toward the final polymer and the genes of precursor biosynthesis were overexpressed (Schmid et al., 2015). Overexpression of genes involved in the EPS assembly such as GTs, Wzx, Wzy, both as single genes or whole cluster, resulted in enhanced yields and precursor conversion rates while in other cases it had a negative effect presumably due to distorting the multidomain protein complex involved in polymerization and secretion (van Kranenburg et al., 1999). Another approach is to increase the EPS productivity by increasing transcription of the operons, which encode the EPS biosynthesis proteins. Single gene knock-outs were also described to enhance yield and in the alteration of EPS chemical structure, as showed in *Azotobacter vinelandii* (Gaytán et al., 2012). However, the strategy to enhance productivity based on genetic engineering might be interesting for EPS with reduced viscosity, for example due to lower molecular weight. The optimization of manufacturing process parameters might be more promising than engineering EPS biosynthesis for many established industrial EPS producers. The highest titers of highly viscous EPS such as xanthan are around 30–50 g/L and represent the current maximum amount, which is manageable by existing bioprocess units (Hublik, 2012).

Another strategy of engineering EPS biosynthesis is to alter the molecular structure and therefore the chemical characteristics and behaviour of the final biopolymer. These modifications can be based on deleting substituents or monomeric sugars from the side chain, or the binding new substituents: in these case a change of the ratio of decoration, such as, the degree of acetylation and pyruvylation, occurred (Donati and Paoletti, 2009). The degree of acetylation and pyruvylation has opposite effects on viscosity. A high degree of pyruvylation resulted in higher viscosity, whereas the presence of more acetyl groups, decreased viscosity of EPS. This finding is a general rule for polysaccharides and can be used in tailoring the EPS viscosity.

Another engineering approaches with respect to the production of xanthan variants included the length of the side chain. A truncated tetramer xanthan version, obtained by deletion of the terminal mannose via inactivation of the glycosil transferase (GT GumI), resulted in a much lower viscosity. The further removal of the glucuronic acid from the side chain by inactivation of GT GumK, resulted in enhanced viscosity compared to the wild type (Schmid et al., 2015). The molecular weight of xanthan was synthetically adjusted by controlling the expression level of the Wzy polymerase Gum E (Galván et al., 2013), while for alginate a similar effect was observed by an over expression of alginate polymerase alg8/alg44 in *Azotobacter vinelandii* producing a high molecular weight alginate type (Díaz-Barrera et al., 2012).

In some EPS, overexpression or mutation of genes involved in the polymerization/ degradation process (e.g., synthase, Wzy, PCP/lyases, glucosidases) represented another possibility to change the rheological properties of the polymers (Galván et al., 2013).

As concerning the haloarchaeal EPS, although the chemical structures have been solved, little is known about their biosynthesis. The EPS from *Haloferax mediterranei* ATCC 33500 was identified to be a heteropolysaccharide containing mannose as the major component. The repeat unit of EPS in *H. mediterranei* contains one mannosyl and two N-acetyl-glucosaminuronyl moieties, and one N-acetyl-glucosaminuronyl group is modified by a sulfonic group. Based on the complete genome sequence of *H. mediterranei*, a gene cluster involved in EPS biosynthesis was identified. Deletion of the gene cluster eliminated EPS synthesis. The mutant strain deficient of EPS biosynthesis showed a remarkable decrease in viscosity and foaming propensity of culture broth, increase in content of dissolved oxygen, and enhanced production of polyhydroxyalkanoate (Zhao et al., 2013). Lü et al. (2017) purified an acidic exopolysaccharide from an extremely halophilic archaeon *Haloarcula hispanica* ATCC 33960, which mainly composed of mannose and galactose with a small amount of glucose in a molar ratio of 55.9: 43.2: 0.9. The authors reported the identification of two glycosyltransferase genes (HAH_1662 and HAH_1667), responsible for the synthesis of EPS. Deletion of either HAH_1662 or HAH_1667 led to loss of the EPS production. In addition, the mutants of *H. hispanica* displayed a different cell surface morphology, retarded growth in low salty environment, an increased adhesion, and swimming ability, suggesting that its biosynthesis might act as an adaptable mechanism to protect the cells against harsh environments.

In conclusion, as emerged from this overview, the genes involved in the different biosynthesis pathways encode various types of GTs, polymerizing and branching enzymes, but also enzymes responsible for addition of substituents or modifications of sugar chain. However, several steps are currently understood, and sometimes the differences between the pathways become less defined. Genome or plasmids contain the genes encoding these enzymes in most of the EPS producing bacteria (Rehm, 2015). The clustering of several GTs and polymerizing as well as secreting enzymes facilitates the identification of EPS operons. Even if many gene clusters involving in the EPS biosynthesis have been known for several years, the function and mode of action of most of the genes and proteins is not completely clarified.

Moreover, the identification of novel EPS clusters by next generation sequencing approaches will enhance our understanding of EPS synthesis pathway variation and modification. By using different tools, such as bioinformatics, structural information of proteins and EPS, it will be enable the implementation of synthetic biology approaches for tailoring microbial EPS. The insights in Wzx and Wzy topology and mechanism might open up the opportunity for incorporation of desired sugars or sugar derivatives resulting in modified EPS structures with hitherto unknown material properties (Rehm, 2015). Recently an innovative bi-enzymatic process was reported:

starting from sucrose, it was possible to obtain the production of short chain fructooligosaccharides and oligolevans. This system was based on an immobilized levansucrase and an endo-inulase, resulting in a highly efficient synthesis system with a yield of more than 65% and a productivity of 96 g/L/h (Tian et al., 2014). The utilization and combination of several carbohydrate modifying enzymes create the potential for industrial production of different low molecular weight oligo- or polysaccharides with applications as food additives (prebiotics) or in medicine.

Chemical composition and properties of halophiles' EPS

Halophilic microorganisms produce EPS as a protection to survive in the extreme saline environments. Thanks to the wide variety of their chemical composition and structural features, the halophiles' EPS possess different properties that make them potentially suitable for applications in several fields from the pharmaceutical, to the biotechnological and the pollution remediation sector (Table 2).

Table 2. Examples of exopolysaccharides from halophiles.

Species	Physico-chemical features	Monosaccharide composition	Properties/Activities	Reference
Alteromonas hispanica F32^T	average MW: 1.9 × 10^7	Man/ Glc/Xyl 62.75/18.15/12.25 (%w/w)	-pseudoplastic behaviour -emulsifying activity	Mata et al., 2008
Halomonas almeriensis	Carbohydrates/ Proteins/Acetyl residues/Sulfate residues 30.5/ 1.1 /0.8/ 1.4 (% w/ w)	Glc/ Man/ Rha 27.5/ 72/ 0.5 (% w/w)	-pseudoplastic behaviour -high capacity for binding metal cations	Llamas et al. 2012
Halomonas anticariensis strain FP35^T	average MW: 2.0 × 10^4 Da	Man/ GalA/Glc 45.0/37.0/15.0 molar ratios	-pseudo-plastic behaviour -high capacity for binding metal cations	Mata et al., 2006
Halomonas anticariensis strain FP36	average MW: 4.6 × 10^4 Da	Man/ GalA/Glc 42.9/37.3/17.0 molar ratios	-pseudo-plastic behaviour -high capacity for binding metal cations	
Halomonas maura	Carbohydrates/ Proteins/Acetyl residues/ Sulfate residues 65/ 2.5/ 0.18/ 6.5 (% w/ w) average MW 4.7 ×·10^6 Da	Man/Glc/GlcA/Gal 34.8/29.3/21.9/14 (% w/w)	-high capacity for binding metal cations -antihemolytic and antithrombogenic nature -nanofibers for biomedical applications	Arias et al. 2003; Raveendran et al., 2013a; Raveendran et al., 2013b

Halomonas nitroreducens strain WB1		EPS A: Man/Glc/Gal 64/28/6 (% w/w) EPS B: Man/Glc/ Rha/Ara 44/18.5/tr/tr (% w/w) EPS C: Man/Glc/Gal/GalA 56.5/ 19/ 14.2/tr (% w/w)	-pseudo-plastic behaviour -capacity for binding metal cations -emulsifying and anti-oxidation properties	Chikkanna et al., 2018
Halomonas smyrnensis		β-(2,6)-d-fructan	-anticoagulation activity -biocompatible and heparin-like activity	Poli et al., 2010; Ergines et al., 2016
Halomonas stenophila strain B100	Carbohydrates/ Proteins/Acetyl residues/Sulfate residues 33.2/1.65/0.25/7.90 (% w/w)	Glc/Gal/Man 44.5/40.5/15 (% w/w)	-proapoptotic effect on T leukaemia cells for the sulfated derivatives	Ruiz-ruiz et al., 2011
Halomonas stenophila strain N12^T	Carbohydrates/ Proteins/Acetyl Residues/Sulfate residues 30.4/1.05/1.95/2.45 (% w/w)	Glc/Man 48.8/25.5 (% w/w).		
Halomonas stenophila strain HK30	Carbohydrates/Acetyl residues/Sulfate residues/Uronic acids 44/2.35/2/1.2 (% w/v).	Glc/GlcA/Man/Fuc/Gal/Rha 24/7.5/5.5/4.5/1.2/1 (% w/ w).	-pseudo-plastic behaviour -emulsifying and flocculating activity	Amjres et al. 2015
Halomonas ventosae strain A112^T	average MW: 5.3 × 10^4 Da	Man/ Glc /Gal 55.9/23.8/13.8 molar ratios	-pseudo-plastic behaviour -high capacity for binding metal cations	Mata et al., 2006
Halomonas ventosae strain A116	average MW: 5.2 × 10^4 Da	Man/ Glc /Gal 56.7/23.7/12.6 molar ratios	-pseudo-plastic behaviour -high capacity for binding metal cations	
Haloarcula japonica strain T5	Carbohydrates/Proteins 70/4.0 (% w/w)	Man/Gal/GlcA 20.0/10.0/30.0 molar ratios	-inhibition of virus penetration and of retroviral reverse transcriptases	Nicolaus et al., 2004
Haloferax mediterranei ATCC 33500		Man/2-amino-2-deoxy-GlcA 1.0/1.1 (% w/w)	-candidate in oil recovery	Lu et al., 2017
Haloterrigena turkmenica	Carbohydrates/ Proteins/ Sulfate residues 91/1.4/2.8 (% w/w)	Glc/GlcN/GlcA/Gal/GalN 1:0.65:0.24:0.22:0.02 molar ratios	-emulsifying activity -antioxidant properties -moisture-retention ability	Squillaci et al., 2016
Idiomarina fontislapidosi F23^T	EPS 1: average MW: 1.5 × 10^6 EPS 2: average MW: 1.5 × 10^4	EPS 1: Man/ Glc /Gal 46.35/28.5/14.85 (%w/w) EPS 2: Man/ Glc /Gal 40.0/40.0/20.0 (%w/w)	-emulsifying activity -capacity for binding metal cations	Mata et al., 2008
Idiomarina ramblicola R22^T	EPS 1: average MW: 5.5 × 10^5 EPS 2: average MW: 2.0 × 10^4	EPS 1: Man/ Glc 68.2/25.0 (%w/w) EPS 2: Man/ Glc 53.6/18.9 (%w/w)		
Salipiger mucosus strain A3^T	Sulfate and phosphate groups	Glc/Man/Gal/Fuc 1.5/2.5/2.5/1 molar ratios	-pseudo-plastic behaviour -capacity for binding metal cations -emulsifying activity	Llamas et al., 2010

Volcaniella eurihalina	Carbohydrates/ Proteins/Acetyl residues/ Sulfate residues/ uronic acids/ hexosaminnes 34.5/14.1/0.6/2.7/1.5/2.4 (% w/w)	Glc/Man/Rha 2.9/1.5/1.0 (% w/w)	-pseudo-plastic behaviour -ability to produce gel-like systems	Béjar et al., 1995; Calvo et al., 1995

Legend: Man mannose, Glc glucose, Xyl xylose, Gal galactose, Rha rhamnose, GalA galacturonic acid, GlcA glucuronic acid, Ara arabinose, Fuc fucose, GlcN glucosamine, GalN galactosamine.

Many halophilic bacteria belonging to *Halomonas* genus, and haloarchaea such as *Haloferax, Haloarcula* and *Haloterrigena,* have been described as EPS producers: some remarkable examples of exopolysaccharides producers and of their related polymers are listed in table 2. The species *Alteromonas hispanica* F32^{T} synthesize an anionic EPS made up of glucose, mannose and xylose: this polymer possesses a pseudoplastic behaviour and a good emulsifying activity, evidenced by dissolution in different solvents. A pseudoplastic behaviour, besides a high capacity for binding metal cations, is shown also by the EPS produced by several species belonging to the genus *Halomonas. Halomonas almeriensis* produces an EPS made up of glucose, mannose and rhamnose with a 1.4% of sulfate content; *Halomonas anticariensis* strain FP35^{T} produces a high molecular weight polymer made of mannose, glucose and galacturonic acid; the polysaccharide produced by *Halomonas anticariensis* strain FP36, is also a high molecular weight polymer made of mannose, glucose and galacturonic acid in similar amounts to those of the FP35^{T} strain. The same properties are shown by the polysaccharides produced by *Halomonas ventosae* strain A112^{T}, *Halomonas ventosae* strain A116, *Halomonas stenophila* strain HK30, *Halomonas nitroreducens* strain WB1. The EPS from *H. ventosae* strain A112^{T} and strain A116 are two high molecular weight polymers both made of mannose, glucose and galactose. The EPS from *H. stenophila* strain HK30 is made of glucose as main monosaccharide with lower percentages of glucuronic acid, mannose, fucose, galactose and rhamnose: this polymer also contains acetyl and sulfate residues and exerts interesting emulsifying and flocculating activities. *H. nitroreducens* strain WB1 produces three main EPS that besides the ability to bind metals and to emulsify oils are also antioxidant agents.

The ability to bind metal ions is also exploited by the EPS mauran, synthesized by *Halomonas maura*. Mauran is a high molecular weight polyanionic EPS composed by four monosaccharide units i.e. mannose, glucose, galacturonic acid and galactose; it also presents as substituting groups acetyl and sulfate residues, accounting for 0.18 and 6.5 % w/w of total weight, respectively. Mauran possesses interesting biological properties since it is a good antihemolytic and antithrombogenic agent; moreover, mauran showed to act against oxidative stress in a mouse fibroblast cell line and in mice liver homogenates. Interestingly, thanks to is biocompatibility, mauran can also be useful for the preparation of nanofibers for biomedical applications like tissue engineering and drug delivery.

Other *Halomonas* species that have been reported as EPS producers include *H. smyrnensis, H. stenophila* strain B100 and *H. stenophila* strain N12^T. The EPS from *H. smyrnensis* is a beta-(2,6)-d-fructan, or levan polysaccharide, whose sulfated derivatives possess a promising anticoagulant activity. Indeed, *H. smyrnensis*' sulfated levan exhibited a high heparin equivalent activity that proceed via thrombin inhibition. Thanks to this activity and to its high biocompatibility, *H. smyrnensis*' sulfated levan has been proposed for the production of biologically active and functionalized films and as building block in scaffolds for cardiac tissue engineering applications.

H. stenophila strain B100 produces an EPS, labelled as B100, that is made of glucose, galactose and mannose, with acetyl and sulfate as main substituting groups; the EPS from *H. stenophila* strain N12^T, labelled as N12, has a similar chemical composition and presents only glucose and mannose as monomer sugars. The native EPS B100 and N12, after deprivation of their constitutive sulfate groups, were oversulfated to demonstrate that their proapoptotic effect on T leukaemia cells is strongly dependent on the presence of these substituent groups.

Other halophilic species belonging to other genera have been studied for their chemical composition and properties. Some examples are offered by the species *Haloarcula japonica* strain T5, *Haloferax mediterranei* ATCC 33500, *Haloterrigena turkmenica*, *Idiomarina fontislapidosi* F23^T and *Idiomarina ramblicola* R22^T, *Salipiger mucosus* strain A3^T and *Volcaniella eurihalina*.

Haloarcula japonica strain T5 produces a sulfated exopolysaccharide made of mannose, galactose and glucuronic acid that has been shown to interfere with the absorption and penetration of viruses into host cells and to inhibit some retroviral reverse transcriptases.

The EPS produced by *Haloferax mediterranei* ATCC 33500 is a sulfated polysaccharide made of mannose and 2-amino-2-deoxy-glucuronic acid, present in similar molar ratios. Thanks to its high viscosity at low concentrations, rheological properties, resistance to extremes values of pH and temperature, this polymer has been proposed for oil recovery, especially in oil deposits with high salinity concentrations.

The EPS produced by the species *Haloterrigena turkmenica* is a sulfated polysaccharide made of glucose and glucosamine as main monomers, with low amounts of galactose, galactosamine and galacturonic acid. This EPS possesses a significant emulsifying activity towards vegetable oils, a moderate antioxidant effect, besides a moisture-retention ability comparable to that of the hyaluronic acid.

The two species *Idiomarina fontislapidosi* F23^T and *Idiomarina ramblicola* R22^T, belonging to the family *Alteromonadaceae*, both produce two high molecular weight EPS that share the common character of being sulfated heteropolymers. *I. fontislapidosi* F23^T produces two polymers both made of mannose, glucose and galactose in different molar ratios; *I. ramblicola* R22^T EPS are instead made of only mannose and glucose. All these polymers display an emulsifying activity and have a good capacity for binding metal cations.

The EPS produced by *Salipiger mucosus* strain A3^T and *Volcaniella eurihalina* are both characterized by a pseudo-plastic behaviour. The polysaccharide from *S. mucosus* strain A3^T is a heteropolymer made of glucose, mannose, galactose and fucose with sulfate and phosphate groups; this EPS has also capacity for binding metal cations and exerts an interesting emulsifying activity. The polysaccharide from *V. eurihalina* is a sulfated and acetylated acidic heteropolysaccharide, whose monomer constituents are rhamnose, glucose, mannose, amino sugars and uronic acids. Besides its pseudo-plastic behaviour, this EPS is able to produce gel-like systems.

Industrial applications of bacterial exopolysaccharides

The microbial exopolysaccharides are a class of high value polymers that have many industrial applications (Wang et al., 2018). Research in this area has led to the production of several microbial polysaccharides on an industrial scale like xanthan, dextran, alginate, gellan, curdlan. Xanthan and gellan are commercially most appreciated on the global market for hydrocolloids with xanthan gum being the only significant bacterial exopolysaccharide, which accounted for 6% of the total market value (Imeson, 2010).

Extremely large molecule of xanthan gives very viscous solutions at relatively low polysaccharide concentrations reflecting in very high thickening and suspending potential, viscosity does not change greatly on raising the temperature in the range of 0-100ºC, emulsions are stable. Despite its high viscosity, xanthan suspensions exhibit high shear thinning, which means that they also flow easily. When shear stress is applied beyond a low minimal value, the viscosity decreases sharply with shear rate. The presence of anionic side chains of the xanthan gum molecules enhances hydration and makes xanthan gum soluble in cold water. Although many researchers purport to have found polysaccharides 'of superior viscosity to xanthan', in reality few match and can maintain their physical properties in the presence of salts, at higher temperatures or extremes of pH (Sutherland, 2002).

For gel forming, the addition of trivalent salts such as chromium or aluminium is required to shield the negative charges on the branched trisaccharide and minimise electrostatic repulsion.

Gellan has rheological properties similar to those of xanthan. Compared with other polysaccharides, gellan has many advantages such as an excellent thermal and acid stability, adjustable gel elasticity and rigidity, high transparency, water solubility, compatibility with other polymers, low concentration requirements and good flavour release. In its native state the polysaccharide does not form a gel. Deacetylation of native gellan results in a change from soft, elastic thermoreversible gels to harder, more brittle gels with higher thermal stability. It is manufactured in four different forms: Kelcogel (food-grade gellans), Gerlite (substitute of agar), Phytagel and Gel-Gro (solidifying agents). Gellan related polysaccharides do yield highly viscous aqueous solutions although none of the others forms gels, and have been proposed for various industrial applications. Whilst this polymer is interesting for many commercial food applications, still the costly downstream processing steps impair the cheap viability of gellan production and it is the most expensive among food gums (Giavasis et al., 2000). Unlike most polysaccharides, levan does not swell in water. The use of levan has not yet received industrial application due to scarce information on its biological safety and polymeric properties and a lack of economic means for large scale production. Presently, it is approved as a food additive in only a few countries. The weak chemical stability of levan in solution is another reason for its limited application.

Bacterial alginate appears to be the most promising in terms of industrial production because of its stable output of alginate. Although commercial alginates derive largely from algal sources there is a large potential for producing tailor-made alginates from bacterial sources, especially if advantage is taken of the genetic tools for controlling the production of the enzymes that are responsible for the synthesis and epimerization. Alginate absorbs water quickly which makes it useful as an additive in dehydrated products (Kumar and Modi, 2009).

Conclusion

Extremophilic microorganisms provide a valuable resource, not only for the elucidation of bioprocesses in extreme environments, but also for the exopolysaccharides they produce, which have a valuable range of physicochemical properties and highly promising commercial applications. The search for new high EPS producing strains is an ongoing process. The improvement of optimized fermentation conditions and polymer down-stream procedures, together with the genetic and metabolic engineering tools, are giving rise the production of microbial polymers with fine-tuned properties for commercial scale manufacturing. It can be predicted that in the near future the production of EPS from extremophiles will undergo fast and effective improvements and become economically more feasible thanks to the use of waste materials as growth media. Moreover, applications of extremophilic EPS in biomedical, food and pollution-mitigation fields can be foreseen based on their known biodegradability, biocompatibility, non-toxicity, and chemical functionality.

Acknowledgement

The authors thank to projects PNRA16_00020 and PNRA16_00274.

References

Amjres, H., Béjar, V., Quesada, E., Abrini, J., and Llamas, I. (2011). *Halomonas rifensis* sp. nov., an exopolysaccharide-producing, halophilic bacterium isolated from a solar saltern. IJSEM, *61*(11), 2600-2605. https://doi.org/10.1099/ijs.0.027268-0

Amjres, H., Béjar, V., Quesada, E., Carranza, D., Abrini, J., Sinquin, C., and Llamas, I. (2015). Characterization of haloglycan, an exopolysaccharide produced by *Halomonas stenophila* HK30. Int. J. Biol. Macromol. 72, 117-124. https://doi.org/10.1016/j.ijbiomac.2014.07.052

Anton, J., Meseguer, I., and Rodriguez-Valera, F. (1988). Production of an extracellular polysaccharide by *Haloferax mediterranei*. Appl. Environ. Microbiol. 54(10), 2381–2386.

Arias, S., del Moral, A., Ferrer, M. R., Tallon, R., Quesada, E., and Béjar, V. (2003). Mauran, an exopolysaccharide produced by the halophilic bacterium *Halomonas maura*, with a novel composition and interesting properties for biotechnology. Extremophiles, *7*(4), 319-326. https://doi.org/10.1007/s00792-003-0325-8

Ateş, Ö., Toksoy Öner, E., and Arga K. Y. (2011). Genome-scale reconstruction of metabolic network for a halophilic extremophile, *Chromohalobacter salexigens* DSM 3043. BMC Systems Biology 5:12, https://doi.org/10.1186/1752-0509-5-12

Ateş, Ö., Arga K.Y., and Toksoy Öner, E. (2013). The stimulatory effect of mannitol on levan biosynthesis: lessons from metabolic systems analysis of *Halomonas smyrnensis* AAD6(T). Biotechnol. Prog. 29, 1386-1397, https://doi.org/10.1002/btpr.1823

Béjar, V., Llamas, I., Calvo, C., and Quesada, E. (1998). Characterization of exopolysaccharides produced by 19 halophilic strains of the species *Halomonas eurihalina*. J. Biotechnol. 61(2), 135-141. http://dx.doi.org/10.1016/S0168-1656(98)00024-8

Bouchotroch, S., Quesada, E., del Moral, A., Llamas, I., and Béjar, V. (2001). *Halomonas maura* sp. nov., a novel moderately halophilic, exopolysaccharide-producing bacterium. IJSEM 51(5), 1625-1632. http://dx.doi.org/10.1099/00207713-51-5-1625

Calvo, C., Ferrer, M.R., Martinez-Checa, F., Béjar, V. and Quesada, E. (1995). Some Rheological Properties of the Extracellular Polysaccharide Produced by *Volcaniella eurihalina* F2-7. Appl. Biochem. Biotechnol. 55, 45. https://doi.org/10.1007/BF02788747

Casillo, A., Parrilli, E., Sannino, F., Mitchell, D. E., Gibson, M. I., Marino, G., and Corsaro, M. M. (2017). Structure-activity relationship of the exopolysaccharide from a psychrophilic bacterium: a strategy for cryoprotection. Carboh. Polym. 156, 364-371. https://doi.org/10.1016/j.carbpol.2016.09.037

Caumette, P. (1993). Ecology and physiology o phototrophic bacteria and sulphate reducing bacteria in marine salterns. Experientia 49, 473–481. https://doi.org/10.1007/BF01955148

Chikkanna, A., Ghosh, D., and Kishore, A. (2018). Expression and characterization of a potential exopolysaccharide from a newly isolated halophilic thermotolerant bacteria *Halomonas nitroreducens* strain WB1. Peer J., 6, e4684. https://doi.org/10.7717/peerj.4684

Dai, X., Shi, X., Gao, X., Liang, J., and Zhang, X-H. (2015). *Salipiger nanhaiensis* sp. nov., a bacterium isolated from deep sea water. IJSEM, 65, 1122-1126, https://doi.org/10.1099/ijs.0.000066

DasSarma, S., and DasSarma, P. (2001). Halophiles *eLS*: John Wiley & Sons, Ltd.

DasSarma, S., and DasSarma, P. (2012). Halophiles. In Encyclopedia of Life Sciences, John Wiley & Sons, Ltd. (Chichester, UK: Wiley Online Library), pp. 691–701. https://doi.org/10.1002/9780470015902.a0000394.pub3

Deming, J.W., Somers, L.K., Straube,W.L., Swartz, D.G., and Macdonell, M.T. (1988). Isolation of an obligately barophilic bacterium and description of a new genus, *Colwellia* gen. nov. Syst. Appl. Microbiol.10(2), 152-160 https://doi.org/10.1016/S0723-2020(88)80030-4

Díaz-Barrera, A., Soto, E., and Altamirano, C. (2012). Alginate production and alg8 gene expression by *Azotobacter vinelandii* in continuous cultures. J. Ind. Microb. Biotechnol. 39 (4), 613-621. https://doi.org/10.1007/s10295-011-1055-z

Donati, I., and Paoletti, S. (2009). Material Properties of Alginates. In: Rehm B.H.A. (ed.) Alginates: Biology and Applications. Springer Berlin Heidelberg, Berlin, Heidelberg, pp 1-53. https://doi.org/10.1007/978-3-540-92679-5_1

Dong, S., Yang, J., Zhang, X.Y., Shi, M., Song, X.Y., Chen, X.L., and Zhang, Y.Z. (2012). Cultivable alginate lyase-excreting bacteria associated with the arctic brown alga Laminaria. Mar. Drugs. DOI https://doi.org/10.3390/md10112481

Erginer, M., Akcay, A., Coskunkan, B., Morova, T., Rende, D., Bucak, S., Baysal, N., Ozisik, R., Eroglu, M.S., Agirbasli, M. and Toksoy Oner, E. (2016). Sulfated levan from *Halomonas smyrnensis* as a bioactive, heparin-mimetic glycan for cardiac tissue engineering applications. Carboh. Polym.149, 289–296. http://dx.doi.org/10.1016/j.carbpol.2016.04.092

Finore, I., Di Donato, P., Mastascusa, V., Nicolaus, B., and Poli, A. (2014). Fermentation Technologies for the Optimization of Marine Microbial Exopolysaccharide Production. Marine Drugs, 12(5), 3005-3024. http://dx.doi.org/10.3390/md12053005

Franklin, M. J., Nivens, D. E., Weadge, J. T. and Howell, P. L. (2011). Biosynthesis of the *Pseudomonas aeruginosa* extracellular polysaccharides, alginate, Pel, and Psl. Front. Microbiol. 2, 167. https://doi.org/10.3389/fmicb.2011.00167

Freitas, F., Alves, V.D., and Reis, M.A.M. (2011). Advances in bacterial exopolysaccharides: from production to biotechnological applications. Trends Biotechnol. 29, 388–398. https://doi.org/10.1016/j.tibtech.2011.03.008

Galván, E.M., Ielmini, M.V., Patel, Y.N., Bianco, M.I., Franceschini, E.A., Schneider, J.C., and Ielpi, L. (2013). Xanthan chain length is modulated by increasing the availability of the polysaccharide copolymerase protein GumC and the outer membrane polysaccharide export protein GumB. Glycobiol. 23 (2), 259-272. https://doi.org/10.1093/glycob/cws146

Gaytán, I., Peña, C., Núñez, C., Córdova, M.S., Espín, G., and Galindo, E. (2012). *Azotobacter vinelandii* lacking the Na+-NQR activity: a potential source for producing alginates with improved properties and at high yield. World J. Microbiol. Biotechnol. 28 (8), 2731-2740. https://doi.org/10.1007/s11274-012-1084-4

Giavasis, I., Harvey, L. M., and McNeil B. (2000). Gellan Gum. Crit. Rev. Biotechnol. 20, 177-211. https://doi.org/10.1080/07388550008984169

González-Domenech, C. M., Martínez-Checa, F., Quesada, E., and Béjar, V. (2008). *Halomonas cerina* sp. nov., a moderately halophilic, denitrifying, exopolysaccharide-producing bacterium. IJSEM 58(4), 803-809. https://doi.org/10.1099/ijs.0.65322-0

Hay, I. D., Ur Rehman, Z., Ghafoor, A., and Rehm, B. H. A. (2010). Bacterial biosynthesis of alginates. J. Chem. Technol. Biotechnol. 85, 752-759. https://doi.org/10.1002/jctb.2372

Hublik, G. (2012). Xanthan. In Polymer Science: A Comprehensive Reference, Matyjaszewski and Möller, eds. (Amsterdam: Elsevier), pp 221-229. https://doi.org/10.1016/B978-0-444-53349-4.00262-4

Imeson, A. (2010). Food stabilisers, thickeners and gelling agents. Blackwell Publishing Ltd, Oxford.

Islam S.T., and Lam, J.S. (2014). Synthesis of bacterial polysaccharides via the Wzx/Wzy-dependent pathway. Can. J. Microbiol. 60 (11), 697-716. https://doi.org/10.1139/cjm-2014-0595

Juez, G., Rodriguez-Valera, F., Ventosa, A., and Kushner, D. J. (1986). *Haloarcula hispanica* sp. nov. and *Haloferax gibbonsii* sp. nov., two new species of extremely halophilic archaebacteria. Syst. Appl. Microbiol. 8, 75-79, https://doi.org/10.1016/S0723-2020(86)80152-7

Kumar, A. S., and Modi, K. (2009). Microbial exopolysaccharides: Variety and potential application. In Microbial Production of Biopolymers and Polymer Precursors: Applications and Perspectives, H. Bernd, and A. Rehm, Eds. (Norfolk, UK: Caister Academic Press), pp. 229-254.

Kucukasik, F., Kazak, H., Guney, D., Finore, I., Poli, A., Yenigun, O., and Oner, E. T. (2011). Molasses as fermentation substrate for levan production by *Halomonas* sp. Appl. Microbiol. Biotechnol. 89(6), 1729-1740. https://doi.org/10.1007/s00253-010-3055-8

Lee, H.K., Chun, J., Moon, E.Y., Ko, S.H., Lee, D.S., Lee, H.S., and Bae, K.S. (2001). *Hahella chejuensis* gen. nov., sp. nov., an extracellular-polysaccharide-producing marine bacterium. Int. J. Syst. Evol. Microbiol. https://doi.org/10.1099/00207713-51-2-661

Li, P., Liu, Z., and Xu, R. (2001). Chemical characterisation of the released polysaccharide from the cyanobacterium *Aphanothece halophytica* GR02. J. Appl. Phycol. 13(1), 71-77 https://doi.org/10.1023/a:1008109501066

Llamas, I., Amjres, H., Mata, J. A., Quesada, E., and Béjar, V. (2012). The Potential Biotechnological Applications of the Exopolysaccharide Produced by the Halophilic Bacterium *Halomonas almeriensis*. Molecules, 17(6), 7103.

Llamas, I., Mata, J. A., Tallon, R., Bressollier, P., Urdaci, M. C., Quesada, E., and Béjar, V. (2010). Characterization of the Exopolysaccharide Produced by *Salipiger mucosus* A3(T), a Halophilic Species Belonging to the *Alphaproteobacteria*, Isolated on the Spanish Mediterranean Seaboard. Marine Drugs, 8(8), 2240-2251. https://doi.org/10.3390/md8082240

Lü, Y., Lu, H., Wang, S., Han, J., Xiang, H., and Jin, C. (2017). An Acidic Exopolysaccharide from *Haloarcula hispanica* ATCC33960 and Two Genes Responsible for Its Synthesis. Archaea, 5842958. https://doi.org/10.1155/2017/5842958

Madhuri, K.V., and Prabhakar, K.V. (2014). Microbial Exopolysaccharides: Biosynthesis and Potential Applications. Orien. J. Chem. 30 (3), 1401-1410. http://doi.org/10.13005/ojc/300362

Martínez-Cánovas, M. J., Quesada, E., Llamas, I., and Béjar, V. (2004a). *Halomonas ventosae* sp. nov., a moderately halophilic, denitrifying, exopolysaccharide-producing bacterium. IJSEM, 54(3), 733-737. http://doi.org/10.1099/ijs.0.02942-0

Martínez-Cánovas, M. J., Quesada, E., Martínez-Checa, F., Moral, A. D., and Béjar, V. (2004b). *Salipiger mucescens* gen. nov., sp. nov., a moderately halophilic, exopolysaccharide-producing bacterium isolated from hypersaline soil, belonging to the α-Proteobacteria. IJSEM, 54(5), 1735-1740. http://doi.org/10.1099/ijs.0.63166-0

Martínez-Cánovas, M.J., Béjar, V., Martínez-Checa, F., and Quesada, E. (2004c). *Halomonas anticariensis* sp. nov., from Fuente de Piedra, a saline-wetland wildfowl reserve in Malaga, southern Spain. IJSEM, 54, 1329-1332, http://doi.org/10.1099/ijs.0.63108-0

Martínez-Cánovas, M.J., Béjar, V., Martínez -Checa, F., Paez, R. and Quesada, E. (2004d). *Idiomarina fontislapidosi* sp. nov., and *Idiomarina ramblicola* sp. nov., isolated from inland hypersaline habitats in Spain. IJSEM, 54, 1793–1797, http://doi.org/10.1099/ ijs.0.63172-0

Martínez-Checa, F., Bejar, V., Llamas, I., del Moral, A. and Quesada, E. (2005). *Alteromonas hispanica* sp. nov., a polyunsaturated- fatty-acid-producing, halophilic becatrium isolated from Fuente de Piedra, southern Spain. IJSEM, 55(Pt 6), 2385-90, https://doi.org/10.1099/ijs.0.63809-0

Marx, J. G., Carpenter, S. D., and Deming, J. W. (2009). Production of cryoprotectant extracellular polysaccharide substances (EPS) by the marine psychrophilic bacterium *Colwellia psychrerythraea* strain 34H under extreme conditions. Canad. J. Microbiol. 55(1), 63-72. https://doi.org/10.1139/w08-130

Mata, J. A., Béjar, V., Bressollier, P., Tallon, R., Urdaci, M. C., Quesada, E., and Llamas, I. (2008). Characterization of exopolysaccharides produced by three moderately halophilic bacteria belonging to the family *Alteromonadaceae*. J. Appl. Microbiol. 105(2), 521-528. https://doi.org/10.1111/j.1365-2672.2008.03789.x

Mata, J. A., Béjar, V., Llamas, I., Arias, S., Bressollier, P., Tallon, R., and Quesada, E. (2006). Exopolysaccharides produced by the recently described halophilic bacteria *Halomonas ventosae* and *Halomonas anticariensis*. Res. Microbiol. 157(9), 827-835. https://doi.org/10.1016/j.resmic.2006.06.004

Maugeri, T. L., Gugliandolo, C., Caccamo, D., Panico, A., Lama, L., Gambacorta, A., and Nicolaus, B. (2002). A halophilic thermotolerant *Bacillus* isolated from a marine hot spring able to produce a new exopolysaccharide. Biotechnol. Lett. 24(7), 515-519. https://doi.org/10.1023/a:1014891431233

Morris, G.A., Li, P., Puaud, M., Liu, Z, Mitchell, J.R., and Harding, S.E. (2000). Hydrodynamic characterisation of the exopolysaccharide from the halophilic cyanobacterium *Aphanothece halophytica* GR02: a comparison with xanthan. Carbohydrate Polymers. DOI

Nicolaus, B., Lama, L., Esposito, E., Manca, M. C., Improta, R., Bellitti, M. R., and Gambacorta, A. (1999). *Haloarcula* spp. able to biosynthesize exo- and endopolymers. J. Ind. Microbiol. Biotechnol. 23(6), 489-496. https://doi.org/10.1038/sj.jim.2900738

Nicolaus, B., Schiano Moriello, V., Lama, L., Poli, A. and Gambacorta, A. (2004). Polysaccharides from extremophilic microorganisms. OLEB, 34, 159–169. https://doi.org/10.1023/B:ORIG.0000009837.37412.d3

Nicolaus, B., Kambourova, M., and Oner, E. T. (2010). Exopolysaccharides from extremophiles: from fundamentals to biotechnology. Environ. Technol. 31(10), 1145-1158. https://doi.org/10.1080/09593330903552094

Oren, A. (2006). Life at high salt conditions. In The Prokaryotes. A Handbook on the Biology of Bacteria: Ecophysiology and Biochemistry; Dworkin, M., Falkow, S., Rosenberg, E., Schleifer, K.H., Stackebrandt, E., Eds. (New York, NY, USA: Springer), Volume 2, pp. 263–282. ISBN: 978-0-387-30742-8.

Oren, A. (2010). Industrial and environmental applications of halophilic microorganisms. Environ. Techn. 31, 825–834. https://doi.org/10.1080/09593330903370026

Oren, A. (2013). Life at high salt concentrations, intracellular KCl concentrations, and acidic proteomes. Front. Microbiol. 4, 315. https://doi.org/10.3389/fmicb.2013.00315

Parolis, H., Parolis, L. A. S., Boán, I. F., Rodríguez-Valera, F., Widmalm, G., Manca, M. C., and Sutherland, I. W. (1996). The structure of the exopolysaccharide produced by the halophilic Archaeon *Haloferax mediterranei* strain R4 (ATCC 33500). Carbohyd. Res. 295(Supplement C), 147-156. https://doi.org/10.1016/S0008-6215(96)90134-2

Poli, A., Moriello, V. S., Esposito, E., Lama, L., Gambacorta, A., and Nicolaus, B. (2004). Exopolysaccharide production by a new *Halomonas* strain CRSS isolated from saline lake Cape Russell in Antarctica growing on complex and defined media. Biotechnol. Lett. 26(21), 1635-1638. http://dx.doi.org/10.1007/s10529-004-3187-y

Poli, A., Esposito, E., Orlando, P., Lama, L., Giordano, A., de Appolonia, F., and Gambacorta, A. (2007). *Halomonas alkaliantarctica* sp. nov., isolated from saline lake Cape Russell in Antarctica, an alkalophilic moderately halophilic, exopolysaccharide-producing bacterium. Syst. Appl. Microbiol. 30(1), 31-38. http://dx.doi.org/10.1016/j.syapm.2006.03.003

Poli, A., Anzelmo, G., and Nicolaus, B. (2010). Bacterial Exopolysaccharides from Extreme Marine Habitats: Production, Characterization and Biological Activities. Marine Drugs, 8(6), 1779. https://doi.org/10.3390/md8061779

Poli, A., Nicolaus, B., Denizci, A. A., Yavuzturk, B., and Kazan, D. (2013). *Halomonas smyrnensis* sp. nov., a moderately halophilic, exopolysaccharide-producing bacterium. IJSEM 63(1), 10-18. http://dx.doi.org/10.1099/ijs.0.037036-0

Poli, A., Finore, I., Romano, I., Gioiello, A., Lama, L., and Nicolaus, B. (2017). Microbial diversity in extreme marine habitats and their biomolecules. Microorganisms. 5(2), 25; https://doi.org/10.3390/microorganisms5020025

Raveendran, S., Palaninathan, V., Chauhan, N., Sakamoto, Y., Yoshida, Y., Maekawa, T., Mohanan, P.V. and Kumar, D.S. (2013a). In vitro evaluation of antioxidant defense mechanism and hemocompatibility of mauran. Carboh. Polym. 98, 108– 115. http://dx.doi.org/10.1016/j.carbpol.2013.04.050

Raveendran, S., Dhandayuthapani, B., Nagaoka, Y., Yoshida, Y. and Kumar, D.S. (2013b). Biocompatible nanofibers based on extremophilic bacterial polysaccharide, Mauran from *Halomonas maura*. Carboh. Polym. 92, 1225– 1233. https://doi.org/10.1016/j.carbpol.2012.10.033

Rehm, B.H.A. (2015). Synthetic biology towards the synthesis of custom-made polysaccharides. Microb. Biotechnol. 8 (1),19-20. https://doi.org/10.1111/1751-7915.12241

Ruiz-Ruiz, C., Srivastava, G.K., Carranza, D., Mata, J.A., Llamas, I., Santamaría, M., Quesada, E. and Molina, I.J. (2011) An exopolysaccharide produced by the novel halophilic bacterium Halomonas stenophila strain B100 selectively induces apoptosis in human T leukaemia cells. Appl Microbiol Biotechnol 89, 345–355. https://doi.org/10.1007/s00253-010-2886-7

Ruffing, A. M., and Chen, R. R. (2012). Transcriptome profiling of a curdlan producing Agrobacterium reveals conserved regulatory mechanisms of exopolysaccharide biosynthesis. Microb. Cell Fact. 11, 17. https://doi.org/10.1186/1475-2859-11-17

Romano, I., Lama, L., Nicolaus, B., Poli, A., Gambacorta, A., and Giordano, A. (2006a). *Halomonas alkaliphila* sp. nov., a novel halotolerant alkaliphilic bacterium isolated from a salt pool in Campania (Italy). J. Gen. Appl. Microbiol. 52(6), 339-348. https://doi.org/10.2323/jgam.52.339

Romano, I., Lama, L., Nicolaus, B., Poli, A., Gambacorta, A., and Giordano, A. (2006b). *Oceanobacillus oncorhynchi* subsp. *incaldanensis* subsp. nov., an alkalitolerant halophile isolated from an algal mat collected from a sulfurous spring in Campania (Italy), and emended description of *Oceanobacillus oncorhynchi*. IJSEM, 56(4), 805-810. https://doi.org/10.1099/ijs.0.63575-0

Schmid, J., Sieber, V., and Rehm, B. (2015). Bacterial exopolysaccharides: biosynthesis pathways and engineering strategies. Front. Microbiol. 6, 496. https://doi.org/10.3389/fmicb.2015.00496

Silvi, S., Barghini, P., Aquilanti, A., Juarez-Jimenez, B., and Fenice, M. (2013). Physiologic and metabolic characterization of a new marine isolate (BM39) of *Pantoea* sp. producing high levels of exopolysaccharide. Microbial Cell Fact. 12(1), 10. https://doi.org/10.1186/1475-2859-12-10

Squillaci, G., Finamore, R., Diana, P., Restaino, O. F., Schiraldi, C., Arbucci, S., and Morana, A. (2016). Production and properties of an exopolysaccharide synthesized by the extreme halophilic archaeon *Haloterrigena turkmenica*. Appl. Microbiol. Biotechnol. 100(2), 613-623. https://doi.org/10.1007/s00253-015-6991-5

Sudo, H., Burgess, J. G., Takemasa, H., Nakamura, N., and Matsunaga, T. (1995). Sulfated exopolysaccharide production by the halophilic cyanobacterium *Aphanocapsa halophytia*. Curr. Microbiol. 30(4), 219-222. https://doi.org/10.1007/bf00293636

Sutherland, I. (2002). A sticky business. Microbial polysaccharides: current products and future trends. Microbiol. Today 29, 70-71.

Tian, F., Karboune, S., and Hill, A. (2014). Synthesis of fructooligosaccharides and oligolevans by the combined use of levansucrase and endo-inulinase in one-step bi-enzymatic system. IFSET 22 (Supplement C), 230-238. https://doi.org/10.1016/j.ifset.2013.12.004

Umezawa, H., Okami, Y., Kurasawa, S., Ohnuki, T., Ishizuka, M., Takeuchi, T.U., Shiio, T., and Yugari, Y. (1983). Marinactan, antitumor polysaccharide produced by marine bacteria. J. Antibiot. 36(5), 471-7.

van Kranenburg, R., Boels, I.C., Kleerebezem, M., and de Vos, W.M. (1999). Genetics and engineering of microbial exopolysaccharides for food: approaches for the production of existing and novel polysaccharides. Curr. Opinion Biotechnol 10 (5), 498-504. https://doi.org/10.1016/S0958-1669(99)00017-8

Ventosa, A., Gutierrez, M.C., Kamekura, M., Dyall-Smith, M.L. (1999). Proposal to transfer *Halococcus turkmenicus, Halobacterium trapanicum* JCM 9743 and strain GSL-11 to *Haloterrigena turkmenica* gen. nov., comb. nov. Int J Syst Bacteriol. DOI 10.1007/s00253-015-6991-5

Vorhölter, F. J., Schneiker, S., Goesmann, A., Krause, L., Bekel, T., Kaiser, O., Linke, B., Patschkowski, T., Rückert, C., Schmid, J., Sidhu, V. K., Sieber, V., Tauch, A., Watt, S. A., Weisshaar, B., Becker, A., Niehaus, K., and Pühler, A. (2008). The genome of *Xanthomonas campestris* pv. *campestris* B100 and its use for the reconstruction of metabolic pathways involved in xanthan biosynthesis. J. Biotechnol. 134, 33-45. https://doi.org/10.1016/j.jbiotec.2007.12.013

Yang, Lu, H., Wang, S., Han, J., Xiang, H., and Jin, C. (2017). An Acidic Exopolysaccharide from *Haloarcula hispanica* ATCC33960 and Two Genes Responsible for Its Synthesis. Archaea, 12. https://doi.org/10.1155/2017/5842958

Wang, J., Salem, D.R., and Sani, R.K. (2018). Extremophilic Exopolysaccharides: A Review and New Perspectives on Engineering Strategies and Applications, Carbohyd. Polym. 205, 8-26. https://doi.org/10.1016/j.carbpol.2018.10.011

Whitney, J.C., and Howell, P.L. (2013). Synthase-dependent exopolysaccharide secretion in Gram-negative bacteria. Trends Microbiol. 21 (2), 63-72. https://doi.org/10.1016/j.tim.2012.10.001

Willis, L. M., and Whitfield, C. (2013). Structure, biosynthesis, and function of bacterial capsular polysaccharides synthesized by ABC transporter-dependent pathways. Carbohydr. Res. 378, 35–44. https://doi.org/10.1016/j.carres.2013.05.007

Zhao, D., Cai, L., Wu, J., Li, M., Liu, H., Han, J., Zhou, J., and Xiang, H. (2013). Improving polyhydroxyalkanoate production by knocking out the genes involved in exopolysaccharide biosynthesis in *Haloferax mediterranei*. Appl. Microbiol. Biotechnol. 97 (7), 3027-3036. https://doi.org/10.1007/s00253-012-4415-3

Exopolysaccharide Production by Thermophilic Microorganisms

3

Songul Yasar Yildiz*

*Department of Bioengineering, Istanbul Medeniyet University, Istanbul, Turkey

DOI: https://doi.org/10.21775/9781912530267.03

Abstract

In recent years, the demand for natural polymers for various industrial applications led to increased interest in microbial exopolysaccharide (EPS) production. Several microorganisms have the ability to synthesize extracellular polysaccharides either as soluble or insoluble polymers. Among those, EPS from thermophilic microorganisms are of special interest owing to its advantages such as short fermentation time, better oxygen mass transfer, decreased viscosity of synthesized polymer and the corresponding culture and the non-toxic nature of the polymer allowing applications in food and pharmaceutical industries. Despite this interest, there is very limited knowledge on EPS biosynthesis mechanisms by thermophilic bacteria, as opposed to the vast amount of reports on mesophilic EPS producers. In this chapter, a brief overview of life under extreme environmental conditions especially high temperatures is presented. This is followed by a discussion of thermophilic microorganisms and their adaptation mechanisms, and specifically focuses on the production of EPSs and their ecological and physiological functions. The application areas of industrially important EPSs from various thermophilic producer strains are also mentioned.

Introduction

The first microbial exopolysaccharide (EPS) was discovered in the mid-19th century in wine. Later, the discovered EPS was identified as dextran synthesized by *Leuconostoc mesenteriode* (Linker and Jones, 1966; Rehm, 2010). In the following years, other EPSs reported such as xhantan, cellulose, alginate etc.. Microbial EPSs are natural, non-toxic, biocompatible and biodegradable biopolymers with many interesting physicochemical and rheological properties. These polymers find wide range of applications in many industrial sectors including textiles, detergents, adhesives, microbial enhanced oil recovery (MEOR), wastewater treatment, dredging, brewing, downstream processing, cosmetology, pharmacology, and food additives (Özcan and Toksoy Oner, 2015). Although, several different types of EPSs have been reported up to date, few of them have been indicated to have significant commercial value in industry or medicine (Costerton et al., 1999). High cost of production compared to their commercial value is the main limitation factor for the use of these microbial polysaccharides. However, using cheaper substrates, improving product yield by optimizing fermentation conditions, or developing higher yielding strains via mutagenesis, and/or genetic and metabolic manipulations, and optimizing downstream processing are the potential solutions to address this problem (Nwodo et al., 2012). On the other hand, the possession of novel properties by the EPSs, which may not be found in polysaccharides produced by plant and algae would invariably translate to high-value applications; thus, product quality wholly surpasses production cost (Costerton et al., 1999; Nwodo et al., 2012). At present, polymers from microbial origin represent only a small fraction of the current global polymer market, but the number and applications of them are gradually increasing. Xanthan, dextran and pullulan are examples of microbial polysaccharides with a considerable market due to their exceptional properties. Biopolymers produced by diverse groups of microorganisms are rapidly emerging as a new and industrially important source of polymeric materials and they are gradually becoming economically competitive with natural gums already presented on the market. Mesophilic biopolymers with properties superior to the commercial ones cannot preserve their functions under industrial process conditions like extremes of temperature, salinity or pH. Hence, most research is focused on the identification of EPS-producing extremophiles with the idea that as these microorganisms survive environmental extremes of desiccation, temperature, pressure, salinity, acidity, heavy metals, and radiation, it is expected that their biopolymers will also have some unique properties to adapt to such extreme conditions. Moreover, almost all commercial EPSs like xanthan and dextran are produced by pathogenic mesophiles; therefore finding non-pathogenic extremophilic strains with high level EPS production ability holds significant industrial importance (Kazak et al., 2010). Extremophiles are organisms that can survive in a variety of extreme environments and they can be belong to Archaea or Bacteria. These extremophiles are classified into several groups such as thermophiles, psychrophiles, halophiles, acidophiles, alkaliphiles, barophiles, radiophiles, etc. (Nicolaus et al., 2004). Since 1991, EPS producing extremophilic microorganisms have been isolated from deep-sea hydrothermal vents characterized by extreme pressure and temperature, high concentrations of H_2S and heavy metals. These include some bacteria such as *Alteromonas macleodii* sub spec *fijiensis*, *Vibrio diabolicus*, *Alteromonas infernos* (Raguénès et al., 1997; Roger et al., 2004) and *Thermotoga maritima* (Nicolaus et al., 2010; Rinker and Kelly, 2000) and archaea such as *Thermococcus litoralis* (Rinker and Kelly, 2000). EPS producing bacteria from marine Mediterranean shallow vents have also been described (Moriello et al., 2003). The thermophilic bacilli *Bacillus thermoantarcticus*, *Geobacillus thermodenitrificans* and *B. licheniformis* were isolated from hot marine shallow vents as producers of large amounts of EPS (Manca et al., 1996).

Extremophiles

Human life is sustained under moderate environments which are generally described by conditions with temperatures between 4 and 40°C, pH near neutral, pressure around 1 atm, water, nutrients and salts at adequate levels, and low ionizing radiations and hydrostatic pressure. Thus ecological systems such as hot springs, salt and soda lakes, deserts and ocean beds and deep thermal vents that are not compatible with the growth and survival of human beings are considered as being "*extreme*" (Brock, 2012; Satyanarayana et al., 2005). Extreme conditions can refer to physical extremes such as temperature, radiation or pressure, but also to geochemical extremes such as salinity, oxygen tension, desiccation and pH.

A variety of microorganisms, known as extremophiles, not only tolerate such extreme conditions, but usually require environmental extremes for their survival and growth. MacElroy coined the term "*extremophile*" in 1974, and from the mid 1970's onwards, a growing number of extreme environments, formerly inaccessible due to technical limitations, are now being explored, which has led to the isolation and identification of a wealth of new organisms (Grant and Horikoshi, 1998; Rainey and Oren, 2006). Extremophiles are classified and named according to the conditions under which they exist (Table 1).

Table 1. Classification of extremophiles.

Environmental Parameter	Phenotype	Condition
Temperature	Hyperthermophile	Growth >80 °C
	Thermophile	Growth 60-80 °C
	Mesophile	15-60 °C
	Psychrophile	<15 °C
Radiation	Radiophiles	
Pressure	Barophile	Weight-loving
	Piezophile	Pressure-lowing
Desiccation	Xerophiles	Anhydrobiotic
Salinity	Halophile	Salt-loving (2-5 M)
pH	Alkalophile	pH>9
	Acidophile	Low pH-loving
Osmotic pressure	Osmophile	Survive in high sugar environments
Chemicals	Metallophile	Able to tolerate high concentrations of metal
	Toxitolerant	Tolerates toxic and xenobiotic chemicals

Due to the higher organisms' cellular complexity and compartmentalization, they were believed to be unable to withstand extreme conditions. Therefore, research on extremophiles has focused mainly on organisms belonging to the Archaea and Bacteria kingdoms. Rather as an exception, number of eukaryotes, both unicellular and multicellular, have evolved to live and thrive in extreme environments (e.g. *Alvinella pompejana, Tetrahymena thermophila* and *Dunaliella salina*). Hence, extremophiles span all three domains of life: Archaea, Bacteria and Eukarya (Rothschild and Mancinelli, 2001). As a result of adaptation to their environment, many extremophilic microorganisms have evolved unique properties of considerable biotechnological importance and, therefore, commercial significance. It is therefore widely accepted that extremophiles provide a valuable resource for exploitation in novel biotechnological processes (Podar and Reysenbach, 2006). The applications of extremophiles, their enzymes (extremozymes) and biopolymers are as varied as their environments. Oil degrading extremophiles are used as oil recovery surfactants, methanogens in methane production, psychrophiles in plant frost protection and sulfur oxidizing ones are used in bioleaching, coal and waste gas desulfurization. The marine microalgae *Dunaliella salina* in whole dried form is taken as a natural dietary supplement for carotenoids and daily nutrients. Whereas S-layer protein and lipids of hyperthermophiles are employed as molecular sieves, as liposomes in drug delivery and cosmetic packaging, bacteriorhodopsin of halophiles are utilized in bioelectronic optical switches and photocurrent generators. Additionally, biodegradable plastics from extremophilic microorganisms have found many applications in waste treatment and in the biomedical sector. Ectoine, an osmoprotectant accumulated by many halophilic bacteria, is used both as a stabilizer of enzyme preparations and as an additive to skin treatment cosmetics (Buenger and Driller, 2004). Industrially important *extremozymes* include DNA polymerases (such as *Taq* DNA polymerase) and alkaline phosphatases that are used in diagnostics procedures, while amylases, pullulanases, proteases and lipases are utilized in cheese manufacturing, dairy processing and detergents. Amylolytic enzymes are used in starch processing, proteases find applications in contact-lens cleaning solutions, in gelatin removal from X-ray films, and in de-hairing animal hides for leather production, and oxidases are used in the preparation of environmental biosensors (Demirjian et al., 2001; Margesin and Schinner, 2001).

Microbial extracellular polysaccharides are known to be involved in protecting cells from numerous external stresses (Nicolaus et al., 2004; Wharton, 2007). EPSs can absorb water and form a highly hydrated matrix. The presence of such a gelled polysaccharide layer around the cell may affect the diffusion of compounds both into and out of the cell, and hence act as a layer of protection to cells against toxic compounds (Bitton and Freihofer, 1977; Jeanthon and Prieur, 1990), very high acidity (Aguilera et. al., 2008), UV radiation (Wang et al., 2007) or digestion by other organisms (Caron, 1987). The specific role of the EPS depends on the natural environment of the microorganism. Changes in pH and salinity have little effect on the viscosity and stability of the EPS layer. Besides their buffering capacity, EPSs may also reduce the rate of water loss during desiccation and assist with water uptake during rehydration. Polymers made up of fructose residues (fructans) are known to be involved in the protection of plants during drought, salt or cold stress by preventing membrane damage (Ritsema and Smeekens, 2003). High concentrations of EPS with high polyhydroxyl content would decrease the freezing point of water in the low temperature, high salinity brine channels and hence act as a cryoprotectant. For example; EPS from a fungal strain, *Phoma herbarum,* isolated from Antarctic soil provides a cryoprotective role in the harsh Antarctic environment where the availability of liquid water and temperatures are extremely low (Selbmann et al., 2002).

Microbial EPSs serve countless ecological functions like stabilizing marine sediments and preventing erosion, improvement of water holding capacity of soil, detoxification of heavy metals and removal of solid matter from water reservoirs (Bender and Phillips, 2004). They form biofilms and aggregates facilitating intracellular interaction and biochemical exchange between organisms (Paerl, 1976; Wotton, 2004). EPSs also contribute to various physiological activities in human beings as anti-tumor, anti-viral, and anti-inflammatory agents, and can act as inducers for interferon, platelet aggregation inhibition, and colony stimulating factor synthesis (Kumar et al., 2004).

Themophilic Microorganisms

Habitats of thermophiles

There are different types of thermal environments on Earth that are populated by microorganisms adapted to life at high temperatures. Many of these environments are associated with volcanic activity. Marine hot springs, geysers, volcanoes and deep-sea hydrothermal vents have all been popular hunting grounds for searching new types of thermophilic microorganisms (Wang et al., 2018). However, thermophiles can also be found in environments such as compost heaps and coal refuse piles, where the high temperature is entirely due to biological activity. Furthermore, thermophiles can survive in man-made environment such as hot water systems of industrial or domestic areas and industrial processes (e.g. paper processing machinery and food processing facilities). Hot springs near to volcanic environments are usually acidic (Urbieta et al., 2014), however, in limestone close regions pH is neutral or slightly alkaline. Thermophiles may also survive under harsh conditions involving extreme pH or high salt concentrations (Urbieta et al., 2015). Generally, microorganisms with an optimal growth temperature between 50 and 80°C are designated as thermophiles, whereas those growing optimally at temperatures higher than 80°C are referred to as hyperthermophiles, which are found in the three domains of life, archaea, bacteria, and eukarya, but the majority are archaea and bacteria (Figure 1)

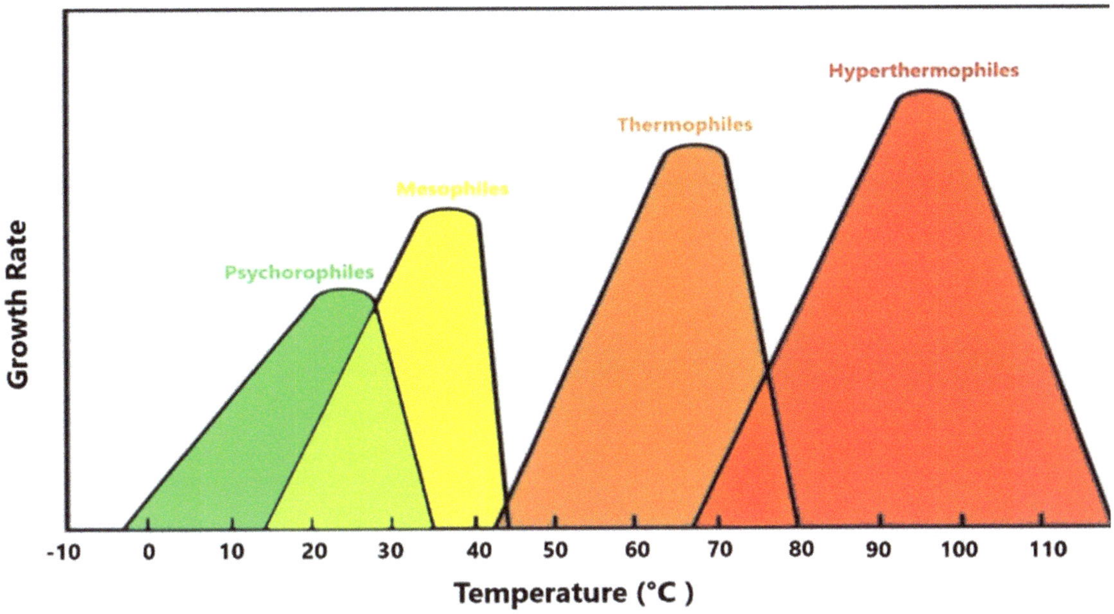

Figure 1: Relation of temperature and growth rates for a typical psychrophilic, mesophilic, thermophilic and hyperthermophilic microorganisms.

EPS Production by Thermohilic Microorganisms

Thermophilic microorganisms proliferated in a wide range of temperature, from 50–60 °C of thermotolerant microorganisms up to 122 °C of hyperthermophile species (Kambourova et al., 2016; Poli et al., 2018). The capability to synthesize exopolysaccharides has been observed for microorganisms belonging to both Archaea and Bacteria domains and in all kind of extremophilic microorganisms. EPSs take role in the survival of the microbial cells as a protector against predators, reservoir of the energy and carbon source and source for the regular nutrient uptake even in environments wherein they would tend to be dispersed (Poli et al., 2018). All EPS producer thermophilic microorganisms are summerized in Table 2.

EPS Production by Bacteria

Aeribacillus pallidus 418 has emerged as an EPS producer among forty-one water, soil, and algobacterial mat samples (Radchenkova et al., 2013). The strain grows at 55°C and uses maltose as a carbon source for the highest amount of EPS. Purification of EPS showed that *Aeribacillus pallidus* 418 synthesize two different

Table 2. Classification of thermophilic EPS producers.

	Microorganism	Carbon source	Growth temperature (°C)	Isolation source	Reference
Bacteria	*Aeribacillus pallidus* 418	Maltose	55	Hot spring, Rupi basin, Bulgaria	(Radchenkova et al., 2014; Radchenkova et al., 2013)
	Bacillus licheniformis B3-15	Glucose	45	Shallow marine hot spring, Vulcano, Italy	(Arena et al., 2006; Maugeri et al., 2002)
	Brevibacillus thermoruber 423	Maltose	55	Hot spring, Blagoevgrad region,Bulgaria	(Yasar Yildiz et al., 2014)
	Geobacillus sp. 4001	Galactose, sucrose	65	Shallow marine hydrothermal vents of Flegrean Ares, Italy	(Nicolaus et al., 2002; Nicolaus et al., 2004)
	Geobacillus sp. 4004	Trehalose, sucrose	65	Shallow marine hydrothermal vents of Flegrean Ares, Italy	(Nicolaus et al., 2002; Nicolaus et al., 2004)
	Geobacillus tepidamans V264	Maltose	60	Hot spring, Velingrad	(Kambourova et al., 2009)
	Geobacillus thermoantarcticus	Mannose	65	Crater of Mount Melbourne, Antarctica	(Manca et al., 1996; Nicolaus et al., 2004)
	Geobacillus thermodenitrificans ArzA-6	Fructose	65	Geothermal spring, Arzakan, Armenian	(Panosyan et al., 2018)
	Geobacillus thermodenitrificans B3-72	Glucose, sucrose	65	Shallow hydrothermal vent, Vulcano island, Italy	(Arena et al., 2009; Nicolaus et al., 2000)
	Geobacillus toebii ArzA-8	Fructose	65	Geothermal spring, Arzakan, Armenian	(Panosyan et al., 2018)
	Rhodothermus marinus DSM 4252^{T}	Glucose, maltose, lactose, sucrose	65	Hot spring, Iceland	(Sardari et al., 2017)
	Rhodothermus marinus MAT 493	Glucose, maltose, lactose, sucrose	65	Hot spring, Iceland	(Sardari et al., 2017)
	Thermotoga maritima	Maltose	88	Geothermal heated marine sediment at Vulcano, Italy	(Rinker and Kelly, 2000)
	Thermus aquaticus YT-1	n.d.	60	Hot spring, Yellowstone, US	(Lin et al., 2011)
Archaea	*Sulfolobus acidocaldarius*	n.d.	76	Hot spring, Yellowstone, US	(Koerdt et al., 2010)
	Sulfolobus solfataricus MT3	Glucose	75	Hot acidic spring, Agnano, Italy	(Nicolaus et al., 1993)
	Sulfolobus solfataricus MT4	Glucose	88	Hot acidic spring, Agnano, Italy	(Nicolaus et al., 1993)
	Sulfolobus tokodaii	n.d.	76	Beppu hot springs, Japan	(Koerdt et al., 2010)
	Thermococcus litoralis	Maltose	88	Shallow marine, thermal spring, Naples, Italy	(Rinker and Kelly, 1996, 2000)

n.d. no data

EPS which constituted by similar sugars, however in different proportions. EPS-1 consists of mannose/glucose/galactosamine/glucosamine/galactose/ribose (69.3/11.2/6.3/5.4/4.7/2.9) and EPS-2 consists of manose/galactose/glucose/ galactosamine/glucosamine/ribose/arabinose (33.9/17.9/15.5/11.7/8.1/5.3/4.9) (Radchenkova et al., 2013). In the further studies, stable oil-in-water emulsions, a low level of separated water phase and high dispersion stability of EPS were reported (Radchenkova et al., 2014).

EPS production by *Bacillus licheniformis* B3-15 from Vulcano Island (Italy) was reported in 2002 (Maugeri et al., 2002). The EPS produced by this bacterium is of new composition which possesses a tetrasaccharide repeating unit essentially constituted by sugars with a manno-pyranosidic configuration. In the study of Arena et al., antiviral and immunomodulatory effects of the EPS were evaluated. It was found that EPS is able to inhibit HSV-2 replication in PBMC by up-regulating expression of peculiar proinflammatory cytokines (Arena et al., 2006).

Brevibacillus thermoruber 423 is reported as a high amount of EPS (863 mg L^{-1}) producer and prove to be a very promising cell factory for microbial EPS production. (Yasar Yildiz et al., 2014). Further studies focused on understanding the polysaccharide production mechanisms in *B. thermoruber* strain 423 and their possible improvement by using whole genome sequence of the bacterium (Yasar Yildiz et al., 2013; Yasar Yildiz et al., 2015).

Nicolaus et al. investigated 14 isolates for EPS production. It is reported that the highest concentrations of EPSs harvested after 48 h of culture were approximately 60 and 50 μg/liter for isolates *Geobacillus sp* 4001 and *Geobacillus sp.* 4004 respectively (Nicolaus et al., 2002). Carbon sources for the highest amount of EPS production are galactose and saccharose for 4001 strain and trehalose and saccharose for 4004 strain (Nicolaus et al., 2004).

Extracellular polysaccharide production from a thermophilic bacterium identified as *Geobacillus tepidamans* V264 and chemical/physical characterization of the purified polymer was reported in 2009 (Kambourova et al., 2009). The reported properties of the polymer such as high viscosity, rigidity and anti-cytotoxicity provide a potential for future application in industrial processes.

The thermophilic bacterium *Geobacillus thermoantarcticus* produces two different EPSs. EPS-1 is a heteropolysaccharide whose repeating unit consists of four different α-D-mannoses and three different β-D-glucoses. EPS-2 is a mannan with four different α-D-mannoses as the repeating unit (Manca et al., 1996).

In a recent study, sixteen thermophilic isolates were investigated for the EPS production. *Geobacillus thermodenitrificans* ArzA-6 and *Geobacillus toebii* ArzA-8 were selected as a two best EPSs producer for the further studies (Panosyan et al., 2018). Fuctose was used as a sole carbon source for both strain and the highest specific EPSs production yields were reported as 0.27 g g^{-1} dry cells and 0.22 g g^{-1} dry cells for strains *G. thermodenitrificans* ArzA-6 and *G. toebii* ArzA-8, respectively. Mannose was found as a major monomer unit of both EPSs (*Geobacillus thermodenitrificans* ArzA-6 and *Geobacillus toebii* ArzA-8 synthesize heteropolymer of Man/Gal/Ara/Fru/Glc (1/0.13/0.1/0.06/0.05) and Man/Gal/Glc/Ara (1/0.5/0.2/0.05) respectively) (Panosyan et al., 2018).

Geobacillus thermodenitrificans B3-72 was reported as a producer of two different EPSs and both of them were based on mannose and glucose in a different ratio (Nicolaus et al., 2000). In the further studies, the effects of EPS-2 on the replication of HSV-2 in human PBMC and WISH cells were investigated. Moreover, it is determined whether the anti-HSV-2 activity observed only in PBMC could be related to its modulatory activity on cytokine production by the same cells (Arena et al., 2009). The results of the research indicate that the effects of this innovative compound may contribute to improve immune surveillance of PBMC toward virus infection, by triggering polarization in favour of the Th1 immune response.

Sardari et al. reported the thermophile *Rhodothermus marinus* as a producer of extracellular EPSs that forms a distinct cellular capsule. EPS from two strains of *R. marinus* namely DSM4252^T and MAT493 are reported and compared. Sulfated EPS with high arabinose and xylose content was synthesized by *R. marinus* (Arena et al., 2009) and the EPSs from *R. marinus* were defined as unique bacterial EPSs since arabinose and xylose are not common sugars in bacterial EPSs (Ahmed et al., 2013; Nichols et al., 2005). Besides, the ratio of monosaccharides also differs between the two strains, indicating that EPS from the respective strains may be useful for different purposes (Arena et al., 2009).

Rinker and Kelly investigated EPS production by *Thermotoga maritima* and effect of carbon and nitrogen sources on the production of EPS (Rinker and Kelly, 2000). Maltose and ammonium chloride were found as the best carbon and nitrogen sources respectively. The maximum yield of EPS production was reported as 120 mg L^{-1}.

Lin et al. isolated a novel extracellular polysaccharide from the biofilm of *Thermus aquaticus* YT-1 and evaluated its structure and immunomodulatory effects.The composition of the polysaccharide was reported as a tetra-saccharide repeating units of galactofuranose, galactopyranose, and N-acetyl-galactosamine (1:1:2) that lacked acidic sugars. The results of the study indicated that TA-1, the major EPS secreted by *T. aquaticus* YT-1, induces murine macrophage and human monocyte cell lines to secrete IL-6, TNF-α, and nitric oxide (Lin et al., 2011).

EPS Production by Archea

Microorganisms belongs to domain of *Archaea* are present in biofilm communities from many different environments (Krüger et al., 2008; Zhang et al., 2008). However, there are a few number of scientific reports on biofilm formation by the members of *Archaea*. The capability of *Sulfolobus solfataricus* to synthesize EPS was first reported more than 25 years ago (Nicolaus et al., 1993). Two strain of *S. solfataricus* namely MT3 and MT4 were investigated for their EPS production capacity. EPSs synthesized by two strains were have same monomers (glucose/mannose/glucosamine/galactose) with different ratio (1.2/1.0/0.18/0.13 and 1.2/1.0/0.77/0.73 respectively) (Nicolaus et al., 1993).

Three closely related hyperthermophilic crenarchaeotes: *Sulfolobus acidocaldarius* and *Sulfolobus tokodaii* together with of *S.solfataricus* were also reported to form biofilm (Koerdt et al., 2010). The effect of multiple environmental conditions on biofilm formation by these three Sulfolobus species was tested and results were indicated presence of extracellular polysaccharides containing glucose, galactose, mannose and N-acetylglucosamine in the structure of the biofilm. Biofilm formation is a probable stress response mechanism of *Archaea* because biofilm formation by *Archaeoglobus profundus*, *Archaeoglobus fulgidus*, *Thermococcus litoralis* were also observed (Kambourova et al., 2016).

The heterotrophic facultative sulfur-dependent hyperthermophile *Thermococcus litoralis* was also reported as a biofilm producer on hydrophilic surfaces under a variety of conditions (Rinker and Kelly, 1996). The EPS produced by *T. litoralis* was a homopolymer of mannose, that is interesting for *Archaea* since mannans are typically produced by eukaryotes, such as plants or yeasts. In the further studies, effect of carbon and nitrogen sources on EPS production were investigated (Rinker and Kelly, 2000).

EPS biosynthesis mechanisms of thermophilic microorganisms

High yield of bacterial EPSs is necessary for the further biotechnological and industrial applications. Traditional methods for yield elevation are strain selection and/or optimization of cultivation conditions. However, it is a known fact that each bacterium has physiological limits and it is difficult to overcome these limits by traditional methods (Freitas et al., 2011a). A number of enzymes and regulatory proteins take role in the regulation of bacterial EPSs biosynthesis (Jaiswal et al., 2014). Four different mechanisms are known for the polymerization of EPSs: extracellular biosynthesis, synthase dependent biosynthesis, ABC-transporter dependent and wzx/wzy- dependent pathways. EPS biosynthesis mechanism of mesophilic organisms including sugar nucleotide synthesis, repeating unit synthesis, and polymerization of the repeating units are studied and well known (De Vuyst et al., 2001; Freitas et al., 2011b). Similar to mesophilic pathway, a thermophilic EPS biosynthesis pathway (ABC-transporter dependent pathway) was reported as a hypothetical mechanism for sugar uptake and EPS biosynthesis by *Brevibacillus thermoruber* 423 (Yasar Yildiz et al., 2015). Furthermore, EPS biosynthesis pathway was also reported for the hyperthermophilic bacterium *Thermotoga maritima* (Johnson et al., 2005). Transcriptions of genes responsible for EPS biosynthesis by *T. maritima* was analyzed in the co-culture of *T. maritima* and *Methanococcus jannaschii*. A polypeptide encoding gene was strongly upregulated and a motif of this polypeptide is similar to that found in peptide signalling molecules in mesophilic bacteria. The complete EPS gene cluster of *Streptococcus thermophiles* was characterized. Products of gene clusters showed similarities with known glycosyltransferases (Delcour et al., 2000). The genes that take role in the biosynthesis of EPS are located on either chromosomes (Harutoshi, 2013; Yasar Yildiz et al., 2015) or plasmid (Dan et al., 2009; Laws et al., 2001). Genetic structure of the microorganism determines the composition and the amount of EPSs. However, type of strain, carbon and nitrogen sources, mineral salts, trace elements, the medium component ratio, fermentation conditions (temperature, pH, agitation and aeration) also affect the composition and the amount of EPSs (Nicolaus et al., 2010). Studies indicated that high C/N ratio increase the EPS production (Radchenkova et al., 2013; Yasar Yildiz et al., 2014). While the amount of nitrogen increases, EPS production decreases and biomass production increases. EPS synthesis can be either growth associated or non-growth associated. However, reports for thermophilic microorganisms indicate presence of growth associated EPS production (Kambourova et al., 2009; Radchenkova et al., 2013; Yasar Yildiz et al., 2014).

When compared with mesophilic producers, thermophiles produce low amount of EPSs usually varied in the range 50–200 mg L^{-1} (Kambourova et al., 2009; Manca et al., 1996; Nicolaus et al., 2003). A higher amount of EPS (366 mg L^{-1}) has been reported for the facultative thermophile *Bacillus licheniformis* (Spanò et al., 2013). Additionally, *B. thermoruber* 423 was reported as microbial cell factory due to its highest EPS production (863 mg L^{-1}) in comparison with other thermophiles (Yasar Yildiz et al., 2014).

The potential of thermophilic EPS in biotechnology

EPSs are natural, nontoxic and biodegradable polymers of sugar residues with high molecular weight. EPSs have a role in various functions such as cell-to-cell interactions, adhesion, biofilm formation and protection of cell against environmental extremes (Toksoy Oner, 2013). As a result of their physicochemical, biological and rheological properties, EPSs like pullulan, gellan, dextran, xanthan, and curdlan have found several applications in sectors including food, medicine, bionanotechnology, cosmetic and environmental industry. Plants or algea are cheaper counterparts of bacteria, hence it is not easy for microbial EPS production process to compete with them (Nicolaus et al., 2010; Toksoy Oner, 2013). Contrary to polysaccharides from plants, literature information on the structural characteristics of bacterial EPSs and especially EPSs produced by thermophiles is limited. New microbial expolysaccharides could come forward in the traditional polysaccharide market if cheap substrates are used and downstream processes with reduced costs, better functional or unique properties were developed (Kambourova et al., 2016). Genera *Bacillus*, *Geobacillus*, *Brevibacillus*, and *Aeribacillus*, that are thermophilic and genus *Thermus* that is extremely thermophilic and genus *Thermotoga* that is hyperthermohilic involves bacteria that are significant EPS producers (Kambourova et al., 2016). EPSs from thermophilic bacteria have several advantages when compared with EPSs from mesophiles. Thermophilic EPSs are thermostable with the highest thermostability reported for the EPS from *Geobacillus tepidamans* V264 (280

°C) (Kambourova et al., 2009). An additional important property of EPSs produced by thermophiles is high molecular weight of these polymers (beginning from several hundred kDa) (Kambourova et al., 2016). Since the growth rate is high at elevated temperature and nutrient components are lower in concentration, fermentation processes of thermophiles are short when compared with mesophiles (Kambourova et al., 2016). Mass transfer is better at high temperature for cultivation (Kumar et al., 2011; Turner et al., 2007). Lower energy consumption can be obtained since viscosity of culture liquid is lower at high temperature (Haki and Rakshit, 2003). Reduced risk of contamination at high temperature (Turner et al., 2007; Xiao et al., 2015). Thermophilic products are non-pathogenic, hence they are applicable in cosmetic and food industry (Nicolaus et al., 2010). On the other hand, EPS production by thermophilic microorganisms are lower in comparison to mesophilic producers (Kambourova et al., 2016). However, *B. thermoruber* 423 has been reported as a microbial cell factory for thermophilic EPS production since it has highest EPS production (863 mg L^{-1}) in comparison with other thermophilic EPS producers (Yasar Yildiz et al., 2014). From this point of view, due to the fast metabolic activities of thermophiles that results in elevated production rates, thermophiles could be considered as efficient cell factories for industrially important EPSs (Nicolaus et al., 2010). Thermophilic bacteria are source of novel biopolymers which have different chemical structure and similarly different physicochemical properties. These properties provide industrial importance for medical and pharmaceutical applications, especially for drug delivery, tissue engineering, as immunostimulatory, immunomodulatory, antitumor, antiviral, anti-inflammatory and antioxidant agents (Arena et al., 2009; Freitas et al., 2014; Lee and Mooney, 2012; Sam et al., 2011). In recent decades, several authors reported anti-tumour activity of microbial exopolysaccharides (Khalikova et al., 2006; Nwodo et al., 2012).

EPSs produced by *Bacillus licheniformis* strain B3-15, *Geobacillus thermodenitrificans* strain B3-72 and *B. licheniformis* strain T14, three thermophilic and thermotolerant bacilli, were investigated for their antiviral and immunomodulatory activities against herpes simplex virus type 2 (HSV-2) (Arena et al., 2009; Arena et al., 2006). All EPSs were non-cytotoxic towards PBMC (peripheral blood mononuclear cells) when concentration of 300 µg mL^{-1} EPSs were used (Poli et al., 2018). EPS1-T14 was able to interfere HSV-2 replication in PBMC by triggering the production of Th1-type cytokines. EPS2-B3-15 showed the best antiviral potential compared with the other EPSs assayed.(Gugliandolo et al., 2014; Marino-Merlo et al., 2017). Additionally, *Bacillus licheniformis* strain T14 isolated from Panarea Island synthesize EPS that has a new fucose monomer and valuable chemical and rheological characteristics were reported for the EPS (Spanò et al., 2013).

The immunomodulatory activity of EPS TA-1 from *Thermus aquaticus* YT-1 was reported in recent years. TA-1 stimulated macrophage cells to produce the cytokines TNF-α and IL-6, which increases the immune response. D-galactofuranose monomers in the composition of EPS TA-1 are probably responsible for reported TA-1 immunoregulatory activity through Toll-like receptor 2 within macrophages, the first line of host defence against bacterial infection (Lin et al., 2011).

EPS synthesized by *Geobacillus tepidamans* V264 inhibited cytotoxic activity of avarol in brine shrimp test. Avarol is a secondary metabolite from the marine sponge *Dysidea avara* and shows strong toxicity (LD50 0.18 g/ml) (Crispino et al., 1989; Minale et al., 1974; Tommonaro et al., 2015). The EPS isolated from *G. tepidamans* V264 increase the value of LD50 of avarol more than twelve-fold from 0.18 μg mL^{-1} up to 2.24 μg mL^{-1} (Kambourova et al., 2009). A possible reason for the activity exerted by EPS is the adhesion of toxic compounds to the surface of the polysaccharide (Kambourova et al., 2016; Poli et al., 2018). The studies indicated that this EPS could be used in pharmacy as anti-cytotoxic drugs (Kambourova et al., 2009).

B. thermoruber 423 was reported as a microbial cell factory for EPS production due to its high amount of EPS production (863 mg L^{-1}) (Yasar Yildiz et al., 2014). The biocompatibility of the EPS was assayed using the monkey kidney fibroblast cell line Cos-7. Results of the study indicated that EPS of *B. thermoruber* 423 is non-pathogenic and highly biocompatible. These results support the probable use of this valuable EPS in biomedical applications as scaffolds or matrices in tissue engineering, drug delivery and wound dressing (Kambourova et al., 2018; Poli et al., 2018).

EPS synthesized by *A. pallidus* 418 has good emulsifying properties. Mixtures of this novel EPS with other biopolymers provide a potential application in cosmetic industry. Synergistic action EPS from *A. pallidus* 418 with xanthan results in a formation of stable emulsion, that is a valuable for its industrial exploration (Radchenkova et al., 2014; Radchenkova et al., 2013).

EPSs synthesized by thermophilic microorganisms are thermostable molecules, that makes thermophilic EPSs significant for industrial applications such as food production process run at elevated temperature (Sajna et al., 2013). These thermophilic EPSs are also good candidates for food and cosmetic creams, emulsions of these molecules can be prepared easily since viscosity is lower and mixing is easier at high temperatures. These molecules also provide long term preservation of the received products even at room temperature. At high temperature of oil drilling fluids, the solutions of the EPSs are able to maintain high viscosity. As a results, they are significant candidates for use as flocculating agents in the thermophilic processes of municipal and wastewater treatment. Research on EPSs synthesized by thermophiles indicated valuable characteristics such as molecular weight suggesting good viscosity; stability of the molecules in harsh industrial conditions, good emulsifying properties and good synergism, biological activity against cytotoxic compounds, antiviral and immunomodulation activities suggesting the potential of EPSs from thermophilic microorganisms in several biotechnological and biomedical processes (Kambourova et al., 2016).

References

Ahmed, Z., Wang, Y., Anjum, N., Ahmad, H., Ahmad, A., and Raza, M. (2013). Characterization of new exopolysaccharides produced by coculturing of L. kefiranofaciens with yoghurt strains. International journal of biological macromolecules *59*, 377-383. https://doi.org/10.1016/j.ijbiomac.2013.04.075

Arena, A., Gugliandolo, C., Stassi, G., Pavone, B., Iannello, D., Bisignano, G., and Maugeri, T.L. (2009). An exopolysaccharide produced by Geobacillus thermodenitrificans strain B3-72: Antiviral activity on immunocompetent cells. Immunology letters *123*, 132-137. https://doi.org/10.1016/j.imlet.2009.03.001

Arena, A., Maugeri, T.L., Pavone, B., Iannello, D., Gugliandolo, C., and Bisignano, G. (2006). Antiviral and immunoregulatory effect of a novel exopolysaccharide from a marine thermotolerant Bacillus licheniformis. International immunopharmacology *6*, 8-13. https://doi.org/10.1016/j.intimp.2005.07.004

Bender, J., and Phillips, P. (2004). Microbial mats for multiple applications in aquaculture and bioremediation. Bioresource Technology *94*, 229-238. https://doi.org/10.1016/j.biortech.2003.12.016

Bitton, G., and Freihofer, V. (1977). Influence of extracellular polysaccharides on the toxicity of copper and cadmium towardKlebsiella aerogenes. Microbial ecology *4*, 119-125.

Brock, T.D. (2012). Thermophilic microorganisms and life at high temperatures (Springer Science & Business Media).

Buenger, J., and Driller, H. (2004). Ectoin: an effective natural substance to prevent UVA-induced premature photoaging. Skin pharmacology and physiology *17*, 232-237. https://doi.org/10.1159/000080216

Caron, D.A. (1987). Grazing of attached bacteria by heterotrophic microflagellates. Microbial Ecology *13*, 203-218.

Costerton, J.W., Stewart, P.S., and Greenberg, E.P. (1999). Bacterial biofilms: a common cause of persistent infections. Science *284*, 1318-1322.

Crispino, A., De Giulio, A., De Rosa, S., and Strazzullo, G. (1989). A new bioactive derivative of avarol from the marine sponge Dysidea avara. Journal of natural products *52*, 646-648.

Dan, T., Fukuda, K., Sugai-Bannai, M., Takakuwa, N., Motoshima, H., and Urashima, T. (2009). Characterization and expression analysis of the exopolysaccharide gene cluster in Lactobacillus fermentum TDS030603. Bioscience, biotechnology, and biochemistry *73*, 2656-2664. https://doi.org/10.1271/bbb.90502

De Vuyst, L., De Vin, F., Vaningelgem, F., and Degeest, B. (2001). Recent developments in the biosynthesis and applications of heteropolysaccharides from lactic acid bacteria. International Dairy Journal *11*, 687-707. https://doi.org/10.1016/S0958-6946(01)00114-5

Delcour, J., Ferain, T., and Hols, P. (2000). Advances in the genetics of thermophilic lactic acid bacteria. Current opinion in biotechnology *11*, 497-504. https://doi.org/10.1016/S0958-1669(00)00134-8

Demirjian, D.C., Morís-Varas, F., and Cassidy, C.S. (2001). Enzymes from extremophiles. Current opinion in chemical biology *5*, 144-151. https://doi.org/10.1016/S1367-5931(00)00183-6

Freitas, F., Alves, V.D., and Reis, M.A. (2011a). Advances in bacterial exopolysaccharides: from production to biotechnological applications. Trends in biotechnology *29*, 388-398. https://doi.org/10.1016/j.tibtech.2011.03.008

Freitas, F., Alves, V.D., Reis, M.A., Crespo, J.G., and Coelhoso, I.M. (2014). Microbial polysaccharide-based membranes: current and future applications. Journal of Applied Polymer Science *131*. https://doi.org/10.1002/app.40047

Freitas, F., Alves, V.D., Torres, C.A., Cruz, M., Sousa, I., Melo, M.J., Ramos, A.M., and Reis, M.A. (2011b). Fucose-containing exopolysaccharide produced by the newly isolated Enterobacter strain A47 DSM 23139. Carbohydrate Polymers *83*, 159-165. https://doi.org/10.1016/j.carbpol.2010.07.034

Grant, W.D., and Horikoshi, K. (1998). Extremophiles: microbial life in extreme environments, Vol 28 (Wiley-Liss New York).

Gugliandolo, C., Spanò, A., Lentini, V., Arena, A., and Maugeri, T. (2014). Antiviral and immunomodulatory effects of a novel bacterial exopolysaccharide of shallow marine vent origin. Journal of applied microbiology *116*, 1028-1034. https://doi.org/10.1111/jam.12422

Haki, G., and Rakshit, S. (2003). Developments in industrially important thermostable enzymes: a review. Bioresource technology *89*, 17-34. https://doi.org/10.1016/S0960-8524(03)00033-6

Harutoshi, T. (2013). Exopolysaccharides of lactic acid bacteria for food and colon health applications. In Lactic Acid Bacteria-R & D for Food, Health and Livestock Purposes (InTech). http://dx.doi.org/10.5772/50839

Jaiswal, P., Sharma, R., Sanodiya, B.S., and Bisen, P.S. (2014). Microbial exopolysaccharides: natural modulators of dairy products. J Appl Pharm Sci *4*, 105-109.

Jeanthon, C., and Prieur, D. (1990). Susceptibility to heavy metals and characterization of heterotrophic bacteria isolated from two hydrothermal vent polychaete annelids, Alvinella pompejana and Alvinella caudata. Applied and environmental microbiology *56*, 3308-3314.

Johnson, M.R., Montero, C.I., Conners, S.B., Shockley, K.R., Bridger, S.L., and Kelly, R.M. (2005). Population density-dependent regulation of exopolysaccharide formation in the hyperthermophilic bacterium Thermotoga maritima. Molecular microbiology *55*, 664-674. https://doi.org/10.1111/j.1365-2958.2004.04419.x

Kambourova, M., Mandeva, R., Dimova, D., Poli, A., Nicolaus, B., and Tommonaro, G. (2009). Production and characterization of a microbial glucan, synthesized by Geobacillus tepidamans V264 isolated from Bulgarian hot spring. Carbohydrate Polymers *77*, 338-343. https://doi.org/10.1016/j.carbpol.2009.01.004

Kambourova, M., Radchenkova, N., Nicolaus, B., Tommonaro, G., Poli, A., Toksoy Oner, E., Yasar Yildiz, S., and Arga, K.Y. (2018). Exopolysaccharides Synthesized by Thermophilic Microorganisms Isolated from Bulgarian Hot Springs. ACTA MICROBIOLOGICA BULGARICA, 68.

Kambourova, M., Radchenkova, N., Tomova, I., and Bojadjieva, I. (2016). Thermophiles as a Promising Source of Exopolysaccharides with Interesting Properties. In Biotechnology of Extremophiles: (Springer), pp. 117-139.

Kazak, H., Toksoy Oner, E., and Dekker, R.F. (2010). Extremophiles as sources of exopolysaccharides. Carbohydrate Polymers: Development, Properties and Applications, 605-619.

Khalikova, T., Korolenko, T., Zhanaeva, S.Y., Kaledin, V., and Kogan, G. (2006). Enhancing effect of new biological response modifier sulfoethylated (1→ 3)-β-D-glucan on antitumor activity of cyclophosphamide in the treatment.

Koerdt, A., Gödeke, J., Berger, J., Thormann, K.M., and Albers, S.-V. (2010). Crenarchaeal biofilm formation under extreme conditions. PLoS One 5, e14104. https://doi.org/10.1371/journal.pone.0014104

Krüger, M., Blumenberg, M., Kasten, S., Wieland, A., Känel, L., Klock, J.H., Michaelis, W., and Seifert, R. (2008). A novel, multi-layered methanotrophic microbial mat system growing on the sediment of the Black Sea. Environmental microbiology *10*, 1934-1947. https://doi.org/10.1111/j.1462-2920.2008.01607.x

Kumar, C.G., Joo, H.-S., Choi, J.-W., Koo, Y.-M., and Chang, C.-S. (2004). Purification and characterization of an extracellular polysaccharide from haloalkalophilic Bacillus sp. I-450. Enzyme and microbial technology *34*, 673-681. https://doi.org/10.1016/j.enzmictec.2004.03.001

Kumar, L., Awasthi, G., and Singh, B. (2011). Extremophiles: a novel source of industrially important enzymes. Biotechnology *10*, 121-135. 10.3923/biotech.2011.121.135

Laws, A., Gu, Y., and Marshall, V. (2001). Biosynthesis, characterisation, and design of bacterial exopolysaccharides from lactic acid bacteria. Biotechnology advances *19*, 597-625. https://doi.org/10.1016/S0734-9750(01)00084-2

Lee, K.Y., and Mooney, D.J. (2012). Alginate: properties and biomedical applications. Progress in polymer science *37*, 106-126. https://doi.org/10.1016/j.progpolymsci.2011.06.003

Lin, M.-H., Yang, Y.-L., Chen, Y.-P., Hua, K.-F., Lu, C.-P., Sheu, F., Lin, G.-H., Tsay, S.-S., Liang, S.-M., and Wu, S.-H. (2011). A novel exopolysaccharide from the biofilm of Thermus aquaticus YT-1 induces the immune response through Toll-like receptor 2. Journal of Biological Chemistry, jbc. M110. 200113.

Linker, A., and Jones, R.S. (1966). A new polysaccharide resembling alginic acid isolated from pseudomonads. Journal of Biological Chemistry *241*, 3845-3851.

Manca, M.C., Lama, L., Improta, R., Esposito, E., Gambacorta, A., and Nicolaus, B. (1996). Chemical composition of two exopolysaccharides from Bacillus thermoantarcticus. Applied and environmental microbiology *62*, 3265-3269.

Margesin, R., and Schinner, F. (2001). Potential of halotolerant and halophilic microorganisms for biotechnology. Extremophiles *5*, 73-83.

Marino-Merlo, F., Papaianni, E., Maugeri, T.L., Zammuto, V., Spanò, A., Nicolaus, B., Poli, A., Di Donato, P., Mosca, C., and Mastino, A. (2017). Anti-herpes simplex virus 1 and immunomodulatory activities of a poly-γ-glutamic acid from Bacillus horneckiae strain APA of shallow vent origin. Applied microbiology and biotechnology *101*, 7487-7496. https://doi.org/10.1007/s00253-017-8472-5

Maugeri, T.L., Gugliandolo, C., Caccamo, D., Panico, A., Lama, L., Gambacorta, A., and Nicolaus, B. (2002). A halophilic thermotolerant Bacillus isolated from a marine hot spring able to produce a new exopolysaccharide. Biotechnology Letters *24*, 515-519. https://doi.org/10.1023/A:1014891431233

Minale, L., Riccio, R., and Sodano, G. (1974). Avarol a novel sesquiterpenoid hydroquinone with a rearranged drimane skeleton from the sponge Disidea avara. Tetrahedron Letters *15*, 3401-3404. https://doi.org/10.1016/S0040-4039(01)91918-5

Moriello, V.S., Lama, L., Poli, A., Gugliandolo, C., Maugeri, T., Gambacorta, A., and Nicolaus, B. (2003). Production of exopolysaccharides from a thermophilic microorganism isolated from a marine hot spring in flegrean areas. Journal of Industrial Microbiology and Biotechnology *30*, 95-101. https://doi.org/10.1007/s10295-002-0019-8

Nichols, C.M., Lardière, S.G., Bowman, J.P., Nichols, P.D., Gibson, J.A., and Guézennec, J. (2005). Chemical characterization of exopolysaccharides from Antarctic marine bacteria. Microbial ecology *49*, 578-589. https://doi.org/10.1007/s00248-004-0093-8

Nicolaus, B., Kambourova, M., and Toksoy Oner, E. (2010). Exopolysaccharides from extremophiles: from fundamentals to biotechnology. Environmental Technology *31*, 1145-1158. https://doi.org/10.1080/09593330903552094

Nicolaus, B., Lama, L., Panico, A., Moriello, V.S., Romano, I., and Gambacorta, A. (2002). Production and characterization of exopolysaccharides excreted by thermophilic bacteria from shallow, marine hydrothermal vents of Flegrean Ares (Italy). Systematic and applied microbiology *25*, 319-325. https://doi.org/10.1078/0723-2020-00128

Nicolaus, B., Manca, M.C., Romano, I., and Lama, L. (1993). Production of an exopolysaccharide from two thermophilic archaea belonging to the genus Sulfolobus. FEMS microbiology letters *109*, 203-206. https://doi.org/10.1111/j.1574-6968.1993.tb06168.x

Nicolaus, B., Moriello, V., Maugeri, T., Gugliandolo, C., and Gambacorta, A. (2003). Bacilli from shallow mediterranean marine vents producers of exopolysaccharides. Recent Research Developments in Microbiology *7*, 197-208.

Nicolaus, B., Moriello, V.S., Lama, L., Poli, A., and Gambacorta, A. (2004). Polysaccharides from extremophilic microorganisms. Origins of Life and Evolution of the Biosphere *34*, 159-169. https://doi.org/10.1023/B:ORIG.0000009837.37412.d3

Nicolaus, B., Panico, A., Manca, M.C., Lama, L., Gambacorta, A., Maugeri, T., Gugliandolo, C., and Caccamo, D. (2000). A Thermophilic Bacillus Isolated From an Eolian Shallow Hydrothermal Vent Able to Produce Exopolysaccharides. Systematic and applied microbiology *23*, 426-432. https://doi.org/10.1016/S0723-2020(00)80074-0

Nwodo, U.U., Green, E., and Okoh, A.I. (2012). Bacterial exopolysaccharides: functionality and prospects. International journal of molecular sciences *13*, 14002-14015. 10.3390/ijms131114002

Toksoy Oner, E. (2013). Microbial Production of Extracellular Polysaccharides from Biomass. In Pretreatment Techniques for Biofuels and Biorefineries (Springer), pp. 35-56. https://doi.org/10.1007/978-3-642-32735-3_2

Özcan, E., and Toksoy Oner, E. (2015). Microbial production of extracellular polysaccharides from biomass sources. Polysaccharides: Bioactivity and Biotechnology, 161-184.
https://doi.org/10.1007/978-3-319-16298-0_51

Paerl, H.W. (1976). Specıfıc associations of the bluegreen algae anabaena and aphanizomenon with bacteria in freshwater blooms 1. Journal of Phycology *12*, 431-435.

Panosyan, H., Di Donato, P., Poli, A., and Nicolaus, B. (2018). Production and characterization of exopolysaccharides by Geobacillus thermodenitrificans ArzA-6 and Geobacillus toebii ArzA-8 strains isolated from an Armenian geothermal spring. Extremophiles, 1-13. https://doi.org/10.1007/s00792-018-1032-9

Podar, M., and Reysenbach, A.-L. (2006). New opportunities revealed by biotechnological explorations of extremophiles. Current opinion in biotechnology *17*, 250-255. https://doi.org/10.1016/j.copbio.2006.05.002

Poli, A., Di Donato, P., Tommonaro, G., Abbamondi, G.R., Finore, I., and Nicolaus, B. (2018). Exopolysaccharide-Producing Microorganisms from Extreme Areas: Chemistry and Application. In Extremophiles in Eurasian Ecosystems: Ecology, Diversity, and Applications (Springer), pp. 405-433. https://doi.org/10.1007/978-981-13-0329-6_15

Radchenkova, N., Vassilev, S., Martinov, M., Kuncheva, M., Panchev, I., Vlaev, S., and Kambourova, M. (2014). Optimization of the aeration and agitation speed of Aeribacillus palidus 418 exopolysaccharide production and the emulsifying properties of the product. Process Biochemistry *49*, 576-582. https://doi.org/10.1016/j.procbio.2014.01.010

Radchenkova, N., Vassilev, S., Panchev, I., Anzelmo, G., Tomova, I., Nicolaus, B., Kuncheva, M., Petrov, K., and Kambourova, M. (2013). Production and properties of two novel exopolysaccharides synthesized by a thermophilic bacterium Aeribacillus pallidus 418. Applied biochemistry and biotechnology *171*, 31-43. https://doi.org/10.1007/s12010-013-0348-2

Raguénès, G., Peres, A., Ruimy, R., Pignet, P., Christen, R., Loaec, M., Rougeaux, H., Barbier, G., and Guezennec, J. (1997). Alteromonas infernus sp. nov., a new polysaccharide-producing bacterium isolated from a deep-sea hydrothermal vent. Journal of applied microbiology *82*, 422-430. https://doi.org/10.1046/j.1365-2672.1997.00125.x

Rainey, F.A., and Oren, A. (2006). 1 Extremophile Microorganisms and the Methods to Handle Them. In Methods in microbiology (Elsevier), pp. 1-25. https://doi.org/10.1016/S0580-9517(08)70004-7

Rehm, B.H. (2010). Bacterial polymers: biosynthesis, modifications and applications. Nature Reviews Microbiology *8*, 578. https://doi.org/10.1038/nrmicro2354

Rinker, K.D., and Kelly, R.M. (1996). Growth physiology of the hyperthermophilic archaeon Thermococcus litoralis: Development of a sulfur-free defined medium, characterization of an exopolysaccharide, and evidence of biofilm formation. Applied and environmental microbiology *62*, 4478-4485.

Rinker, K.D., and Kelly, R.M. (2000). Effect of carbon and nitrogen sources on growth dynamics and exopolysaccharide production for the hyperthermophilic archaeon Thermococcus litoralis and bacterium Thermotoga maritima. Biotechnology and Bioengineering *69*, 537-547. https://doi.org/10.1002/1097-0290(20000905)69:5<537::AID-BIT8>3.0.CO;2-7

Ritsema, T., and Smeekens, S. (2003). Fructans: beneficial for plants and humans. Current opinion in plant biology *6*, 223-230. https://doi.org/10.1016/S1369-5266(03)00034-7

Roger, O., Kervarec, N., Ratiskol, J., Colliec-Jouault, S., and Chevolot, L. (2004). Structural studies of the main exopolysaccharide produced by the deep-sea bacterium Alteromonas infernus. Carbohydrate research *339*, 2371-2380. https://doi.org/10.1016/j.carres.2004.07.021

Rothschild, L.J., and Mancinelli, R.L. (2001). Life in extreme environments. Nature *409*, 1092. https://doi.org/10.1038/35059215

Sajna, K.V., Sukumaran, R.K., Gottumukkala, L.D., Jayamurthy, H., Dhar, K.S., and Pandey, A. (2013). Studies on structural and physical characteristics of a novel exopolysaccharide from Pseudozyma sp. NII 08165. International journal of biological macromolecules *59*, 84-89. https://doi.org/10.1016/j.ijbiomac.2013.04.025

Sam, S., Kucukasik, F., Yenigun, O., Nicolaus, B., Toksoy Oner, E., and Yukselen, M.A. (2011). Flocculating performances of exopolysaccharides produced by a halophilic bacterial strain cultivated on agro-industrial waste. Bioresource technology *102*, 1788-1794. https://doi.org/10.1016/j.ijbiomac.2013.04.025

Satyanarayana, T., Raghukumar, C., and Shivaji, S. (2005). Extremophilic microbes: Diversity and perspectives. Current science, 78-90.

Selbmann, L., Onofri, S., Fenice, M., Federici, F., and Petruccioli, M. (2002). Production and structural characterization of the exopolysaccharide of the Antarctic fungus Phoma herbarum CCFEE 5080. Research in Microbiology *153*, 585-592. https://doi.org/10.1016/S0923-2508(02)01372-4

Spanò, A., Gugliandolo, C., Lentini, V., Maugeri, T.L., Anzelmo, G., Poli, A., and Nicolaus, B. (2013). A novel EPS-producing strain of Bacillus licheniformis isolated from a shallow vent off Panarea Island (Italy). Current microbiology *67*, 21-29. https://doi.org/10.1007/s00284-013-0327-4

Tommonaro, G., Iodice, C., AbdEl-Hady, F.K., Guerriero, G., and Pejin, B. (2015). The Mediterranean sponge Dysidea avara as a 40 year inspiration of marine natural product chemists. Journal of Biodiversity & Endangered Species *2015*. http://dx.doi.org/10.4172/2332-2543.S1.001

Turner, P., Mamo, G., and Karlsson, E.N. (2007). Potential and utilization of thermophiles and thermostable enzymes in biorefining. Microbial Cell Factories *6*, 1. https://doi.org/10.1186/1475-2859-6-9

Urbieta, M.S., Donati, E.R., Chan, K.-G., Shahar, S., Sin, L.L., and Goh, K.M. (2015). Thermophiles in the genomic era: biodiversity, science, and applications. Biotechnology Advances *33*, 633-647. https://doi.org/10.1016/j.biotechadv.2015.04.007

Urbieta, M.S., Toril, E.G., Giaveno, M.A., Bazán, Á.A., and Donati, E.R. (2014). Archaeal and bacterial diversity in five different hydrothermal ponds in the Copahue region in Argentina. Systematic and applied microbiology *37*, 429-441. https://doi.org/10.1016/j.syapm.2014.05.012

Wang, J., Salem, D.R., and Sani, R.K. (2018). Extremophilic Exopolysaccharides: A Review and New Perspectives on Engineering Strategies and Applications. Carbohydrate Polymers. https://doi.org/10.1016/j.carbpol.2018.10.011

Wang, Z.-W., Liu, Y., and Tay, J.-H. (2007). Biodegradability of extracellular polymeric substances produced by aerobic granules. Applied microbiology and biotechnology *74*, 462-466. https://doi.org/10.1007/s00253-006-0686-x

Wharton, D.A. (2007). Life at the limits: organisms in extreme environments (Cambridge University Press).

Wotton, R.S. (2004). The utiquity and many roles of exopolymers (EPS) in aquatic systems. Scientia marina *68*, 13-21. https://doi.org/10.3989/scimar.2004.68s113

Xiao, Z., Zhang, Y., Xi, L., Huo, F., Zhao, J.y., and Li, J. (2015). Thermophilic production of polyhydroxyalkanoates by a novel Aneurinibacillus strain isolated from Gudao oilfield, China. Journal of basic microbiology *55*, 1125-1133. https://doi.org/10.1002/jobm.201400843

Yasar Yildiz, S., Anzelmo, G., Ozer, T., Radchenkova, N., Genc, S., Di Donato, P., Nicolaus, B., Toksoy Oner, E., and Kambourova, M. (2014). Brevibacillus themoruber: a promising microbial cell factory for exopolysaccharide production. Journal of applied microbiology *116*, 314-324. https://doi.org/10.1111/jam.12362

Yasar Yildiz, S., Kambourova, M., Arga, K.Y., and Toksoy Oner, E. (2013). Draft genome sequence of exopolysaccharide-producing thermophilic bacterium

Brevibacillus thermoruber strain 423. Genome announcements *1*, e00774-00713. 10.1128/genomeA.00774-13

Yasar Yildiz, S., Radchenkova, N., Arga, K.Y., Kambourova, M., and Toksoy Oner, E. (2015). Genomic analysis of Brevibacillus thermoruber 423 reveals its biotechnological and industrial potential. Applied microbiology and biotechnology *99*, 2277-2289. https://doi.org/10.1007/s00253-015-6388-5

Zhang, C.L., Ye, Q., Huang, Z., Li, W., Chen, J., Song, Z., Zhao, W., Bagwell, C., Inskeep, W.P., and Ross, C. (2008). Global occurrence of archaeal amoA genes in terrestrial hot springs. Applied and Environmental Microbiology *74*, 6417-6426.

Engineering of Microbial Polysaccharide Structures

4

Jochen Schmid

Norwegian University of Science and Technology, Department of Biotechnology and Food Science, Sem Sælands vei 6-8, 7491 Trondheim, Norway

DOI: https://doi.org/10.21775/9781912530267.04

Abstract

This review will deal with the structural diversity of bacterial (exo)polysaccharides based on the biosynthesis pathways and the highly specific glycosyltransferase proteins. Especially the so called Wzx/Wzy biosynthesis pathway will be explained in detail and the challenges and prospects of tailoring the chemical structures and thus polysaccharide properties will be highlighted. The differences of chemical modification and *in-vivo* modification will be discussed. Finally, genetic engineering approaches for altered chemical structures and enhanced productivities will be displayed and discussed to highlight challenges and perspectives for engineering hetero- as well as homopolysaccharides.

Structural Diversity of Microbial Exopolysaccharides

The class of microbial, especially bacterial exopolysaccharides is defined by a highly structural diversity. This structural diversity is based on several aspects which includes the type of sugars and sugar derivatives which are assembled, the type of bonds present in the polymer, additional substituents such as acetyl, pyruvyl or glycerol residues and branching of the polymer by additional side chains. For many exopolysaccharides (EPS) the complete chemical structure is described and was analyzed by determination of the monomer composition, methylation experiments to define the chemical bonds and additional information such as NMR spectra or mass spectroscopy (Altman and Li, 2010; Cui, 2005; Mischnick, 2011). But NMR analysis of highly viscous polysaccharides is complex and only very rare specialists are available which can interpret the spectra of novel identified polysaccharides (Bubb, 2003).

Within the last years also image analysis by for example atomic force microscopy became more important for analysis of the secondary and tertiary structures (Wang and Nie, 2018). For the class of homopolysaccharides, just consisting of one type of sugar monomers, such as Dextrans and Levans, the type and degree of branching as well as the molecular weight will result in highly differing functionalities as defined by the structure. The different homopolysaccharides show a high potential for medical applications due to their antiviral and antitumoral properties (Moscovici, 2015), or act in the form of oligosaccharides as prebiotics (Mitmesser and Combs, 2017). The class of heteropolysaccharides offers an even higher degree of structural variety due to the presence of many different sugar monomers and sugar derivatives. Based on the various sugar monomers and the high variety of possible combinations, including different chemical bonds as well as decoration with typical substituents and branching, the degree of variations seems to be unlimited. The carbohydrate structure database is one of the most current databases which summarizes the structural diversity of polysaccharides and thus shows impressively the great amount of EPS (Toukach and Egorova, 2016). But our knowledge is still limited due to the problems of polysaccharide structure analysis as well as defined identification procedures of novel microbial exopolysaccharide producers (Rühmann et al., 2015). Very interestingly the heteropolysaccharides are more responsible for cell-cell interactions and pathogenicity compared to the homopolysaccharides (Cress et al., 2014). Besides some medical applications, the main commercial field of applications is highly defined by their rheological characteristics and properties. Table 1 gives an overview of the different commercially available polysaccharides based on their chemical composition and applications.

Xanthan as the main commercial representative of the bacterial heteropolysaccharides shows very well thickening behavior and is used to adjust the texture of many foods all over the world (Hublik, 2012). Additionally, it is applied in many sanitary products to enhance cleaning effects by increasing the viscosity to remain longer at the place to be applied. Xanthan consists of 3 different types of sugars and derivatives out of them, namely glucose (Glc), mannose (Man) and glucuronic acid (GlcA). The ratio is 2:2:1 respectively and depending on the cultivation conditions, or production strain a variable degree of acetyl and pyruvyl residues is present at both mannose monomers of the side chain (Figure 1). It has been identified relatively early, that the degree and presence of these substituents massively influences the rheological behavior of Xanthan (Hassler and Doherty, 1990; Tako and Nakamura, 1984).

Thus, it is clearly seen that the degree and localization of the substituents massively influences the polysaccharide characteristics. This fact massively enhances the structural diversity combined with the possibility to tailor the polymer characteristics via chemical or genetically driven approaches. Up to now there exist just a very low number of engineered polysaccharide structures what is mainly based on the differences in chemical and genetic engineering approaches and possibilities. Main engineering is done in the field of deacetylation and depyruvylation, due to the relatively easy way to perform this. Chemical deacetylation and depyruvylation can be done by alkaline conditions under slightly enhanced temperature conditions.

Table 1. An overview of the different types of polysaccharides based on their chemical structure

Polysaccharide	Production strain	Chemical Composition	Field of Application
Alginate	*Pseudomonas* sp.	GulA, ManA, Ace	Food, Feed, Medicine
Cellulose	*Gluconoacetobacter sp.*, *Agrobacterium sp.*	Glc	Food, Medicine, Materials
Curdlan	*Agrobacterium sp.*, *Bacillus sp.*, *Cellulomonas sp.*, *Rhizobium sp.*	Glc	Food, Cosmetics, Medicine, Construction Industry
Dextran	*Leuconostoc sp.*, *Streptococcus sp.*	Glc	Construction Chemistry, Food, Feed
Diutan	*Sphingomonas sp.*	Glc, Rha, GlcA, Ace	Construction Chemistry
Gellan	*Sphingomonas sp.*	Glc, Rha, GlcA, Ace, Gly	Construction Chemistry, Food, Feed, Research

Hyaluronic acid	*Streptococcus sp.*, recombinant systems	GlcA, GlcNAc	Medicine, Cosmetics
Levan	*Pseudomonas sp. Bacillus sp.*, *Zymomonas sp.*, *Halomonas sp.*, recombinant systems	Fru, Glc*	Food (prebiotic), Feed, Medicine, Cosmetics, Glue
Succinoglycan	*Rhizobium* sp.	Glc, Gal, Ace, Pyr, Succ	Oil Industry, Cosmetics
Welan	*Sphingomonas sp.*	Glc, Rha, GlcA, Man, Ace	Construction Chemistry
Xanthan	*Xanthomonas sp.*	Glc, Man, GluA Ace, Pyr	Food, Feed, Technical Applications, Oil Drilling

Glc, glucose; Rha, rhamnose; Fru, fructose; Gal, galactose; Man, mannose; GlcA, glucuronic acid; ManA, mannuronic acid; GulA, guluronic acid; GalA, galacturonic acid; GlcNAc, *N*-acetyl-glucosamine; Pyr, pyruvate; Ace, acetate; Gly, glycerate; Suc, succinate

* Levan in general is a fructopolymer but contains a single glucose monomer as priming sugar of the growing polymer strand.

There are commercial variants of deacetylated Xanthan and Gellan available, which are used in tailored applications (Giavasis et al., 2000). Additional effects of depyruvylated Xanthan exist, but no commercial applications are described (Tako and Nakamura, 1988). Furthermore, chemically modified variants of plant-based polysaccharides such as cellulose are very prominent in many products and fulfill needs such as high water retention and pH stability over a broad range (Li et al., 2016). Further chemical modifications can be performed such as etherification, oxidation or sulfonation as nicely reviewed by (Cumpstey, 2013). The main representatives of this class of chemically modified polysaccharides are hydroxyethyl methylcellulose and hydroxypropyl methylcellulose. They are mainly applied in the construction industry due to their function to improve product characteristics like crack-resistance, adhesive strength, workability and water retention capacity (Jérémie Pourchez, 2010; Peschard et al., 2004). But, the chemical modification of cellulose, mainly coming from cotton wool, the most water consuming plant breeding in combination with a high amount of pesticides, necessary for its successful production, is in the need of very harsh and not ecological conditions (Chen and Burns, 2006). Thus, this kind of modified biopolymers shows a high degree of suitability for many applications but is anything but green and sustainable (Laursen et al., 1997). To circumvent classical chemical modification based on harsh and non-ecological conditions, genetic engineering of production strains represents a very promising alternative. But this alternative shows also some limitation, which will be discussed in the following section.

Polysaccharide biosynthesis pathways

For a better understanding of the limitations and prospective of genetically engineering of microbial exopolysaccharides the biosynthesis pathways must be elucidated. The following section will focus on the different biosynthesis pathways as present in bacteria due to the limited knowledge and lower diversity in higher organisms such as fungi.

Homo- and heteropolysaccharides

The main difference of the biosynthesis pathways for bacterial exopolysaccharide production can be found in-between hetero- and homopolysaccharides. For homopolysaccharides such as Dextran and Levan, there is only one single enzyme involved in the biosynthesis, the so-called dextrose sucrase or Levan sucrase (Öner et al., 2016; Robyt et al., 2008). These enzymes are located at the outer cell wall or can even be secreted and need di- or trisaccharide's as substrates for the polymerization. The sucrases uses the energy of the glycosidic bonds of the substrates to transfer the cleaved products to the growing polymer chain. In the case of Levan sucrases, just the cleaved fructose of sucrose is transferred towards the growing polymer chain, which often starts with an initial glucose monomer. Thus, only one enzyme is involved in the biosynthesis. Additional enzymes can be responsible for Levan degradation or the production of fructooligosaccharides from Levan as it is the case for the levanases which can act as endo- or exo-levanases (Miasnikov, 1997). The next, a bit more complex pathway is the so-called synthase based biosynthesis, which results in the biosynthesis of homopolysaccharides such as bacterial cellulose or Curdlan (McIntosh et al., 2005). The synthase-based biosynthesis is widespread amongst different strains such as Gluconoacetobacter and Agrobacterium strains to make glucan like polysaccharides (Benziman et al., 1980). This pathway is encoded by the main *bscABC* genes which are highly conserved in most strains, but also can differ in size and organization (Römling and Galperin, 2015). Cellulose biosynthesis was remarkably characterized within the last years on the molecular level, and the mechanism of UDP-Glucose monomer incorporation into the polymer was described in detail (McNamara et al., 2015).

Another very well characterized synthase-based biosynthesis pathway exists for bacterial Alginates. It comprises several genes which are clustered in the genome and encode for the polymerization of mannuronic acid monomers (Rehm and Valla, 1997). Very interestingly the mannuronic acid (M) monomers can be epimerized in the later stage of the secretion process of the polymer towards guluronic acid (G) monomers (Ertesvag, 2015). This enables the production of a heteropolysaccharide which is produced via a synthase-based biosynthesis. The Alginate biosynthesis cluster consists of 11 *alg* genes which encode polymerization, transport and secretion as well as for epimerization of M to G (Rehm, 2010). This synthase-based exopolysaccharide biosynthesis pathway is very well characterized towards assembly of the nucleotide activated sugar monomers, the *in-vivo* epimerization process as well as transportation and even regulation. It has to be mentioned here, that nearly all commercial Alginates are derived from macro algae and differ highly in their M:G ratio and pattern. These differences result from harvesting them at different seasons, regions as well as different treatment methods to extract them from the algal biomass. These non-uniform M:G ratios and patterns can highly influence the rheological behavior as well as chelation properties, since chelation mainly depends on the G content and G-block formation within the polymer, which can form highly rigid "crystalline" regions in the presence of divalent ions such as Ca^{2+} (Andriamanantoanina and Rinaudo, 2010).

Heteropolysaccharides

For heteropolysaccharides the biosynthesis pathway is more complex and thus comprises more enzymatic steps (Rehm, 2010; Schmid and Sieber, 2015). These steps comprise the assembly of the so-called repeating unit, which is assembled by the action of several glycosyltransferases using a lipid anchor at the cytosolic side of the inner membrane as starting point (Schmid et al., 2015). Sequentially the different carbohydrates monomers of the so-called repeating unit are transferred towards this linker in a highly specific way, including decoration with substituents such as acetyl- or pyruvyl- groups. The highly specific repeating unit is then flipped towards the periplasmic space by the so-called Wzx flippase and following polymerized by the action of the polymerase protein Wzy (Schmid, 2018). Further membrane spanning proteins realize the secretion to the environmental space. One of the best characterized biosynthesis pathways for a heteropolysaccharide is the biosynthesis of Xanthan, which is also the microbial polysaccharide of main commercial interest (Hublik, 2012). Production titers of Xanthan can reach up to 60 g/L and are mainly limited by the high viscosity of the fermentation broth, thus limiting the oxygen and mass transfer which are essential for cell growth and polysaccharide production (Flores et al., 1994). The biosynthesis of Xanthan is encoded by the so-called *gum*-cluster and includes 12 genes which are well described in literature (Katzen et al., 1998; Vorhölter et al., 2008). The Xanthan biosynthesis, its chemical structure and the encoding *gum*-cluster is given in Figure 1.

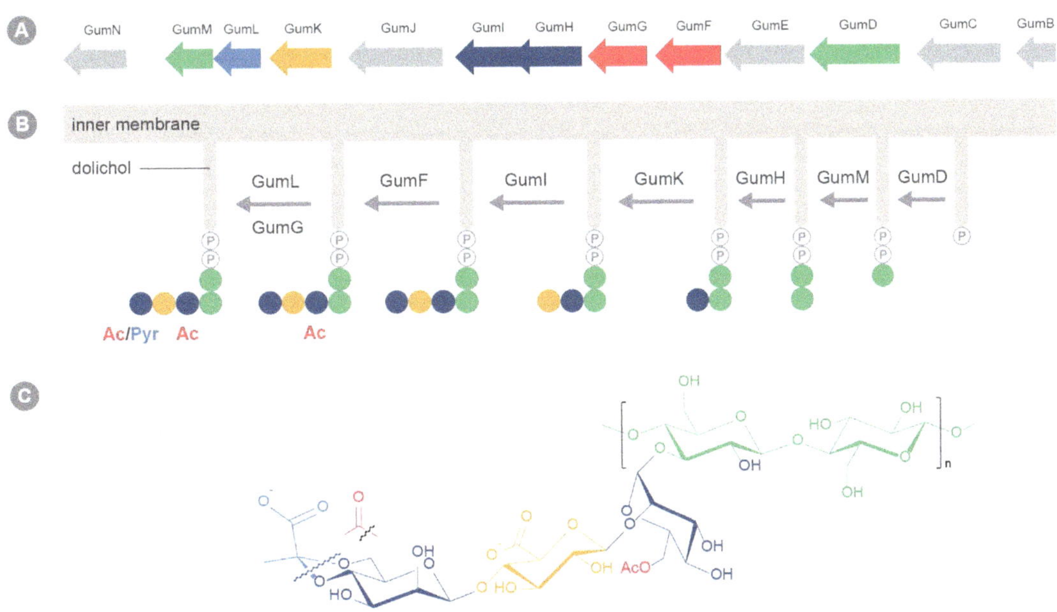

Figure 1. The *gum*-cluster encoding the whole biosynthesis machinery of Xanthan by *Xanthomonas campestris* is given in A. The schematic assembly of the repeating unit is presented in B, including the dolichol lipid anchor at the cytosolic side of the inner membrane. The action of the sugar monomer transferring proteins is depicted by the arrows and the corresponding Protein names. The acetylation and pyruvylation is performed by the three transferases GumF, GumG and GumL respectively. It is not clear if the order of the steps for adding the substituents is the same as given here. The terminal mannose of the repeating unit can be either acetylated or pyruvylated. The final chemical structure of Xanthan is given in C. The color code determines the corresponding genes, which are responsible for adding the sugars and substituents which are given in the same colors.

In brief, most of the *gum* genes encode for (glycosyl)transferases which transfer highly selective sugar monomers and substituents towards the growing repeating unit which is bound to a dolichol anchor at the cytosolic side of the inner membrane of the cell. Thus, the repeating unit of Xanthan as given in Figure 1 (b) is assembled by sequential action of the glycosyltransferases GumD, GumM, GumH and GumK, until the repeating unit is completed by the action of GumI. It has to be mentioned that by the definition for glycosyltransferases, GumD is not a glycosyltransferase, it is more a kinase, since it transfers not a nucleotide activated sugar, but a phosphorylated sugar. But in general, the enzymes which transfer the initial monosaccharide onto the dolichol anchor are named initial glycosyltransferases.

Once the repeating unit is finished, the complete dolichol anchored repeating unit is flipped towards the periplasmic space by action of the flippase encoded by the *gumJ* gene. This is one of the processes which is not fully understood up to now. In a next step the various repeating units as flipped into the periplasm are released from the dolichol anchor and polymerized by action of the polymerase GumE. Finally, the growing polymer chain is transferred through the cell wall by assistance of the proteins encoded by *gumC* and *gumB*. Whereas the action and mechanisms of the different glycosyltransferases which assemble the repeating units are very well characterized, relative low information is available on the mechanisms and structures of the following proteins in the biosynthesis pathways. More and more information get available but still many research has to be performed to obtain clear insights into this highly complex process (Islam and Lam, 2013). Additionally, it is not completely clear at what point or at what location the substituent groups are assembled to the repeating units, this may occur in the periplasm or the cytosol. The only thing for sure is that the acetyl- and pyruvyl- transferases are membrane bound proteins (Vorhölter et al., 2008).

The biosynthesis machinery for Xanthan is described to be highly specific, but *in-vivo* modifications of the chemical structure are also described. Main reports are available on the deletion of the acetyl- and pyruvyl- groups and result in the same product titers as for the wildtype polymer, but have distinctive different rheological behavior, thus extending the portfolio and field of applications for Xanthan and its derivatives thereof (Hassler and Doherty, 1990). Xanthan in its natural appearance shows a high viscosity with pseudoplastic behavior, which is decreased by deacetylation and especially depyruvylation (Wu et al., 2019). Additionally the degree of substitution also influences the temperature stability of the viscosity (Gansbiller et al., 2019).

In the case of Succinoglycan, one of the EPS with a long repeating unit of 9 monomer units, in combination with acetylation, pyruvylation and succinylation, the pyruvylation of the terminal glucose residue of the terminal glucose of the repeating unit seems to be essential for flipping towards the periplasmic space or polymerization (Becker et al., 1993; Reuber and Walker, 1993). For Xanthan also further modifications of the sidechain are described, which includes the deletion of *gumI* for example, resulting in a truncated version of Xanthan, lacking the terminal mannose (Hassler and Doherty, 1990). Based on complementation experiments of *wzx* genes within different strains and species, it was found that there are highly specific Wzx variants existent, which solely accept their native repeating units (Hong et al., 2012; Hong and Reeves, 2014; Wang et al., 2012). But also, the translocation of non-native repeating units is described which often results in lowered product titers, and seems to be highly dependent on the initial dolichol-linked sugar (Hong and Reeves, 2014; Islam et al., 2010; Marolda et al., 2006; Rütering et al., 2017). The lowered product titers of *in-vivo* truncations of the side chain might be balanced by overexpression of the native flippase gene (Marolda et al., 2006). The main reports of side chain modification describe the truncation of one sugar moiety from the terminal end, only limited reports are available which describe the complete removal of the side chain, and most of them seems to end up in non-EPS production or might be lethal deletions (Katzen et al., 1998; Wang et al., 2012). But for this experimental data it seems to be obvious that more the presence of a side chain is essential to be efficiently recognized by the flippase and polymerase proteins than the nature or composition of the side chain. No information is available on the effects of synthetically *in vivo* acetylated or pyruvylated polysaccharides on the secretion efficiency.

In addition, no reports can be found on alteration of the native backbone of heteropolysaccharides, what might be based on the high specificity of the glycosyltransferase proteins. In sum, for the future design of synthetically exopolysaccharide structures which can be produced under highly reproducible and efficient process conditions, mainly the understanding of the mechanisms of the flippase and polymerase proteins is essential.

Modification of homopolysaccharides

Modification of homopolysaccharides such as Levans and Dextrans can be done by protein engineering of the different sucrase enzymes, what can mainly influence the substrate specificity as well as the grade of polymerization and branching of the final products (Gangoiti et al., 2018; Meng et al., 2016). Thus, various variants were generated which accept a vast number of substrates resulting in more heterogenous oligosaccharides and polysaccharides with other monomers than only fructose or glucose. Just recently the recombinant expression of bacterial cellulose biosynthesis in *E. coli* strains was reported and might enhance the chance for *in-vivo* characterization and modification of this highly promising glucan polymer (Buldum et al., 2018). For the Alginate biosynthesis several approaches already exist which includes the targeted modification of the G:M ratio and sequence within the Alginate variants by *in-vivo* or *in-vitro* modification via the highly specific epimerase enzymes (de Reese et al., 2015; Ertesvag, 2015). Therefore, the *in-vivo* modification of tailored microbial homopolysaccharides seems to be more progressed than in the field of heteropolysaccharides.

Conclusion

Just recently the latest insights in heterologous expression of native as well as synthetically designed biosynthesis pathways were reviewed for capsular polysaccharides (Williams et al., 2018), exopolysaccharides (Schmid, 2018) as well as for EPS from extremophiles (Wang et al., 2018). All these articles describe the high potential of metabolic engineering, application of omics technologies as well as heterologous expression by use of well characterized so-called chassis organisms and agree that the *in-vivo* modification will result in synthetic polysaccharide structures. Based on that insights the future development of tailored EPS production seems to be close and will be massively advanced by systems- and synthetic-biology approaches.

References

Altman, E., and Li, J. (2010). Characterization of Polysaccharides Using Mass Spectrometry for Bacterial Serotyping. In Functional Glycomics: Methods and Protocols, J. Li, ed. (Totowa, NJ: Humana Press), pp. 245-257. https://doi: 10.1007/978-1-60761-454-8_17.

Andriamanantoanina, H., and Rinaudo, M. (2010). Relationship between the molecular structure of alginates and their gelation in acidic conditions. Polymer International *59*, 1531-1541. https://doi.org/10.1002/pi.2943

Becker, A., Kleickmann, A., Küster, H., Keller, M., Arnold, W., and Pühler, A. (1993). Analysis of the *Rhizobium meliloti* genes *exoU*, *exoV*, *exoW*, *exoT*, and *exoI* involved in exopolysaccharide biosynthesis and nodule invasion: *exoU* and *exoW* probably encode glucosyltransferases. Molecular plant-microbe interactions: MPMI *6*, 735-744. https://doi: 10.1094/MPMI-6-735

Benziman, M., Haigler, C.H., Brown, R.M., White, A.R., and Cooper, K.M. (1980). Cellulose biogenesis: Polymerization and crystallization are coupled processes in *Acetobacter xylinum*. Proceedings of the National Academy of Sciences of the United States of America *77*, 6678-6682. https://doi: 10.1073/pnas.77.11.6678

Bubb, W.A. (2003). NMR spectroscopy in the study of carbohydrates: Characterizing the structural complexity. Concepts in Magnetic Resonance Part A *19A*, 1-19. https://doi.org/10.1002/cmr.a.10080

Buldum, G., Bismarck, A., and Mantalaris, A. (2018). Recombinant biosynthesis of bacterial cellulose in genetically modified *Escherichia coli*. Bioprocess and Biosystems Engineering *41*, 265-279. doi: 10.1007/s00449-017-1864-1

Chen, H.-L., and Burns, L. (2006). Environmental Analysis of Textile Products, Vol 24. https://doi.org/10.1177/0887302X06293065

Cress, B.F., Englaender, J.A., He, W., Kasper, D., Linhardt, R.J., and Koffas, M.A.G. (2014). Masquerading microbial pathogens: Capsular polysaccharides mimic host-tissue molecules. FEMS microbiology reviews *38*, 660-697. https://doi.org/ 10.1111/1574-6976.12056

Cui, S. (2005). Structural Analysis of Polysaccharides.

Cumpstey, I. (2013). Chemical Modification of Polysaccharides. ISRN Organic Chemistry *2013*, 27. http://dx.doi.org/10.1155/2013/417672

de Reese, J., Sperl, N., Schmid, J., Sieber, V., and Plank, J. (2015). Effect of biotechnologically modified alginates on LDH structures. Bioinspired, Biomimetic and Nanobiomaterials *4*, 174-186. https://doi.org/10.1680/jbibn.14.00032

Ertesvag, H. (2015). Alginate-modifying enzymes: biological roles and biotechnological uses. Front Microbiol *6*, 523. https://doi.org/10.3389/fmicb.2015.00523

Flores, F., Torres, L.G., and Galindo, E. (1994). Effect of the dissolved oxygen tension during cultivation of *X. campestris* on the production and quality of xanthan gum. Journal of Biotechnology *34*, 165-173. https://doi.org/10.1016/0168-1656(94)90086-8

Gangoiti, J., Pijning, T., and Dijkhuizen, L. (2018). Biotechnological potential of novel glycoside hydrolase family 70 enzymes synthesizing α-glucans from starch and sucrose. Biotechnology Advances *36*, 196-207. https://doi.org/10.1016/j.biotechadv.2017.11.001

Gansbiller, M., Schmid, J., and Sieber, V. (2019). In-depth rheological characterization of genetically modified Xanthan-variants Carbohydrate Polymers *213*,236-246. https://doi.org/10.1016/j.carbpol.2019.02.055

Giavasis, I., Harvey, L.M., and McNeil, B. (2000). Gellan Gum. Critical Reviews in Biotechnology *20*, 177-211. https://doi.org/10.1080/07388550008984169

Hassler, R.A., and Doherty, D.H. (1990). Genetic engineering of polysaccharide structure: production of variants of xanthan gum in *Xanthomonas campestris*. Biotechnol Prog *6*, 182-187. https://doi.org/10.1021/bp00003a003

Hong, Y., Cunneen, M.M., and Reeves, P.R. (2012). The Wzx translocases for *Salmonella enterica* O-antigen processing have unexpected serotype specificity. Molecular microbiology *84*, 620-630. doi:10.1111/j.1365-2958.2012.08048.x

Hong, Y., and Reeves, P.R. (2014). Diversity of O-antigen repeat-unit structures can account for the substantial sequence variation in Wzx translocases. Journal of bacteriology, JB. 01323-01313. https://doi: 10.1128/JB.01323-13

Hublik, G. (2012). Xanthan. In Polymer Science: A Comprehensive Reference, M. K, and M. M, eds. (Elsevier), pp. 221-229.

Islam, S.T., and Lam, J.S. (2013). Wzx flippase-mediated membrane translocation of sugar polymer precursors in bacteria. Environ Microbiol *15*, 1001-1015. https://doi.org/10.1111/j.1462-2920.2012.02890.x

Islam, S.T., Taylor, V.L., Qi, M., and Lam, J.S. (2010). Membrane topology mapping of the O-antigen flippase (Wzx), polymerase (Wzy), and ligase (WaaL) from *Pseudomonas aeruginosa* PAO1 reveals novel domain architectures. MBio *1*. https://doi: 10.1128/mBio.00189-10

Jérémie Pourchez, A.P., Philippe Grosseau, René Guyonnet, Bernard Guilhot, F. Vallée (2010). HPMC and HEMC influence on cement hydration. Cement and ConcreteResearch*36*,288-294. https://doi.org/10.1016/j.cemconres.2005.08.003

Katzen, F., Ferreiro, D.U., Oddo, C.G., Ielmini, M.V., Becker, A., Pühler, A., and Ielpi, L. (1998). *Xanthomonas campestris* pv. *campestris* gum Mutants: Effects on Xanthan Biosynthesis and Plant Virulence. Journal of Bacteriology *180*, 1607-1617.

Laursen, S., Bagh, J., Hansen, J., K. Jensen, O., and Werther, I. (1997). Environmental assessment of textiles. Life cycle screening of textiles containing cotton, wool, viscose, polyester or acrylic fibres. MILJOPROJEKT, 369.

Li, S., Xiong, Q., Lai, X., Li, X., Wan, M., Zhang, J., Yan, Y., Cao, M., Lu, L., Guan, J., *et al.* (2016). Molecular Modification of Polysaccharides and Resulting Bioactivities. Comprehensive Reviews in Food Science and Food Safety *15*, 237-250. https://doi.org/10.1111/1541-4337.12161

Marolda, C.L., Tatar, L.D., Alaimo, C., Aebi, M., and Valvano, M.A. (2006). Interplay of the Wzx translocase and the corresponding polymerase and chain length regulator proteins in the translocation and periplasmic assembly of lipopolysaccharide O antigen. Journal of bacteriology *188*, 5124-5135. https://doi: 10.1128/JB.00461-06

McIntosh, M., Stone, B.A., and Stanisich, V.A. (2005). Curdlan and other bacterial (1→3)-β-d-glucans. Applied Microbiology and Biotechnology *68*, 163-173. https://doi:10.1007/s00253-005-1959-5

McNamara, J.T., Morgan, J.L.W., and Zimmer, J. (2015). A Molecular Description of Cellulose Biosynthesis. Annual review of biochemistry *84*, 895-921. https://doi: 10.1146/annurev-biochem-060614-033930.

Meng, X., Gangoiti, J., Bai, Y., Pijning, T., Van Leeuwen, S.S., and Dijkhuizen, L. (2016). Structure–function relationships of family GH70 glucansucrase and 4,6-α-glucanotransferase enzymes, and their evolutionary relationships with family GH13 enzymes. Cellular and Molecular Life Sciences *73*, 2681-2706. https://doi: 10.1007/s00018-016-2245-7.

Miasnikov, A.N. (1997). Characterization of a novel endo-levanase and its gene from *Bacillus* sp. L7. FEMS Microbiology Letters *154*, 23-28. https://doi.org/10.1111/j.1574-6968.1997.tb12619.x

Mischnick, P. (2011). Structural analysis of polysaccharides and polysaccharide derivatives. Macromolecular Symposia *99*, 3-13. https://doi.org/10.1002/masy.19950990103

Mitmesser, S., and Combs, M. (2017). Chapter 23 - Prebiotics: Inulin and Other Oligosaccharides. In The Microbiota in Gastrointestinal Pathophysiology, M.H. Floch, Y. Ringel, and W. Allan Walker, eds. (Boston: Academic Press), pp. 201-208. https://doi.org/10.1016/B978-0-12-804024-9.00023-9

Moscovici, M. (2015). Present and future medical applications of microbial exopolysaccharides. Frontiers in Microbiology *6*, 1012. https://doi.org/10.3389/fmicb.2015.01012

Öner, E.T., Hernández, L., and Combie, J. (2016). Review of Levan polysaccharide: From a century of past experiences to future prospects. Biotechnology Advances *34*, 827-844. https://doi.org/10.1016/j.biotechadv.2016.05.002

Peschard, A., Govin, A., Grosseau, P., Guilhot, B., and Guyonnet, R. (2004). Effect of Polysaccharides on the Hydration of Cement Paste at Early Ages, Vol 34. https://doi.org/10.1016/j.cemconres.2004.04.001

Rehm, B.H. (2010). Bacterial polymers: biosynthesis, modifications and applications. Nat Rev Microbiol *8*, 578-592. doi: 10.1038/nrmicro2354

Rehm, B.H., and Valla, S. (1997). Bacterial alginates: biosynthesis and applications. Appl Microbiol Biotechnol *48*, 281-288. https://doi.org/10.1007/s002530051051

Reuber, T.L., and Walker, G.C. (1993). Biosynthesis of succinoglycan, a symbiotically important exopolysaccharide of *Rhizobium meliloti*. Cell *74*, 269-280. https://doi.org/10.1016/0092-8674(93)90418-P

Robyt, J.F., Yoon, S.-H., and Mukerjea, R. (2008). Dextransucrase and the mechanism for dextran biosynthesis. Carbohydrate Research *343*, 3039-3048. https://doi.org/10.1016/j.carres.2008.09.012

Römling, U., and Galperin, M.Y. (2015). Bacterial cellulose biosynthesis: diversity of operons, subunits, products and functions. Trends in microbiology *23*, 545-557. https://doi.org/10.1016/j.tim.2015.05.005

Rühmann, B., Schmid, J., and Sieber, V. (2015). Methods to identify the unexplored diversity of microbial exopolysaccharides. Front Microbiol *6*, 565. https://doi.org/10.3389/fmicb.2015.00565

Rütering, M., Cress, B.F., Schilling, M., Rühmann, B., Koffas, M.A.G., Sieber, V., and Schmid, J. (2017). Tailor-made exopolysaccharides—CRISPR-Cas9 mediated genome editing in *Paenibacillus polymyxa*. Synthetic Biology *2*, ysx007-ysx007. https://doi.org/10.1093/synbio/ysx007

Schmid, J. (2018). Recent insights in microbial exopolysaccharide biosynthesis and engineering strategies. Current Opinion in Biotechnology *53*, 130-136. https://doi.org/10.1016/j.copbio.2018.01.005

Schmid, J., and Sieber, V. (2015). Enzymatic transformations involved in the biosynthesis of microbial exo-polysaccharides based on the assembly of repeat units. Chembiochem *16*, 1141-1147. https://doi.org/10.1002/cbic.201500035

Schmid, J., Sieber, V., and Rehm, B. (2015). Bacterial exopolysaccharides: biosynthesis pathways and engineering strategies. Front Microbiol *6*, 496. https://doi.org/10.3389/fmicb.2015.00496

Tako, M., and Nakamura, S. (1984). Rheological Properties of Deacetylated Xanthan in Aqueous Media. Agricultural and Biological Chemistry *48*, 2987-2993. https://doi.org/10.1271/bbb1961.48.2987

Tako, M., and Nakamura, S. (1988). Rheological Properties of Depyruvated Xanthan in Aqueous Media1. Agricultural and Biological Chemistry *52*, 1585-1586. https://doi.org/10.1271/bbb1961.52.1585

Toukach, P.V., and Egorova, K.S. (2016). Carbohydrate structure database merged from bacterial, archaeal, plant and fungal parts. Nucleic Acids Research *44*, D1229-D1236. https://doi.org/10.1093/nar/gkv840.

Vorhölter, F.-J., Schneiker, S., Goesmann, A., Krause, L., Bekel, T., Kaiser, O., Linke, B., Patschkowski, T., Rückert, C., Schmid, J., *et al.* (2008). The genome of *Xanthomonas campestris* pv. *campestris* B100 and its use for the reconstruction of metabolic pathways involved in xanthan biosynthesis. Journal of Biotechnology *134*, 33-45. https://doi.org/10.1016/j.jbiotec.2007.12.013

Wang, J., and Nie, S. (2018). Application of atomic force microscopy in microscopic analysis of polysaccharide. Trends in Food Science & Technology 87, 35-46. https://doi.org/10.1016/j.tifs.2018.02.005

Wang, J., Salem, D., and Sani, R. (2018). Extremophilic Exopolysaccharides: A Review and New Perspectives on Engineering Strategies and Applications. Carbohydrate Polymers, 8-26. https://doi.org/10.1016/j.carbpol.2018.10.011

Wang, X., Yang, F., and von Bodman, S.B. (2012). The genetic and structural basis of two distinct terminal side branch residues in stewartan and amylovoran exopolysaccharides and their potential role in host adaptation. Molecular microbiology *83*, 195-207. https://doi.org/10.1111/j.1365-2958.2011.07926.x

Williams, A., Linhardt, R.J., and Koffas, M.A.G. (2018). Metabolic engineering of capsular polysaccharides. Emerging Topics in Life Sciences. https://doi.org. 10.1042/ETLS20180003

Wu, M., Qu, J., Shen, Y., Dai, X., Wei, W., Shi, Z., Li, G., and Ma, T. (2019). Gel properties of xanthan containing a single repeating unit with saturated pyruvate produced by an engineered *Xanthomonas campestris* CGMCC 15155. Food Hydrocolloids *87*, 747-757. https://doi.org/10.1016/j.foodhyd.2018.09.002

Microbial Exopolysaccharides Based Drug Delivery Systems

5

Aynur Muduroglu Kirmizibekmez[1], Songul Yasar Yildiz[2], Ozlem Ates Duru[1*]

[1]Department of Medical Services and Techniques, Nisantasi University, Istanbul Turkey
[2]Department of Bioengineering, Istanbul Medeniyet University, Istanbul, Turkey
*corresponding author: ozlem.ates@nisantasi.edu.tr

DOI: https://doi.org/10.21775/9781912530267.05

Abstract
Microbial exopolysaccharides (EPSs), one of the main groups of microbially produced biopolymers, have long been accepted as significantly important biomaterials in industrial areas. Therefore there is an increased interest in microbial EPSs that show great diversity and functions with unique and commercially relevant material properties. These biodegradable, biocompatible, both human and environmental friendly polysaccharides have many uses as drug delivery vehicles, controlled drug carriers or scaffolds for tissue engineering, implantable biomaterials, etc. In this review, after a brief description of main principles and routes of drug delivery and microbial exopolysaccaharides, the microbial EPSs xanthan, dextran, levan, gellan, and curdlan are discussed in more detail with a special focus on their uses indrug delivery systems

Introduction

Drug administration is a method applied for the treatment of diseases by formulating various drug delivery systems. (Tiwari et al., 2012). In recent years, intensive studies have been conducted to increase the efficiency of medications.

A drug delivery system is defined as a formulation or vehicle allowing the utilization of a therapeutic agent in the body. Such a system improves the effectiveness and safety of drugs by controlling the rate, timing and location of their release *in vivo*. Drug delivery systems are generally stratified into two groups; namely, macromolecular drug delivery systems and particle delivery systems, such as microspheres (Ram,2018).

Design of a drug is important with regard to drug administration. Physical characteristics such as low lipophilicity, high toxicity and chemical instability are undesired features in therapeutic design. Absorption and degradation of a drug should meet the desired levels (Arnab De, 2013). Various polymeric materials such as microbial exopolysaccharides are often used to design convenient drug profiles and to ensure consistent manufacturing. Polymeric micelles have drawn considerable attention among drug delivery systems. Filling type and level of polymers used in preferred hydrophilic matrix systems, particle size of polymers and matrix porosity are important aspects (Deepak et al., 2007). Unlike non-degradable polymers, biodegradable polymers as well as self-regulating polymeric hydrogel systems have been developed and manufactured in recent years, providing convenience for drug delivery systems (Bhatia, 2016).

In this chapter, first section briefly discussed main topics such as drug delivery methods, drug formulations and targeted drug systems. In the next part a brief summary of microbial EPSs, their main properties, functions and industrial applications were presented. Lastly microbial EPSs such as dextran, xanthan, levan and pullulan based drug delivery systems were evaluated.

Drug Delivery Routes

Systemic or local drug administration is employed for the treatment of diseases. Certain routes of drug delivery such as oral, parenteral, transdermal, transmucosal, nasal and pulmonary drug delivery systems are briefly described below.

Oral Drug Delivery

Oral drug administration is the most convenient and most commonly preferred method for patients and clinicians. There are three routes of oral drug delivery: sublingual delivery, administration via buccal mucosa (mouthwashes, aerosol sprays, chewing gums, bio-adhesive tablets etc.) and administration via oral cavity (Shojaei, 1998, Hamid et al., 2018).

Oral drug administration has various disadvantages such as dissolution, permeability and solubility as well as metabolic and enzymatic degradation. Attempts are made to overcome majority of these challenges by means of certain polymer-based nanocarriers (Pragasam et al., 2017). Compared to micro-devices or macroscopic drug delivery systems, some polymeric nanocarriers offer a number of different advantages such as the ability to reach specific areas and improve intracellular distribution (Bhatia, 2016).

Parenteral Drug Delivery

Parenteral drug delivery refers to the injection of a drug under the skin (subcutaneous), into a muscle (intramuscular), into a vein (intravenous) or through vascular access (infusion). In this modality, drugs are introduced to the body through non-gastrointestinal route in order to maintain therapeutic effectiveness. This method is convenient not only in terms of bioavailability but also for use in patients with cognitive impairment who are unable to receive their medication orally (Gulati and Gupta, 2017, Martinez, 2011).

Several important drugs are formulated for parenteral use only. Bioactive macromolecules have certain advantages such as controlled release, convenient administration, less complex manufacturing and less stressful manufacturing conditions for sensitive drug molecules through (semi) solid in situ formation systems. Therefore, a number of polymeric drug delivery systems with the capability of generating a drug chamber in the injection site are currently investigated (Packhaeuser et al., 2004).

Transdermal Drug Delivery

Starting from ancient times until the present day, humans have tried treating diseases by placing various substances on the skin. In the current era, topical formulations are developed for this type of administration. Transdermal applications create an alternative particularly for oral drugs (Mark and Robert, 2008). Adhesive patches and transdermal drug delivery systems are used for the treatment and prevention of systemic diseases (Richard and Jonathan, 2003). Water-insoluble active compounds constitute a particular challenge in pharmaceutical technology. Therefore, transdermal administration is required to control and determine the amount of active components that reach skin layers (Desfrançois et al., 2018).

Transmucosal Drug Delivery

Transmucosal drug delivery may be defined as drug administration via body orifices such as oral and nasal administration. It is a preferred route of drug delivery due to being a convenient and painless method. This modality exerts its effect upon bypassing certain defensive organs such as gastric environment. The onset of action is rapid since transmucosal membranes are rich in blood flow, allowing therapeutics to enter circulation rapidly (Motwani and Lipworth, 1991). Bioadhesion is an important aspect in drug delivery systems and refers to the adhesion of a drug carrier system to a biological location. In recent years, highly adhesive polymers have been discovered, translating into important steps in transmucosal drug delivery. Further concentrated use of polymers allow improved adhesion (Roy and Prabhakar, 2010).

Nasal Drug Delivery

Nasal drug delivery is a preferred method among patients owing to its convenience. Nasal cavity is directly linked to the brain through olfactory nerves. Therefore, this method is considered as a good alternative for the treatment of certain brain disorders. However, drugs delivered by this route may also enter systemic circulation due to the rich web of blood vessels (Alexander and Saraf, 2018).

Pulmonary Drug Delivery

Aerosols that can be breathed in through inhalation have been utilized for the treatment of inflammation of the respiratory tract. While aerosols in various forms have been in use for the treatment of respiratory conditions starting from mid-twentieth century, employing the pulmonary route to administer systemic drugs is a relatively new area of interest. This method is commonly used as it is non-invasive, readily available and convenient. Slow-release therapeutic particles, prolonged presence in the respiratory tract and the limited rate of presence in blood circulation also contribute to the preference for this delivery method (Edwards et al., 1997, 1998). Pulmonary drug delivery offers various advantages compared to certain other routes of administration (oral, buccal, transdermal etc.). These include a very limited number of side effects, rapid clinical response and being effective at smaller amounts compared to systemic administration (Labiris and Dolovich 2003). Nanoparticles, especially those comprised of biodegradable polymers, are preferred as they are able to reach target regions (such as a population of lung cells), offer biocompatibility and degrade in an appropriate period of time (Tiwari et al., 2012).

Drug Delivery to the Central Nervous System

Central nervous system is the most sensitive structure in human body that is under the protection of various structures, mainly by a strong skull. Unfortunately, the structures that protect the brain may interfere with therapeutic interventions. The structure referred to as the blood-brain barrier interferes with the access of drugs to the brain, sometimes making treatment ineffective in neurodegenerative diseases such as cerebrovascular disorders which cause great harm to the central nervous system (Misra et al., 2003). Recently, research on drugs targeted to cross the blood-brain barrier has intensified. In addition to the blood-brain barrier, blood-cerebrospinal fluid barrier (BCB) represents another interference limiting the access through central nervous system. BCB is located in choroid plexus epithelium to limit the entry of molecules into cerebrospinal fluid (Yuan and Taft, 2007).

Several strategies have been studied to overcome the barriers that interfere with the cerebral penetration of therapeutic agents (Dwibhashyam and Nagappa, 2008). Drug delivery to the central nervous system may be invasive or non-invasive. However, invasive methods are not preferred due to potential severe inflammation and occurrence of serious complications such as neuron damage.

Nanoparticles developed by loading drugs to nanospheres is the most interesting approach among non-invasive methods (Kasinathan et al., 2014). Nanoparticles pass the blood-brain barrier owing to their small size and offer an effective solution for the treatment of central nervous system disorders. This method is used for the treatment of several conditions including cardiovascular disease as well as cerebral disorders (Kreuter, 2014).

Intra-osseous Infusion

Despite the increasing popularity of intra-osseous infusion for adults in emergency medicine, a limited number of studies on pharmacokinetics of intra-osseous drugs in humans have been conducted to date (Von Hoff et al., 2008). Intra-osseous intervention is preferred in the event of small veins in young children, obese patients for whom vascular access poses a challenge and in patients with vasoconstriction (Calkins et al., 2000). This method is of particular importance in conditions where medical intervention should not be delayed such as cardiac arrest and shock (Bielski et al., 2017). Intra-osseous administration may be performed by means of manual or semi-automated devices. Manual devices are more affordable; however, they require vigilance and proper training (Neuhaus, 2014). The administration may be performed in one of the three regions: proximal tibia, malleolus and radius. Utilized under compulsory circumstances, this administration into long bones allows the administered agents to pass into the bone marrow and systemic venous system (Bosomworth, 2008; Petitpas et al., 2016).

Drug Formulations

Environmental factors such as pH, temperature and enzymes, are also important in drug release mechanisms in addition to drug delivery aspects such as drug diffusion and erosion principles. Apart from bioavailability problems, drugs may not be soluble both in water and under organic conditions (Müller et al.,2001; Hamid et al., 2018). Drugs with different modifications are manufactured for this reason.

Sustained Release

Several drugs have been formulated successfully for sustained release, also referred to as long-acting formulations. Sustained-release drugs offer various therapeutic advantages. These include advantages such as minimizing systemic toxicity and maximizing effectiveness. Since prolonged presence of a drug in the body may result in toxic effects, drugs should be used with caution (Radomsky, 2000). Polymeric materials are often used to design the most convenient profile and to ensure appropriate, consistent manufacturing. Designing a successful profile requires rigorous controls and comprehensive studies on protein degradation during manufacturing, burst release, efficiency of encapsulation, with particular attention on quality of polymers utilized in hydrophilic matrix systems (Wu and Jin, 2008; Chauhan,2018).

Controlled Release

Controlled release mechanisms aim to deliver a given drug with a particular rate over a specified duration of time. The control on release of drugs is developed by multidisciplinary branches such as the science of polymers and engineering technologies. Polymeric layers in particular are commonly used in modified-release coated tablets and materials (Mustafa et al.,2016). Polymeric micelles are preferred owing to having small particles, being partially hydrophobic and effective in long-term release of drugs (Deepak et al., 2007).

Prodrugs

High-quality studies have been conducted in prodrug development since these agents are able to regulate the pharmacokinetics, pharmacodynamics and toxic effects of drugs (Jornada et al., 2016). The goal of designing a drug is mainly minimizing side effects while increasing the effectiveness. Functions and molecular structures of targeted prodrugs, enzymes and carrier systems should be adequately understood (Redasani and Bari, 2015). Prodrugs should have good access to the relevant receptor and be responsible for the activation of chemicals or enzymes in the target region (Jain, 2008).

Prodrugs may be classified into two main groups, namely carrier-linked prodrugs and bioprecursors.

A bioprecursor is an inactive compound without a carrier that is converted into active drug through a chain of reactions. Polymeric skeletons are often used as carriers in carrier-linked prodrugs (macromolecular prodrugs) (Jornada et al., 2016). Currently, different characteristics of polymeric drug formulations have been discovered and their immune protection effects have become an area of interest (Hoste et al., 2004).

Capsules and Tablets

Capsules are created by placing a drug formulation (which may be in powder, semi-solid or liquid form) inside a shell. Capsules may easily adapt to units containing different drugs or to different release modes (Wilson and Crowley, 2011).

There are currently two main types of capsules in use:

- Soft gelatin capsules
- Hard capsules

Enteric coating of hard gelatin capsules may be obtained from cellulose-based polymers. Soft gelatin capsules contain liquid or semi-solid drugs and cannot be opened without destroying the capsule (Ram, 2018b).

Tablets represent another alternative. Due to patient convenience, tablets are the most commonly preferred form. Especially in heart diseases, sublingual tablets serve as an effective way of drug delivery owing to their rapid onset of action and being readily available.

Drug Delivery Devices

Drug delivery devices (DDDs) require a multidisciplinary approach. Nanoparticle drug delivery systems (consisting of polymers, lipids etc.) are commonly manufactured and used. In recent years, there have been studies on metal and silica-based inorganic structures for DDD systems. These structures offer a wide range of use as they have magnetic, optical and electronic characteristics. Shells of these inorganic structures are usually coated with biological polymers (Wang et al., 2016).

Continuous-release drugs are implanted to the required and most convenient part of the body by means of a device. The polymers used in these implants should not cause any irritation or inflammation in the implantation area. (Jain, 2008).

Targeted Delivery Systems

This refers to targeted drug delivery over a specified duration and at a specified amount to administer a therapeutic to a cell, to intracellular structures, tissues or a specific site of an organ. Targeted drug delivery is most often used for cancer tissues (Ram, 2018a). New and emerging technologies such as liposomes, proliposomes, microspheres as well as nanotechnology and molecular mechanisms are commonly employed in targeted drug delivery systems (Tiwari et al., 2012).

Nanotechnology-Based Drug Delivery

Nanotechnology-based products have been introduced for use in several fields. There are challenges with regard to the delivery of therapeutics such as susceptibility to degradation, rapid clearance and certain biological barriers. Nanoparticles facilitate overcoming these difficulties (Ako-adounvo et al., 2017). Nano-carriers are designed to deliver localized effects. The wide surface area of nanoparticles allows transferring high amounts of drugs, reducing the need for high-dose carriers (Hamid et al., 2018). Nanoparticle technology paves the way for important advances such as early diagnosis of several cancers and identification of pathogens. Nanoparticles are divided into two main groups: non-rigid nanoparticles (such as liposomes) and rigid nanoparticles (such as degradable polymeric nanoparticles). There have been extensive studies on polymeric nanoparticles recently. For instance, rapid degradation of acrylic polymers prevents accumulation in the target organ, thereby keeping side effects at a minimum level (Ray and Mitra, 2017).

Microbial Polysaccharides

Polysaccharides can be defined as “compounds consisting of a large number of monosaccharides linked glycosidically” (Panitch et al., 2014). Moreover, the term polysaccharide can be used for the sugar part of a glycoconjugate (glycolipid, glycoprotein and proteopolysaccharide), even though the carbohydrate is only an oligosaccharide (Dwek, 1996). Polysaccharides can be homopolysaccharides or heteropolysaccharides of monomeric sugar residues, and can be either branched or linear.

Polysaccharides have biological roles that contribute to physical/structural integrity, formation of extracellular matrix, protein folding, signal transduction, and data transport between cells. Most of the polysaccharides are present on the outer surface of the cell or secreted to the extracellular environment. As a result, they take role in variable interactions between cell–cell, cell–matrix, and cell–molecule that are critical to the development and function of a complex multicellular organism. Polysaccharides can also be involved in the interactions between different organisms (Varki and Sharon, 2009). Polysaccharides can function as adhesion molecules, recognition molecules and signaling molecules due to their variable complex structure on the cell surface (Ofek et al., 2003a; Ofek et al., 2003b; Sharon and Lis,

1989). Likewise, cell surface polysaccharides take a role in many physiologically significant functions such as embryonic development, cell differentiation, growth of organism, cell–cell recognition, cell signaling, host–pathogen interaction during infection, host immune response, development of disease, metastasis, intracellular trafficking and localization, rate of degradation and membrane rigidity (Ghazarian et al., 2011). Furthermore glycoconjugates i.e. polysaccharides covalently bound to other macromolecules, like proteins or lipids (glycoproteins and glycolipids respectively) play important structural roles in cell communication and signaling events (von der Lieth et al., 2009). Two types of linkages between oligosasaccharides and proteins are present. The first type, called N-linked polysaccharides include the binding of N-acetylglucosamine to the amide side chain of asparagine. The second type, O-linked polysaccharides contain the binding of C-1 of N-acetylgalactosamine to the hydroxyl function of serine or threonine (Gorelik et al., 2001).

Despite polysaccharide is one of the four major components of living cells, their functions are less understood compared to protein, lipid or DNA. Nonetheless, polysaccharides are the most abundant and structurally varied biopolymers present in nature. Glycosylation (polysaccharide biosynthesis) consist of the many pathways and enzymatic activities, that generate the variability of the secretory pathway (Lowe and Marth, 2003).

Microbial exopolysaccharides (EPS)

Exopolysaccharide (EPS) is a term first used by Sutherland in 1972 to describe high molecular weight carbohydrate polymers produced by many marine bacteria (Sutherland, 1972). EPSs are known as polysaccharides found in extracellular matrix of the microbial cell (Rodríguez-Carvajal et al., 2008). The biosynthesis of most EPSs is very similar to synthesis of bacterial wall polymers such as peptidoglycan and lipopolysaccharide (Suresh Kumar et al., 2007). various microorganisms (gram-positive and gram-negative bacteria, fungi and also some algae) synthesize EPSs and secrete into the extracellular environment either as soluble or insoluble polymers. As a result, together with composition and functions of EPSs, chemical and physical properties that establish primary conformation of them show differences from one bacterial species to another.

Composition of EPSs contains majorly sugar residues (a wide range of sugar residues) and some non-carbohydrate substituents (such as acetate, pyruvate, succinate and phosphate) in minor amount (Llamas et al., 2012; Nicolaus et al., 2010; Staudt et al., 2012; Vu et al., 2009).

EPSs, which are surrounding the environment of microbial cells as a layer with no apparent association to any particular cell, are found in two basic forms as discrete capsular material or excreted as slime material unattached to the cell surface. A capsule form (capsular polysaccharide, CPS, K-antigens) is eventually linked with the cell surface at a limited number of discrete sites and slime form is bound to the cell surface with larger and smaller molecules forming the slime (Kumar and Mody, 2009; Llamas et al., 2012; Mata et al., 2006; Nicolaus et al., 2010).

EPSs can be categorized according to their monomeric composition as homopolysaccharides which are composed of a one type of monosaccharide, like dextran, levan or mutan and heteropolysaccharides which have more complex structures than homopolysaccharides and are made up of numerous types of monosaccharides and also varying in size from disaccharides to heptasaccharides like xanthans or gellans.

Heteropolysaccharides are usually produced inside the cell in the form of repeating units. Most of the homopolysaccharides are neutral glucans, whereas heteropolysaccharides are mainly polyanionic owing to the occurrence of uronic acids (D-glucuronic, D-galacturonic and D-mannuronic acids) or ketal-linked pyruvate (Donot et al., 2012; Llamas et al., 2012; Nicolaus et al., 2010; Nwodo et al., 2012).

Most of EPSs producer bacteria have been reported to synthesize either homopolysaccharide or heteropolysaccharide (Suresh Kumar et al., 2007). Besides, it is reported that some bacteria (*Aeromonas salominicida, Serratia marcescens* and *Pseudomonas* sp. strain NCIB 2021) can synthesize both of homo- and heteropolysaccharides (Kwon et al., 1994). Based on monomeric units and linkage bonds between these monomers homopolysaccharides can be classified as α-D-glucans/β-D-glucans, fructans and polygalactans. On the other hand, heteropolysaccharides can be composed of sugar residues such as D-glucose, D-galactose, L-rhamnose and N-acetylglucosamine (GlcNAc), N-acetylgalactosamine (GalNAc) or glucuronic acid (GlcA) and non-sugar components such as phosphate, acetyl and glycerol. Homopolysaccharides and heteropolysaccharides are synthesized by different enzymes at different sites of the cell. Sucrose is the specific substrate required for the homopolysaccharide biosynthesis and homopolysaccharides are synthesized extracellularly. On the other hand, heteropolysaccharide are synthesized intracellularly and monomers are located across the membrane by isoprenoid glycosyl carrier lipids for extracellular polymerization (Nwodo et al., 2012).

EPSs have also been categorized in seven classes (Flemming and Wingender, 2010) based on their functionality such as:

- constructive or structural (serve in the matrix help water retention and cell protection)
- sorptive (composed of charged polymers)
- surface-active (including molecules with amphiphilic behaviour)
- active
- informative
- redox-active
- nutritive

The industrial applications of glycans was based until recently on materials produced by plants (e.g. starch, cellulose, pectins, galactomanans, gums) or algae (e.g. carragenan, alginates, agar) (Ruas-Madiedo and De Los Reyes-Gavilán, 2005). In the last three decades, biotechnologies for the microbial production of extracellular polysaccharides were developed. Most of them are industrially valuable polymers that can be an alternative to with plant and algal EPSs as well as synthetic polysaccharides (Singh et al., 2009; Zhao et al., 2009). Although the first commercial polysaccharide that is approvable for food industry is the dextran, a number of polymers gain usability for commercial purposes. The success is because of their properties and to the diversity of producing microorganisms and synthesized polysaccharides (Sutherland, 2002). Because of their versatile and fascinating physico-chemical and rheological properties, EPSs from microorganisms serve as new bioproducts and find themselves broad areas of applications in numerous industries like food (Dlamini et al., 2009), medicine (Miranda et al., 2008; Novak et al., 2009) textiles (Fijan et al., 2009), detergents (Friedemann et al., 2009), adhesives (Hoffmann et al., 2009), microbial enhanced oil recovery (MEOR) (Singh et al., 2007), wastewater treatment (Chen et al., 2008; Pant and Adholeya, 2007), dredging (Yim et al., 2007), brewing (Alsteens et al., 2008), downstream processing (Maina et al., 2008), cosmetology (Im et al., 2009), pharmacology (Kumar et al., 2004; Yan et al., 2009).

EPS accumulation on the surface of the cell is an adaptation strategy of microorganisms. Glycan coat on the surface of the cell is generally more tick compared to dimension of the cell. Though, the biological function of EPSs in microorganism is most likely more complicated and diverse than currently known (Kambourova et al., 2016). Although some of the EPSs have same function, most of them perform different functions according to their environmental conditions and taxonomic classification of microorganism (Nwodo et al., 2012; Poli et al., 2011). Some of the functions of EPSs are given below:

- EPS is a general physical barrier that protects microorganism from environmental stress including pH, temperature, atmospheric pressure osmotic stress and light intensity (Otero and Vincenzini, 2003)
- For the maintenance of hydration in the environment EPS on the cell surface provides water retention in environments that are water-deficient (Kambourova et al., 2016).
- In extreme habitats, EPSs hold materials that can be used as nutrient on cell surface. In the bulk water phase, biodegradable compounds and cations are bound or accumulated by the film formed by EPS (Kambourova et al., 2016).
- EPSs adhere organisms to surfaces and they can serve as flocculants or emulsifiers. EPS synthesis is used strategically in order to grow and adhere to solid surfaces via hydrophobic site coverage on the cell envelope in bacteria (Kambourova et al., 2016).
- Effects of EPS film surrounded the cell on the diffusion of harmful materials, such as antibiotics or toxic compounds are also known (Nicolaus et al., 2010).
- It is known that EPSs bind and neutralize bacteriophage and play role in cell protection (Vu et al., 2009).

- According to the studies up to date, EPSs are not degraded by producer microorganism or utilized as energy source (Donot et al., 2012). However, in the case of low organic source in extreme habitats EPSs from hyperthermophilic species (*Thermotoga, Thermococcus* and *Sulfolobus*) can take a role as extracellular storage polymers (Nicolaus et al., 1993; Rinker and Kelly, 2000).

Microbial EPS based drug delivery systems

Microbial EPSs, that have critically important parameters of biocompatibility and biodegradability, are emerging as ideal candidates for the development of smart drug delivery systems such as hydrogels, microgels etc. (Zykwinska et al.,2016). Furthermore, they have desirable properties as stability, no toxicity, safety and easily reproducibility which also important to gain importance for synthesis of nanoparticles used in drug delivery, tissue engineering, cancer treatment etc. (Sivasankar et al., 2017).

Natural polysaccharides are biodegradable and easily modified substitutes for synthetic polymers since the polysaccharides have biocompatibility, high stability, low immunogenicity, bioactivity, targeting ability and also they can exist in positive, negative, or neutral charge states. Therefore, microbial exopolysaccharides, which is a subgroup of natural polysaccharides, encounter the requirements to develop ideal micelles (nanosized colloidal particles composed of a hydrophobic core and hydrophilic shell) that are appropriate carriers for poor water soluble or insoluble drugs (Zhang et al., 2013).

For controlled drug delivery systems hydrogels which are hydrophilic polymeric networks have an excessive attention with their responsive characteristics to physical and chemical factors. Since there is a requirement of chemical modifications to induce hydrogel formation for drug delivery systems, the toxicity of cross-linking molecules may be appeared as limiting factors in biological applications. Therefore, the research for new molecules is still important. One of these new molecules is an innovative polysaccharide HE800 exopolysaccharide which is secreted by deep-sea hydrothermal bacteria *Vibrio diabolicus*, demonstrates remarkable biological activities as a result of its unique structure. It has been used to generate microgels that could be used as carriers for a model protein, Bovine Serum Albumin (BSA). The results of the study showed the high potential of HE800 EPS as innovative protein microcarriers for biomedical applications (Zykwinska et al.,2016). Cardosa et al. (2017) also prepared hydrogels from gellan exopolysaccharide and starch retrogradation blends. Therefore, they obtained hydrogels that can be modulated and specified according to the purpose of uses of drug delivery systems by changing polymer concentrations and cross-linking process.

EPSs can be used as reducing and stabilizing agents for the synthesis of nanoparticles for drug delivery systems since they have hydroxyl and hemiacetal groups, which plays a vital role in reduction and stabilization. The utilization of EPSs for nanoparticle synthesis enhances the eco-friendly characteristics of nanoparticles by preventing the usage of toxic chemicals (Saravanan et al., 2017). For instance, the exopolysaccharide from *Streptomyces violaceus* were used to synthesize silver nanoparticles that showed promising antibacterial and antioxidant activity. These findings suggested the development of novel nanoparticles using EPS that could be used in future biomedical applications (Sivasankar et al., 2018). Sirajunnisa and Surendhiran (2014) reported the silver nanoparticles (SNP) using EPS from *Pseudomonas fluorescens* revealing the proteins of the polysaccharide served as a capping agent. The SNP molecules showed stability more than 6 months and the antimicrobial and antioxidant activities were also investigated. The findings showed that SNP molecules synthesized in that study could be utilized as a broad spectrum antimicrobial agent and have potential to be an antioxidant. Dhanka et al. (2017) reported preparation of methotrexate (MTX) loaded the EPS gellan microparticles by water-in-oil (W/O) emulsion solvent diffusion method and they developed microparticles a rough surface morphology. These gellan microparticles were found to be biocompatible and hemocompatible. MTX release experiments resulted in 84% MTX release within 1440 minutes. The findings demonstrated the injectable applicability of MTX loaded gellan gum microparticles.

Complexes consisted of polysaccharide and protein were widely used in medical fields due to the properties of hydration and structuration, forming hydrogen or covalent bonding easily, hydrophobic interaction and ability to form self-assemble nanoparticles (Arjama et al.,2018). Arjama and coworkers (2018) encapsulated the high amounts of Doxorubicin into sericin-albumin-gellan nanocomposites to analyze the drug release in acidic tumor environment and they obtained results that showed the potential of this type nanocomposites in drug delivery. They found that 84% of doxorubicin was released *in vitro* condition after 120 h and of 42 % of cancer cells were survived after treatment with nanocomposites.

Electrospun nanofibers with microbial EPSs mostly produced by electrospinning offer an alternative solution for drug delivery systems due to the their properties like high surface-to-volume ratio, high porosity, good stability, gelling capacity, biocompatibility, etc. Kefiran, a glucogalactan heteropolysaccharide, is a suitable candidate for development of nanostructured materials with its unique properties and antibacterial, anti-inflammatory, and protective effects. For instance, kefiran based nanofiber mat for doxycycline delivery was prepared and reducing the side effects of drug including cytotoxicity by kefiran nanofibers was obtained (Dadashi et al., 2019).

A specific dose of drugs at a controlled rate is delivered through the skin and into the blood stream by transdermal drug delivery system (TDDS) which is more useful than the oral and parenteral routes. This system has advantages such as promotion of drug bioavailability and avoiding degradation by pH associated deactivation and enzymes. TDDS is also a possible solution for the disabled or unconscious patients that cannot swallow the drugs. Microbial exopolysaacharides like gellan, xanthan and cellulose are able to be used in TDDS systems (Saidin et al., 2018). Gellan has been used as a gelling agent for transdermal delivery of diclofenac sodium (Carmona-Moran et al., 2016) and for hydrogel patch coated with gold nanorods for transdermal delivery of fluorescein isothiocyanate-modified ovalbumin (Haine et al., 2017). Grafted xanthan was utilized as a matrix film for transdermal delivery of atenolol (Mundargi et al., 2007). Bacterial cellulose was used as a matrix patch in TDDS establishing a drug permeation rate of diclofenac sodium which is similar to the commercial patch of Olfen® (Silva et al., 2014).

Microbial EPSs drug delivery researches focused on dextran, levan, pullulan and xanthan were also discussed in details in the following section.

Dextran

Dextran is a homopolysaccharides composed of α-1,6 glycosidic linkages, is produced by various strains of bacteria such as *Leuconostoc, Streptococcus* and *Acetobacter* in the presence of sucrose (Patel et al., 2011; Casettari et al., 2015). Due to its flexible structure caused by free rotation of glycosidic bond and it has important properties like high solubility in water, biocompatibility and biodegradability it becomes a functional hydrocolloid having commercial applications in food, pharmaceutical and chemical industries as adjuvant, emulsifier, thickener, carrier, stabilizer therapeutic agent to restore blood volume, and for the matrix preparation of chromatography (Ahmad et al., 2015, Ates, 2015). Dextran has also ability to be chemically modified by different reagents which makes it a suitable drug carrier (Yadav and Karthiyekan, 2019).

Acryloyl crosslinked dextran dialdehyde (ACDD) nanoparticles were used to synthesize a novel insulin delivery system. This system is composed of glucose oxidase immobilized ACDD which is a novel glucose responsive and *in vitro* pH triggered delivery system. The *in vitro* insulin release studies were performed to investigate release behavior and kinetics at different pH values also with different glucose concentrations. A 'non-Fickian' behavior of insulin diffusion release and controlled release of about 90% under artificial intestinal fluid conditions were observed presenting the potential of this system for subcutaneous insulin therapy (Shivani et al., 2018). Dextran was also used as a component of nanocomposite scaffolds for preparing anticancer drug carriers. El-Meliegy et al. (2018) reported the design of nanocomposite scaffolds based on dicalcium phosphate, dextran and carboxymethyl cellulose. Cytotoxicity and drug release studies were also carried out and the findings demonstrated the impressive cytotoxicity and 90% drug release (at high dicalcium phosphate concentrations).

For anti-cancer drug delivery systems use of dextran based hydrogels and micelles have been performed in several studies. In one study, the synthesis of polyacrylic acid grafted dextran nanohydrogels was carried and *in vitro* release of doxorubicin, anti-cancer effects *in vivo*, and biosafety for these nanohydrogels were investigated. The pH- redox controlled drug release, reducing toxicity of free doxorubicin and inhibition of tumor cells were obtained (Wang et al., 2017). In another study, dextran microspheres were loaded with vincristine sulfate and findings demonstrated the sustained drug release and increased cytotoxicity for THP-1 human leukemia cells (Thakur et al., 2016).

Levan

Levan is a microbial exopolysaccharide which is composed of β-d-fructofuranose with β(2-6) linkages between fructose rings and it is synthesized by direct conversion of sucrose with the enyzme levansucrase by a wide range of microorganisms such as *Acetobacter*, *Bacillus*, *Erwinia*, *Gluconobacter*, *Pseudomonas*, *Zymomonas*, *Halomonas* (Erkorkmaz et al., 2018). Levan homopolysaccharide has a great potential in foods, feeds, cosmetics, pharmaceutical, and chemical industries with various distinguished properties such as high solubility in oil and water, biocompatibility, biodegradability, adhesivity, and film-forming ability (Kang et al., 2009; Kazak Sarilmiser et al., 2015).

Sezer et al. (2011) investigated the potential of levan as a nanoparticle for drug delivery. They prepared levan nanoparticles with bovine serum albumin (BSA) encapsulated and analyzed the characteristics of these nanoparticles and drug release. The results demonstrated 10% and 23% of BSA released within the first 10 days and affected release behavior by initial loading capacity of protein. Another study reported the encapsulation of vancomycin (VANCO) into levan microparticles. The investigation of release behavior showed that the VANCO release fit to Higuchi kinetic model. The VANCO loaded levan microparticles have shown antibacterial effect against *Bacillus subtilis* ATCC 6633 and also biocompatibility according to the cytotoxicity analysis (Sezer et al., 2015).

Levan was also used with N-isopropyl acrylamide for preparation of biocompatible and temperature responsive hydrogels. Levan/N-isopropyl acrylamide hydrogels were analyzed for 5-aminosalicylic acid release and biocompatibility. The release of the drug reached almost 100 % and the increased biocompatibility was obtained by increasing levan content (Osman et al., 2017). In another study, Patel et al. (2019) synthesized paclitaxel loaded levan micelles that were targeted to Glucose Transporter 5 and investigated their anticancer activity in breast cancer cell line and also performed hemocompatibility studies. A nineteen-fold improvement in the IC50 value was obtained in this study.

Recently, González-Garcinuño and coworkers (2019) produced levan-capped silver nanoparticles to investigate silver release and silver bactericidal effect. The nanoparticles showing bactericide effect, for *E. coli* and *B. subtilis* demonstrated a

29% loading capacity. Additionally, a mathematical model was developed to investigate the potential of the drug delivery system of levan nanoparticles introduced in an alginate gel for designing drug release parameters.

Pullulan

Pullulan is composed of maltotriose reduplicative units *connected by α-1,4*-linkages and is produced by strains such as *Aureobasidium pullulans*, *Remella mesenterica*, *Cryphonectria parasitica*, and *Teloschistes flavicans*. It has outstanding chemical and physical properties, such as high solubility in water, low viscosity, non-toxicity, slow digestibility, high plasticity, and excellent film-forming (Cheng et al., 2011; Sheng et al., 2014). Due to this superior properties pullulan represents a high value profile as a component of hydrogels and nano/micro capsules for biomedical applications such as drug/gene delivery, tissue engineering, imaging, etc. For instance, a hydrogel prepared by a mixture of pullulan and poly(acrylic acid) containing Clotrimazole-loaded nanocapsules was developed for the treatment of vulvovaginal candidiasis (de Lima et al., 2017).

Recently, Choudhury (2019) reported synthesis of a novel composite hydrogel using gellan and pullulan maintaining its elasticity at high temperatures and frequency. This hydrogel demonstrated a high potential to be used as a drug carrier since the characterization studies resulted in fast gelling and stability of elasticity. In addition, Lin et al. (2019) prepared a pullulan based glucose-sensitive hydrogel and encapsulated insulin with this hydrogel to investigate release of insulin. The studies were resulted in smart release of insulin in response to changes in the glucose concentration which proposing a closed-loop systems for diabetes treatment.

Carrigy et al. (2019) reported pullulan trehalose powder, produced either with a monodisperse droplet chain or a spray dryer that could be used as microparticle respiratory drug delivery system. They investigated the characteristics of this compressible and non-cohesive platform and the promising findings demonstrated no morphological change, high dispersibility, optimal size for inhalation, and adequate total lung dose that pointed out the possibility of being a good candidate for respiratory drug delivery.

Doxorubicin-loaded nanodiamonds were encapsulated by pullulan-all-trans-Retinal (pullulan-ATR) self-assembled nanoparticles which were pH-sensitive and showed homogeneous size distribution with good dispersity. Intratumoral penetration of these nanoparticles were performed by ultrasound and enhancement of the intracellular retention of DOX with 17.3% of injected dosage of DOX in the tumor tissue. The novel and potential use of these nanoparticles for clinical cancer therapy was also confirmed with improved anti-tumor effect in the DOX-sensitive HepG2 and DOX-resistant HepG2/ADR tumor models *in vivo*.

Xanthan

Xanthan which is known as the most important commercial microbial EPS is produced by the plant-pathogen-proteobacterium *Xanthomonas campestris pv. campestris* (Xcc) (Frese et al., 2014). This EPS is a heteropolysaccharide composed of repetitive pentasaccharide units consisting of monomeric units of two glucose, two mannose, and one glucuronic acid residue with a backbone chain consisting of (1-4)-β-d-glucan cellulose (Khan et al., 2007). Xanthan has superior properties and rheological characteristics such as high viscosity at low concentrations, stability over a wide range of pH and temperature. High molecular weight xanthan has found a wide range of applications as a thickening or stabilizing agent in food, pharmaceutical, textile, paper, paint, cosmetics and oil drilling industries (Ates 2015, Ates and Toksoy Oner, 2017).

Xanthan has been used in drug delivery systems as excipient in tablets or as supporting hydrogels. To have the desired properties for drug carriers, xanthan is used with other polymers. Aloe or curcumin loaded hydrogels of xanthan and chitosan can find potential usage for skin repairing or skin diseases (Petri, 2015). For instance, effective utilization of xanthan-plant galactomannan hydrogels loaded with curcumin in skin-diseases, as anti-inflammatory and antioxidant drugs demonstrated the potential of these hydrogels as an alternative treatment method for skin-cancer and psoriasis (Koop et al., 2015).

In addition, Sabaa et al., (2019) synthesized xanthan gum (XG)/poly (N–vinyl imidazole (PVI) hydrogel as a potential protein carrier to investigate the controlled release of proteins. For this purpose, BSA was encapsulated with this xanthan based hydrogel, subsequently cytotoxicity and protein release analysis were performed. The BSA release showed increase with increased polymer concentrations. Besides the results confirmed the biocompatibility of hydrogel with no cytotoxic effects and its potential for delivery of therapeutic proteins.

Another study reported the successful preparation and characterization of xanthan based microspheres and hydrogels to investigate colorectal release of omega-3 polyunsaturated fatty acids (Trombino et al., 2019). Furthermore, Hanna and Saad (2019) prepared hydrogel of Ntrimehtyl chitosan (TMC)/sodium carboxymethyl xanthan gum (CMXG) for encapsulation of ciprofloxacin (CFX). The release kinetics of loaded CFX hydrogel followed zero-order kinetics and a successful released with high activity against the gram positive and gram negative bacteria was obtained confirming the potential of non-toxic CMXG for controlled release drug delivery system.

Conclusions

Microbially produced exopolysaccharides are mainly associated with high-value applications in various areas and are rapidly emerging as new and industrially important biomaterials. Therefore, a growing body of research associated with these high-value microbial EPSs makes these natural polysaccharides increasingly essential. For drug carriers systems they are non-toxic alternatives with their potentially harmful characteristics and unique properties. Consequently, the high potential of microbial exopolysaccharides as well-regulated drug delivery systems will enhance the researches to improve therapeutic efficacy and reduce undesired side effects of drugs.

References

Ahmad, N. H., Mustafa, S., and Che Man, Y. B. (2015). Microbial polysaccharides and their modification approaches: a review. Int. J. Food Prop. 18, 332–347. doi: 10.1080/10942912.2012.693561

Ako-adounvo, A. A., Beatriz, M., Rayssa, C. L., Ayuk, P., and Pradeep, K. K. (2017). Drug and Gene Delivery Materials and Devices. In Emergıng Nanotechnologıes For Diagnostıcs, Drug Delıvery, and Medical Devices, Mitra, K. M., Kıshore, C., Abhırup, M, ed. (Washington DC, USA: Elsevier), pp.375-86. http://dx.doi.org/10.1016/B978-0-323-42978-8.00015-2

Alexander, A., Saraf, S. (2018). Nose-to-brain drug delivery approach: A key to easily accessing the brain for the treatment of Alzheimer's disease. Neural Regen Res. 13, 2102-2104. https://doi.org/10.4103/1673-5374.241458.

Alsteens, D., Dupres, V., Mc Evoy, K., Wildling, L., and Gruber, H.J. (2008). Structure, cell wall elasticity and polysaccharide properties of living yeast? cells, as probed by AFM. Nanotechnology 19, 384005. https://doi.org/10.1088/0957-4484/19/38/384005

Arjama, M., Mehnath, S., Rajan, M., Jeyaraj, M. (2018). Sericin/RBA embedded gellan gum based smart nanosystem for pH responsive drug delivery. International journal of biological macromolecules, 120, 1561-1571., doi:10.1016/j.ijbiomac.2018.09.146

Arnab De. (2013). Application of Peptide-Based Prodrug Chemistry in Drug Development.(NW, USA: Springer). https://dx.doi.org/10.1007/978-1-4614-48754

Ates, O. And Oner, E. T. (2017). Microbial Xanthan, Levan, Gellan, and Curdlan as Food Additives. Microbial Functional Foods and Nutraceuticals, 149. https://doi.org/10.1002/9781119048961.ch7

Ates, O., Arga, K. Y., and Toksoy Öner, E. (2013). The stimulatory effect of mannitol on levan biosynthesis: lessons from metabolic systems analysis of *Halomonas smyrnensis* AAD6T. *Biotechnol. Prog.* 29, 1386–1397. doi:10.1002/btpr.1823

Ates, O., Toksoy Oner, E., and Arga, K. Y. (2011). Genome-scale reconstruction of metabolic network for a halophilic extremophile, *Chromohalobacter salexigens* DSM 3043. *BMC Syst. Biol.* 5:12. doi:10.1186/1752-0509-5-12

Bhatia, S. (2016). Systems for Drug Delivery. (Switzerland: Springer International Publishing). https://dx.doi.org/10.1007/978-3-319-41926-8_1

Bielski, K., Szarpak, L., Smereka, j., Ladny, J.R., Leung, S., and Ruetzler, K. (2017). Comparison of four different intraosseous access devices during simulated pediatric resuscitation. A randomized crossover manikin trial. Eur J Pediatr. 176, 865–871. https://doi.org/10.1007/s00431-017-2922-z.

Bosomworth, N.J. (2008). The occasional intraosseous infusion. Can J Rural Med. Spring;13(2):80-3.

Calkins, M.D., Fitzgerald, G., Bentley, T.B, Burris, D. (2000). Intraosseous infusion devices: a comparison for potential use in special operations. J Trauma. 48, 1068-74.

Carmona-Moran, C. A., Zavgorodnya, O., Penman, A. D., Kharlampieva, E., Bridges Jr, S. L., Hergenrother, R. W., ... & Wick, T. M. (2016). Development of gellan gum containing formulations for transdermal drug delivery: Component evaluation and controlled drug release using temperature responsive nanogels. International journal of pharmaceutics, 509(1-2), 465-476. https://doi.org/10.1016/j.ijpharm.2016.05.062

Carrigy, N. B., Ordoubadi, M., Liu, Y., Melhem, O., Barona, D., Wang, H., ... & Vehring, R. (2019). Amorphous pullulan trehalose microparticle platform for respiratory delivery. International Journal of Pharmaceutics, 563, 156-168. https://doi.org/10.1016/j.ijpharm.2019.04.004

Casettari, L., Bonacucina, G., Morris, G. A., Perinelli, D. R., Lucaioli, P., Cespi, M., et al. (2015). Dextran and its potential use as tablet excipient. *Powder Technol.* 273, 125–132. doi:10.1016/j.powtec.2014.12.030

Chauhan, A.S. (2018). Dendrimers for Drug Delivery. Molecules. 23(4). pii: E938. https://doi.org/10.3390/molecules23040938.

Chen, W., Zhao, Z., Chen, S.-F., and Li, Y.-Q. (2008). Optimization for the production of exopolysaccharide from *Fomes fomentarius* in submerged culture and its antitumor effect in vitro. Bioresource technology 99, 3187-3194. https://doi.org/10.1016/j.biortech.2007.05.049

Cheng, K. C., Demirci, A., and Catchmark, J. M. (2011). Pullulan: biosynthesis, production, and applications. Appl. Microbiol. Biotechnol. 92, 29–44. doi:10.1007/s00253-011-3477-y

Choudhury, A. R. (2019). Synthesis and rheological characterization of a novel thermostable quick setting composite hydrogel of gellan and pullulan. International journal of biological macromolecules, 125, 979-988. https://doi.org/10.1016/j.ijbiomac.2018.12.147

Dadashi, S., Boddohi, S., & Soleimani, N. (2019). Preparation, characterization, and antibacterial effect of doxycycline loaded kefiran nanofibers. Journal of Drug Delivery Science and Technology. https://doi.org/10.1016/j.jddst.2019.06.012.

de Oliveira Cardoso, V. M., Cury, B. S. F., Evangelista, R. C., and Gremião, M. P. D. (2017). Development and characterization of cross-linked gellan gum and retrograded starch blend hydrogels for drug delivery applications. Journal of the mechanical behavior of biomedical materials, 65, 317-333. https://doi.org/10.1016/j.jmbbm.2016.08.005

Deepak, T., Michel, D., Yashwant, P. (2007). Nanoparticulate Drug Delivery Systems. In Nanoparticulate Drug-Delivery Systems: An Overview Deepak, T., Michel, D., Yashwant, P., ed. (NW,USA: Informa Healthcare), pp.1-32

Desfrançois, C., Auzély, R., Texier, I. (2018). Lipid Nanoparticles and Their Hydrogel Composites for Drug Delivery: A Review. Pharmaceuticals (Basel). 11 pii E118. https://doi.org/10.3390/ph11040118

Dhanka, M., Shetty, C., & Srivastava, R. (2018). Methotrexate loaded gellan gum microparticles for drug delivery. International journal of biological macromolecules, 110, 346-356. https://doi.org/10.1016/j.ijbiomac.2017.12.026

Dlamini, A.M., Peiris, P.S., Bavor, J.H., and Kailasapathy, K. (2009). Rheological characteristics of an exopolysaccharide produced by a strain of *Klebsiella oxytoca*. Journal of bioscience and bioengineering 107, 272-274. https://doi.org/10.1016/j.jbiosc.2008.11.023

Donot, F., Fontana, A., Baccou, J., and Schorr-Galindo, S. (2012). Microbial exopolysaccharides: main examples of synthesis, excretion, genetics and extraction. Carbohydrate Polymers 87, 951-962. https://doi.org/10.1016/j.carbpol.2011.08.083

Dwek, R.A. (1996). Glycobiology: toward understanding the function of sugars. Chemical Reviews 96, 683-720. https://doi.org/10.1021/cr940283b

Dwibhashyam, V. S. N. M., and Nagappa, A. N. (2008). Strategies for enhanced drug delivery to the central nervous system. Indian J Pharm Sci, 70, 145-153. https://dx.doi.org/10.4103%2F0250-474X.41446

El-Meliegy, E., Mabrouk, M., Kamal, G. M., Awad, S. M., El-Tohamy, A. M., & El Gohary, M. I. (2018). Anticancer drug carriers using dicalcium phosphate/dextran/CMCnanocomposite scaffolds. Journal of Drug Delivery Science and Technology, 45, 315-322. https://doi.org/10.1016/j.jddst.2018.03.026

Edwards, D.A, Hanes, J., Caponetti, G., Hrkach, J., Ben-Jebria, A., Eskew, M.L, Mintzes, J., Deaver, D., Lotan, N., and Langer, R. (1997). Large Porous Particles for Pulmonary Drug Delivery. Science 276, 1868. https://doi.org/10.1126/science.276.5320.1868.

Edwards, D.A., Ben-Jebria, A and Langer, R. (1998). Recent advances in pulmonary drug delivery using large, porous inhaled particles. Journal of Applied Physiology. 85, 279-385 (Review). https://doi.org/10.1152/jappl.1998.85.2.379

Erkorkmaz, B. A., Kırtel, O., Duru, Ö. A., Öner, E. T. (2018). Development of a cost-effective production process for *Halomonas* levan. Bioprocess and biosystems engineering, 41(9), 1247-1259. https://doi.org/10.1007/s00449-018-1952-x

Fijan, R., Basile, M., Šostar-Turk, S., Žagar, E., Žigon, M., and Lapasin, R. (2009). A study of rheological and molecular weight properties of recycled polysaccharides used as thickeners in textile printing. Carbohydrate Polymers 76, 8-16. https://doi.org/10.1016/j.carbpol.2008.09.027

Flemming, H.-C., and Wingender, J. (2010). The biofilm matrix. Nature Reviews Microbiology 8, 623. https://doi.org/10.1038/nrmicro2415

Frese, M., Schatschneider, S., Voss, J., Vorhölter, F. J., and Niehaus, K. (2014). Characterization of the pyrophosphate-dependent 6-phosphofructokinase from *Xanthomonas campestris* pv. campestris. Arch. Biochem. Biophys. 546, 53–63. doi: 10.1016/j.abb.2014.01.023

Friedemann, K., Stallmach, F., and Kärger, J. (2009). Carboxylates and sulfates of polysaccharides for controlled internal water release during cement hydration. Cement and Concrete Composites 31, 244-249. https://doi.org/10.1016/j.cemconcomp.2009.02.010

Ghazarian, H., Idoni, B., and Oppenheimer, S.B. (2011). A glycobiology review: carbohydrates, lectins and implications in cancer therapeutics. Acta histochemica 113, 236-247. https://doi.org/10.1016/j.acthis.2010.02.004

González-Garcinuño, Á., Masa, R., Hernández, M., Domínguez, Á., Tabernero, A., & del Valle, E. M. (2019). Levan-Capped Silver Nanoparticles for Bactericidal Formulations: Release and Activity Modelling. International journal of molecular sciences, 20(6), 1502. https://doi.org/10.3390/ijms20061502.

Gorelik, E., Galili, U., and Raz, A. (2001). On the role of cell surface carbohydrates and their binding proteins (lectins) in tumor metastasis. Cancer and Metastasis Reviews 20, 245-277. https://doi.org/10.1023/A:1015535427597

Gulati, N., Gupta, H. (2011). Parenteral drug delivery: a review. Recent Pat Drug Deliv Formul. 5 :133-45.

Haine AT, Koga Y, Hashimoto Y, et al., Enhancement of transdermal protein delivery by photothermal effect of gold nanorods coated on polysaccharide-based hydrogel. Eur J Pharm Biopharm, 2017; 119:91-95. https://doi.org/10.1016/j.ejpb.2017.06.005

Hamid, R. R., Mohammaaossein, E.,Abolfazl, A. A., Andreas, Ö. (2018). A Review of Biomaterials and Their Applications in Drug Delivery. (NW, USA: Springer). https://doi.org/10.1007/978-981-10-0503-9

Hanna, D. H., Saad, G. R. (2019). Encapsulation of ciprofloxacin within modified xanthan gum-chitosan based hydrogel for drug delivery. Bioorganic chemistry, 84, 115-124. https://doi.org/10.1016/j.bioorg.2018.11.036

Hoffmann, B., Volkmer, E., Kokott, A., Augat, P., Ohnmacht, M., Sedlmayr, N., Schieker, M., Claes, L., Mutschler, W., and Ziegler, G. (2009). Characterisation of a new bioadhesive system based on polysaccharides with the potential to be used as bone glue. Journal of Materials Science: Materials in Medicine 20, 2001-2009. https://link.springer.com/article/10.1007/s10856-009-3782-5

Hoste, K., De Winne, K., and Schacht, E. (2004). Polymeric prodrugs. Int J Pharm. 277, 119-31. https://doi.org/10.1016/j.ijpharm.2003.07.016.

Im, J.-H., Song, J.-M., Kang, J.-H., and Kang, D.-J. (2009). Optimization of medium components for high-molecular-weight hyaluronic acid production by Streptococcus sp. ID9102 via a statistical approach. Journal of Industrial Microbiology & Biotechnology 36, 1337-1344. https://link.springer.com/article/10.1007/s10295-009-0618-8

Jain, K. K.(2008). Drug Delivery Systems. (Totowa, NJ: Human Press).

Jornada, H. D., Fernandes, F.D.S., Chiba, D. E., Melo, T.R.F., Santos, J.L., and Chung, M.C. (2016). The Prodrug Approach: A Successful Tool for Improving Drug Solubility. Moleculs. 21, 42. https://doi.org/10.3390/molecules21010042.

Kambourova, M., Radchenkova, N., Tomova, I., and Bojadjieva, I. (2016). Thermophiles as a Promising Source of Exopolysaccharides with Interesting Properties. In Biotechnology of Extremophiles: (Springer), pp. 117-139. https://doi.org/10.1007/978-3-319-13521-2_4

Kang, S. A., Jang, K. H., Seo, J. W., Kim, K. H., Kim, Y. H., Rairakhwada, D., et al. (2009). "Levan: applications and perspectives," in Microbial Production of Biopolymers and Polymer Precursors: Applications and Perspectives, ed. Rehm B. (Norfolk: Caister Academic Press, Inc.), 145–161.

Kasinathan, N., Volety, S. M., Josyula, V.R. (2014). Application of Experimental Design in Preparation of Nanoliposomes Containing Hyaluronidase. J Drug Deliv. 2014, 948650.https://dx.doi.org/10.1155%2F2014%2F948650

Kazak Sarilmiser, H., Ates, Ö, Özdemir, G., Arga, K. Y., and Toksoy Öner, E. (2015). Effective stimulating factors for microbial levan production by *Halomonas Smyrnensis* AAD6T. J. Biosci. Bioeng. 119, 455–463. doi:10.1016/j.jbiosc.2014.09.019

Khan, T., Park, J. K., and Kwon, J. H. (2007). Functional biopolymers produced by biochemical technology considering applications in food engineering. Korean J. Chem. Eng. 24, 816–826. doi:10.1007/s11814-007-0047-1

Koop, H. S., de Freitas, R. A., de Souza, M. M., Savi, R. Jr., and Silveira, J. L. M. (2015). Topical curcumin-loaded hydrogels obtained using galactomannan from *Schizolobium parahybae* and xanthan. Carbohydr. Polym. 116, 229–236. doi: 10.1016/j.carbpol.2014.07.043

Kreuter, J. (2014). Drug delivery to the central nervous system by polymeric nanoparticles: what do we know? Adv Drug Deliv Rev. 71,2-14. https://doi.org/10.1016/j.addr.2013.08.008.

Kumar, A.S., and Mody, K. (2009). Microbial exopolysaccharides: variety and potential applications. Microbial Production of Biopolymers and Polymer Precursors: Applications and Perspectives, ed. Rehm B. Caister Academic: Norfolk, UK, 2009; pp: 229-253.

Kumar, C.G., Joo, H.-S., Choi, J.-W., Koo, Y.-M., and Chang, C.-S. (2004). Purification and characterization of an extracellular polysaccharide from haloalkalophilic *Bacillus* sp. I-450. Enzyme and microbial technology 34, 673-681. https://doi.org/10.1016/j.enzmictec.2004.03.001

Labiris, N.R., and Dolovich, M.B. (2003). Pulmonary drug delivery. Part I: Physiological factors affecting therapeutic effectiveness of aerosolized medications. Br J Clin Pharmacol. 56, 588–599. https://dx.doi.org/10.1046%2Fj.1365-2125.2003.01892.x

Li, H., Ma, M., Zhang, J., Hou, W., Chen, H., Zeng, D., & Wang, Z. (2019). Ultrasound-enhanced Delivery of Doxorubicin-loaded Nanodiamonds from Pullulan-all-trans-Retinal Nanoparticles for Effective Cancer Therapy. ACS applied materials & interfaces. DOI: 10.1021/acsami.9b03559

Lin, K., Yi, J., Mao, X., Wu, H., Zhang, L. M., & Yang, L. (2019). Glucose-sensitive hydrogels from covalently modified carboxylated pullulan and concanavalin A for smart controlled release of insulin. Reactive and Functional Polymers. 139,112-119. https://doi.org/10.1016/j.reactfunctpolym.2019.01.016

Llamas, I., Amjres, H., Mata, J.A., Quesada, E., and Béjar, V. (2012). The potential biotechnological applications of the exopolysaccharide produced by the halophilic bacterium *Halomonas almeriensis*. Molecules 17, 7103-7120. https://doi.org/10.3390/molecules17067103

Lowe, J.B., and Marth, J.D. (2003). A genetic approach to mammalian glycan function. Annual review of biochemistry 72, 643-691. https://doi.org/10.1146/annurev.biochem.72.121801.161809

Maina, N.H., Tenkanen, M., Maaheimo, H., Juvonen, R., and Virkki, L. (2008). NMR spectroscopic analysis of exopolysaccharides produced by *Leuconostoc citreum* and *Weissella confusa*. Carbohydrate research 343, 1446-1455. https://doi.org/10.1016/j.carres.2008.04.012

Mark, R. P., and Robert, L. (2008). Transdermal drug delivery. Nat. Biotechnol. 26, 1261-1268. https://dx.doi.org/10.1038%2Fnbt.1504

Martinez, M.N. (2011). Factors influencing the use and interpretation of animal models in the development of parenteral drug delivery systems. AAPS J. 13, 632-49. https://doi.org/10.1208/s12248-011-9303-8

Mata, J.A., Béjar, V., Llamas, I., Arias, S., Bressollier, P., Tallon, R., Urdaci, M.C., and Quesada, E. (2006). Exopolysaccharides produced by the recently described halophilic bacteria Halomonas ventosae and Halomonas anticariensis. Research in Microbiology 157, 827-835. https://doi.org/10.1016/j.resmic.2006.06.004

Miranda, C.C., Dekker, R.F., Serpeloni, J.M., Fonseca, E.A., Cólus, I.M., and Barbosa, A.M. (2008). Anticlastogenic activity exhibited by botryosphaeran, a new exopolysaccharide produced by *Botryosphaeria rhodina* MAMB-05. International journal of biological macromolecules 42, 172-177. https://doi.org/10.1016/j.ijbiomac.2007.10.010

Misra, A., Ganesh, S., Shahiwala, A., Shah, S.P. (2003). Drug delivery to the central nervous system: a review. J Pharm Pharm Sci. 6, 252-73.

Motwani, J.G., and Lipworth, B.J. (1991). Clinical pharmacokinetics of drugs administered buccally and sublingually. Clin Pharmacokinet. 21, 83–94. https://doi.org/10.2165/00003088-199121020-00001

Mundargi RC, Patil S a, Agnihotri S a, Aminabhavi TM. Evaluation and controlled release characteristics of modified xanthan films for transdermal delivery of atenolol. Drug Dev Ind Pharm, 2007; 33(1):79- 90. https://doi.org/10.1080/03639040600975030

Mustafa, S., Devi, V.K., and Pai, R,S. (2016). Comparative study of kanamycin sulphate microparticles and nanoparticles for intramuscular administration: preparation in vitro release and preliminary in vivo evaluation. J Microencapsul. 33, 679-688. https://doi.org/10.1080/02652048.2016.1248511.

Müller, R.H., Jacobs, C., Kayser, O. (2001). Nanosuspensions as particulate drug formulations in therapy: Rationale for development and what we can expect for the future. Advanced Drug Delivery Reviews. 47, 3-19. https://doi.org/10.1016/S0169-409X(00)00118-6.

Neuhaus, D. (2014). Intraosseous infusion in elective and emergency pediatric anesthesia: when should we use it? Current Opinion in Anaesthesiology, 27(3): 282-287. https://doi.org/10.1097/ACO.0000000000000069.

Nicolaus, B., Kambourova, M., and Oner, E.T. (2010). Exopolysaccharides from extremophiles: from fundamentals to biotechnology. Environmental Technology 31, 1145-1158. https://doi.org/10.1080/09593330903552094

Nicolaus, B., Manca, M.C., Romano, I., and Lama, L. (1993). Production of an exopolysaccharide from two thermophilic archaea belonging to the genus *Sulfolobus*. FEMS microbiology letters 109, 203-206. https://doi.org/10.1111/j.1574-6968.1993.tb06168.x

Novak, M.T., Bryers, J.D., and Reichert, W.M. (2009). Biomimetic strategies based on viruses and bacteria for the development of immune evasive biomaterials. Biomaterials 30, 1989-2005. https://doi.org/10.1016/j.biomaterials.2008.11.025

Nwodo, U.U., Green, E., and Okoh, A.I. (2012). Bacterial exopolysaccharides: functionality and prospects. International journal of molecular sciences 13, 14002-14015. https://doi.org/10.3390/ijms131114002

Ofek, I., Hasty, D.L., and Doyle, R.J. (2003a). Bacterial adhesion to animal cells and tissues (American Society of Microbiology).

Ofek, I., Hasty, D.L., and Sharon, N. (2003b). Anti-adhesion therapy of bacterial diseases: prospects and problems. FEMS Immunology & Medical Microbiology 38, 181-191. https://doi.org/10.1016/S0928-8244(03)00228-1

Osman, A., Oner, E. T., and Eroglu, M. S., (2017). Novel levan and pNIPA temperature sensitive hydrogels for 5-ASA controlled release, Carbohydrate Polymers. http://dx.doi.org/10.1016/j.carbpol.2017.01.097

Otero, A., and Vincenzini, M. (2003). Extracellular polysaccharide synthesis by Nostoc strains as affected by N source and light intensity. Journal of biotechnology 102, 143-152. https://doi.org/10.1016/S0168-1656(03)00022-1

Packhaeuser, C.B., Schnieders, J., Oster, C.G,. and Kissel, T. (2004). In situ forming parenteral drug delivery systems: an overview. Eur J Pharm Biopharm. 58, 445 55. https://doi.org/10.1016/j.ejpb.2004.03.003

Panitch, A., Paderi, J.E., Sharma, S., Stuart, K.A., and Vazquez-Portalatin, N.M. (2014). Extracellular matrix-binding synthetic peptidoglycans (Google Patents).

Pant, D., and Adholeya, A. (2007). Biological approaches for treatment of distillery wastewater: a review. Bioresource technology 98, 2321-2334. https://doi.org/10.1016/j.biortech.2006.09.027

Patel, S., Majumder, A., and Goyal, A. (2011). Potentials of exopolysaccha-rides from lactic acid bacteria. *Indian J. Microbiol.* 52, 3–12. doi:10.1007/ s12088-011-0148-8

Patel, P., and Agrawal, Y. (2019). Preparation and In-vitro Evaluation of Levan Micelles: A Polyfructan Based Nano-carrier for Breast Cancer Targeted Delivery. Drug Delivery Letters, 9(2), 97-107. https://doi.org/ 10.2174/2210303109666190102115814

Petitpas, F., Guenezan, J., Vendeuvre, T., Scepi, M., Oriot, D., and Mimoz, O. (2016). Use of intra-osseous access in adults: a systematic review.Crit Care. 20, 102. https://doi.org/10.1186/s13054-016-1277-6.

Petri, D. F. (2015). Xanthan gum: A versatile biopolymer for biomedical and technological applications. Journal of Applied Polymer Science, 132(23). DOI: 10.1002/app.42035

Poli, A., Di Donato, P., Abbamondi, G.R., and Nicolaus, B. (2011). Synthesis, production, and biotechnological applications of exopolysaccharides and polyhydroxyalkanoates by archaea. Archaea 2011. https://doi.org/ 10.1155/2011/693253

Pragasam, V., Yuvashree, M., Gokulakannan, R. (2017). Challenges in oral drug delivery: a nano-based strategy to overcome. In Nanostructures for Oral Medicine Micro and Nano Technologies, Ecarterina A., Alexandru, M. G, ed. (Washington DC, USA: Elsevier) pp.173-201. https://doi.org/10.1016/B978-0-323-47720-8. 00008-0

Radomsky, M. (2000). Product Development Principles of Sustained-Release Injectable Formulations. In Sustained-Release Injectable Products. Senior, J., and Radomsky, M, ed. (USA, informa healthcare) pp.13-22.

Ram, I.M. (2018)a. Pharmaceutical Dosage Forms and Drug Delivery. In Drug Delivery System, Ram, I.M., Ajit, S.N, ed. (NW, USA: CRC Press), pp.313-336.

Ram, I.M. (2018)b. Pharmaceutical Dosage Forms and Drug Delivery. In Drug Delivery System, Ram, I.M., Ajit, S.N, ed. (NW, USA: CRC Press), pp.509-531.

Ray, A., and Mitra, K. M (2017). Nanotechnology in Intracellular Trafficking, Imaging, and Delivery of Therapeutic Agents. In Emergıng Nanotechnologıes For Dıagnostıcs, Drug Delivery, And Medical Devices, Mitra, K. M., Kishore, C., Abhirup, M, ed. (Washington DC, USA: Elsevier), pp.375-86. http://dx.doi.org/ 10.1016/B978-0-323-42978-8.00015-2

Redasani, K.V., and Bari, S.B. (2015). Prodrug Design Perspectives, Approaches and Applications in Medicinal Chemistry (USA, Academic Press).

Richard, H.G., and Jonathan, H. (2003) Transdermal Drug Delivery. (NY, USA; Marcel Dekker,).

Rinker, K.D., and Kelly, R.M. (2000). Effect of carbon and nitrogen sources on growth dynamics and exopolysaccharide production for the hyperthermophilic archaeon *Thermococcus litoralis* and bacterium *Thermotoga maritima*. Biotechnology and Bioengineering 69, 537-547. https://doi.org/10.1002/1097-0290(20000905)69:5%3C537::AID-BIT8%3E3.0.CO;2-7

Rodríguez-Carvajal, M.A., Sánchez, J.I., Campelo, A.B., Martínez, B., Rodríguez, A., and Gil-Serrano, A.M. (2008). Structure of the high-molecular weight exopolysaccharide isolated from *Lactobacillus pen*tosus LPS26. Carbohydrate research 343, 3066-3070. https://doi.org/10.1016/j.carres.2008.08.028

Roy, S. K., and Prabhakar, B. R. (2010). Bioadhesive Polymeric Platforms for Transmucosal Drug Delivery Systems – a Review. Tropical Journal of Pharmaceutical Research. 9, 91-104. https://doi.org/10.4314/tipr.v9i1.52043

Ruas-Madiedo, P., and De Los Reyes-Gavilán, C. (2005). Invited review: methods for the screening, isolation, and characterization of exopolysaccharides produced by lactic acid bacteria. Journal of Dairy Science 88, 843-856. https://doi.org/10.3168/jds.S0022-0302(05)72750-8

Sabaa, M. W., Hanna, D. H., Elella, M. H. A., & Mohamed, R. R. (2019). Encapsulation of bovine serum albumin within novel xanthan gum based hydrogel for protein delivery. Materials Science and Engineering: C, 94, 1044-1055. https://doi.org/10.1016/j.msec.2018.10.040

Saidin, N. M., Anuar, N. K., & Affandi, M. M. M. (2018). Roles of Polysaccharides in Transdermal Drug Delivery System and Future Prospects. Journal of Applied Pharmaceutical Science, 8(03), 141-157. doi: 10.7324/JAPS.2018.8320

Sagitha, P., Reshmi, C. R., Sundaran, S. P., Binoy, A., Mishra, N., Sujith, A. (2018). In-vitro evaluation on drug release kinetics and antibacterial activity of dextran modified polyurethane fibrous membrane. International journal of biological macromolecules, 126, 717-730. https://doi.org/10.1016/j.ijbiomac.2018.12.155

Saravanan, C., Rajesh, R., Kaviarasan, T., Muthukumar, K., Kavitake, D., & Shetty, P. H. (2017). Synthesis of silver nanoparticles using bacterial exopolysaccharide and its application for degradation of azo-dyes. Biotechnology reports, 15, 33-40. http://dx.doi.org/10.1016/j.btre.2017.02.006

Sezer, A. D., Kazak Sarılmışer, H., Rayaman, E., Çevikbaş, A., Öner, E. T., Akbuğa, J. (2017). Development and characterization of vancomycin-loaded levan-based microparticular system for drug delivery. Pharmaceutical development and technology, 22(5), 627-634. DOI: 10.3109/10837450.2015.1116564.

Sezer, A. D., Kazak, H., Öner, E. T., Akbuğa, J. (2011). Levan-based nanocarrier system for peptide and protein drug delivery: optimization and influence of experimental parameters on the nanoparticle characteristics. Carbohydrate Polymers, 84(1), 358-363. doi:10.1016/j.carbpol.2010.11.046.

Sharon, N., and Lis, H. (1989). Lectins as cell recognition molecules. Science 246, 227-234. https://doi.org/10.1126/science.2552581

Sheng, L., Zhu, G., and Tong, Q. (2014). Comparative proteomic analysis of *Aureobasidium pullulans* in the presence of high and low levels of nitrogen source. J. Agric. Food Chem. 62, 10529–10534. doi:10.1021/jf503390f

Jamwal, S., Ram, B., Ranote, S., Dharela, R., & Chauhan, G. S. (2018). New glucose oxidase-immobilized stimuli-responsive dextran nanoparticles for insulin delivery. International journal of biological macromolecules, 123, 968-978. https://doi.org/10.1016/ j.ijbiomac.2018.11.147

Shojaei, A.H. (1998). Buccal mucosa as a route for systemic drug delivery: a review. J Pharm Pharm Sci. 1:15-30

Silva NHCS, Rodrigues AF, Almeida IF, et al., Bacterial cellulose membranes as transdermal delivery systems for diclofenac: In vitro dissolution and permeation studies. Carbohydr Polym, 2014; 106:264- 269. https://doi.org/10.1016/j.carbpol.2014.02.014

Singh, A., Van Hamme, J.D., and Ward, O.P. (2007). Surfactants in microbiology and biotechnology: Part 2. Application aspects. Biotechnology advances 25, 99-121. https://doi.org/10.1016/j.biotechadv.2006.10.004

Singh, A.K., Pindi, P.K., Dube, S., Sundareswaran, V., and Shivaji, S. (2009). Importance of trmE for growth of the psychrophile *Pseudomonas syringae* at low temperatures. Applied and environmental microbiology 75, 4419-4426. https://doi.org/10.1128/AEM.01523-08

Sivasankar, P., Seedevi, P., Poongodi, S., Sivakumar, M., Murugan, T., Sivakumar, L., Sivakumar, K. & Balasubramanian, T. (2018). Characterization, antimicrobial and antioxidant property of exopolysaccharide mediated silver nanoparticles synthesized by *Streptomyces violaceus* MM72. Carbohydrate polymers, 181, 752-759. https://doi.org/10.1016/j.carbpol.2017.11.082

Staudt, A.K., Wolfe, L.G., and Shrout, J.D. (2012). Variations in exopolysaccharide production by *Rhizobium tropici*. Archives of microbiology 194, 197-206. https://doi.org/10.1007/s00203-011-0742-5

Suresh Kumar, A., Mody, K., and Jha, B. (2007). Bacterial exopolysaccharides–a perception. Journal of basic microbiology 47, 103-117. https://doi.org/10.1002/jobm.200610203

Sutherland, I. (1972). Bacterial exopolysaccharides. Advances in microbial physiology 8, 143-213.

Sutherland, I. (2002). A sticky business. Microbial polysaccharides: current products and future trends. Microbiology Today 29, 70-71.

Tiwari, G., Tiwari, R., Sriwastawa, B., Bhati, L., Pandey, S., Pandey, P., and Bannerjee., S. K. (2012). Drug delivery systems: An updated review. Int J Pharm Investig. 2: 2–11. 2–11. https://dx.doi.org/10.4103%2F2230-973X.96920

Thakur, V., Kush, P., Pandey, R. S., Jain, U. K., Chandra, R., & Madan, J. (2016). Vincristine sulfate loaded dextran microspheres amalgamated with thermosensitive gel offered sustained release and enhanced cytotoxicity in THP-1, human leukemia cells: in vitro and in vivo study. Materials Science and Engineering: C, 61, 113-122. https://doi.org/10.1016/j.msec.2015.12.015

Trombino, S., Serini, S., Cassano, R., & Calviello, G. (2019). Xanthan gum-based materials for omega-3 PUFA delivery: Preparation, characterization and antineoplastic activity evaluation. Carbohydrate polymers, 208, 431-440. https://doi.org/10.1016/j.carbpol.2019.01.001

Varki, A., and Sharon, N. (2009). Historical background and overview. In Essentials of Glycobiology. 2nd edition. Cold Spring Harbor Laboratory Press.

von der Lieth, C.-W., Lütteke, T., and Frank, M. (2009). Bioinformatics for glycobiology and glycomics: an Introduction (John Wiley & Sons).

Von Hoff, D. D., John G.KuhnPharm D, .BurrisIII, A.H., Miller, L. J. (2008). Does intraosseous equal intravenous? A pharmacokinetic study. The American Journal of Emergency Medicine. 26, 31-38. https://doi.org/10.1016/j.ajem.2007.03.024.

Vu, B., Chen, M., Crawford, R.J., and Ivanova, E.P. (2009). Bacterial extracellular polysaccharides involved in biofilm formation. Molecules 14, 2535-2554. https://doi.org/10.3390/molecules14072535

Wang, B., Hu, L., Siahaan, T. J. (2016). Drug Delivery Principles and Applications. (Hoboken, New Jersey: John Wiley & Sons)

Wang, H., Dai, T., Zhou, S., Huang, X., Li, S., Sun, K., ... & Dou, H. (2017). Self-assembly assisted fabrication of dextran-based nanohydrogels with reduction-cleavable junctions for applications as efficient drug delivery systems. Scientific reports, 7, 40011. DOI: 10.1038/srep40011.

Wilson, C. G., and Crowley, P. J. (2011). Controlled Release in Oral Drug Delivery. . (NW, USA: Springer). https://doi.org/10.1007/978-1-4614-1004-1.

Wu F, Jin T. (2008). Polymer-based sustained-release dosage forms for protein drugs, challenges, and recent advances. AAPS PharmSciTech. 9, 1218-29. https://doi.org/10.1208/s12249-008-9148-3.

Yadav, H., Karthikeyan, C. (2019). Natural polysaccharides: Structural features and properties. Polysaccharide Carriers for Drug D*elivery*, 1.

Yan, J.K., Li, L., Wang, Z.M., Leung, P.H., Wang, W.Q., and Wu, J.Y. (2009). Acidic degradation and enhanced antioxidant activities of exopolysaccharides from *Cordyceps sinensis* mycelial culture. Food Chemistry 117, 641-646. https://doi.org/10.1016/j.foodchem.2009.04.068

Yim, J.H., Kim, S.J., Ahn, S.H., and Lee, H.K. (2007). Characterization of a novel bioflocculant, p-KG03, from a marine dinoflagellate, *Gyrodinium impudicum* KG03. Bioresource technology 98, 361-367. https://doi.org/10.1016/j.biortech.2005.12.021

Yuan, X., and Taft, D. (2007). Advanced Formulation Strategies for Central Nervous System Drug Delivery. In Advanced Drug Formulation Design to Optimize Therapeutic Outcomes. O. Williams III Robert, David R. Taft, Jason T. McConville, ed. (USA; Informa Healthcare).Tropical Journal of Pharmaceutical Research, February 2010; 9 (1): 91-104

Zhang, N., Wardwell, P., Bader, R. (2013). Polysaccharide-based micelles for drug delivery. Pharmaceutics, 5(2), 329-352. https://doi.org/10.3390/pharmaceutics5020329

Zhao, Q., Zhao, M., Yang, B., and Cui, C. (2009). Effect of xanthan gum on the physical properties and textural characteristics of whipped cream. Food Chemistry 116, 624-628. https://doi.org/10.1016/j.foodchem.2009.02.079

Zykwinska, A., Marquis, M., Sinquin, C., Cuenot, S., and Colliec-Jouault, S (2016).Assembly of HE800 exopolysaccharide produced by a deep-sea hydrothermal bacterium into microgels for protein delivery applications, Carbohydrate Polymers, http://dx.doi.org/10.1016/j.carbpol.2016.01.056

Fructan Biosynthesis in Bacteria

6

Lázaro Hernández[1]*, Yamira Quintero[1], Alexis Musacchio[2]

[1]Enzyme Technology Group, Center for Genetic Engineering and Biotechnology (CIGB). PO Box 6162, Havana 10600, Cuba.
[2]System Biology Department, Center for Genetic Engineering and Biotechnology (CIGB). PO Box 6162, Havana 10600, Cuba.
*corresponding author: lazaro.hernandez@cigb.edu.cu

DOI: https://doi.org/10.21775/9781912530267.06

Abstract

In nature, fructans are synthesized from sucrose by a relative low number of plant species and a wide range of microorganisms, including bacteria, archaea, and fungi. Levan, the predominant bacterial fructose polymer, is a biofilm-forming exopolysaccharide that fulfills diverse biological roles. The biosynthetic enzyme levansucrase (EC 2.4.1.10) catalyzes sucrose hydrolysis, sucrose fructosylation, and fructan polymerization, as the three main activities. The hydrolase/transferase ratio, the spectrum of fructan products, and the overall levan yield vary depending on the intrinsic properties of each enzyme and the reaction conditions. This chapter provides a detailed update on the most relevant genetic, biochemical, structural, and evolutionary aspects of levansucrase. Several levan-producing bacteria are enzymatically equipped to utilize the polymer as an alternative carbon source under starving conditions. The last section describes how the strictly controlled transcription of levanase genes ensures the prevalence of levan synthesis over degradation whenever exogenous sucrose is available as an energetic substrate.

Introduction

In nature, fructans are produced from sucrose by certain families of higher plants and a wide range of microorganisms, including bacteria, archaea, and fungi. Plants synthesize fructans in the cell vacuole by the concerted action of two or more fructosyltransferases, which differ in their substrate specificity. The size of plant fructans rarely exceed of 10^2 fructosyl units. Bacterial fructosyltransferases are multifunctional enzymes that synthesize fructans with a degree of polymerization (DP) ranging 10^3–10^4 fructosyl units. Fungal fructosyltransferases most often yield short-chain fructooligosaccharides (FOS). Archaeal fructosyltransferases and fructans remain uncharacterized.

This chapter provides an updated review on the most relevant genetic, biochemical, structural, and evolutionary aspects of fructan biosynthesis in bacteria, making emphasis on levansucrase, the most abundant fructosyltransferase. Levan synthesis and degradation are two exclusive but also complementary habitat-related processes. The last section focuses the biological significance of the finely coordinated transcriptional regulation of the fructan-metabolic genes in a bacterial host.

Bacterial fructosyltransferases enzyme classification

In bacteria, two distinct fructosyltransferases catalyze the conversion of sucrose into fructans. The enzyme is named inulosucrase (EC 2.4.1.9) when it synthesizes inulin [β-(2→1) polyfructan], and levansucrase (EC 2.4.1.10) when the polymerization product is levan [β-(2→6) polyfructan]. Levansucrases can also form β-(2→1) fructosyl–fructose linkages either to synthesize inulin–type FOS or to create branches in the growing β-(2→6) linked chains.

Levansucrase is wide spread in Gram–positive and Gram–negative bacteria of different habitats (Toksoy Öner et al., 2016). Inulosucrase production has been reported so far for the Gram–positive species *Streptococcus mutans* (Rosell and Birkhed, 1974), *Leuconostoc citreum* (Olivares-Illana et al., 2002), *Lactobacillus reuteri* (van Hijum et al., 2002), *Bacillus sp. 217C-11* (Wada et al., 2003), *Lactobacillus johnsonii* (Anwar et al., 2008), *Lactobacillus gasseri* (Anwar et al., 2010), *Weissella confusa* (Malang et al., 2015; Malik et al., 2015), *Streptomyces viridochromogenes* (Frasch et al., 2017), and *Salipaludibacillus agaradhaerens* (Kralj et al., 2018). Some bacteria contain both a levansucrase and an inulosucrase.

Gene expression

Most levan–producing species have a single levansucrase gene, which is often part of a chromosomal operon. Blast searches in sequenced genomes have revealed the presence of putative levansucrase genes also in plasmids. The plant pathogen *Pseudomonas syringae* carries three levansucrase genes (two in the chromosome and one in a plasmid), but only two of them are functionally expressed (Li and Ullrich, 2001).

Sucrose is required to stimulate levansucrase transcription in the genera *Bacillus*, *Paenibacillus* and *Geobacillus* (Choi et al., 2004; Homann et al., 2007; Inthanavong et al., 2013). *Bacillus subtilis* was the first species to have the levansucrase gene sequenced (Steinmetz et al., 1985), and the control of its expression has been thoroughly investigated. The levansucrase gene (*sacB*) initiates a tricistronic operon with the second and third genes encoding an endo–levanase (LevB) and a transmembrane protein (YveA), respectively. A stem–loop structure functioning as a conditional transcription terminator is placed upstream to each coding sequence provoking partial arrests of the RNA polymerase at the operon leader region and the two intergenic regions (Pereira et al. 2001). The sucrose–inducible transcription of the *sacB–levB–yveA* operon involves an anti–termination mechanism negatively controlled by the phosphotransferase system (Crutz et al., 1990; Declerck et al., 2002). When the sucrose concentration is higher than 10 mM, the operon expression is autogenously activated via a feedback loop present in the *levB* transcript, which interacts with the anti–terminator protein SacY (Daguer et al., 2004).

Constitutive expression is a common feature for levansucrase genes of bacteria that reside in sucrose–containing habitats, for instance, *Erwinia amylovora* (Geier and Geider, 1993), *Streptococcus salivarius* (Rathsam and Jacques, 1998), *Erwinia herbicola* (Cote, 1988), *Actinomyces naeslundii* (Bergeron et al., 2000), *Gluconacetobacter diazotrophicus* (Hernández et al., 2000), and *Zymomonas mobilis* (Senthilkumar et al., 2009). The components of growth media and the culture conditions can influence the yield of constitutively expressed levansucrases. For instance, *G. diazotrophicus* Pal5^{T} produced a higher amount of levan on a solid medium with a high phosphate concentration (> 28 mM) (Idogawa et al., 2014). In liquid cultures, *G. diazotrophicus* SRT4 displayed maximal levansucrase production in mannitol– or sorbitol–supplemented media at initial pH of 8, although cell growth was optimal when using sucrose or glycerol as the carbon source and under acidic conditions (initial pH 5–6) (Hernández et al., 1999b). The stimulatory effect of mannitol on levan biosynthesis has been reported for three other Gram–negative bacteria (Tkachenko et al., 1980; Cote, 1988; Ateş et al., 2013), confirming the regulatory link between carbohydrate metabolism and levansucrase gene expression.

Protein secretion

Levan synthesis is naturally designed to occur outside the cell, where the exopolysaccharide plays its biological function. Levansucrases are either released to the external medium or remain attached to the cell surface. Variations in the enzyme secretion routes have been described for different species.

Levansucrase secretion in Gram–positive bacteria involves the proteolytic cleavage of a signal–peptide–containing precursor protein (Rathsam et al., 1993; Steinmetz et al., 1985; van Hijum et al., 2004; Waldherr et al., 2008). *Bacillus subtilis* secretes levansucrase in a two–step process (Petit-Glatron et al., 1987).

In the first step, the removal of an N–terminal signal peptide of 29 aa results in a transient cell–associated protein form. In the second step, the mature levansucrase is released to the external medium via a membrane active–transport system that is directly linked to the influx of iron into the bacteria (Chambert and Petit-Glatron 1988).

The presence or absence of a cleavable signal peptide in Gram–negative levansucrases is a trait that reflects the enzyme evolutionary history. A typical N –terminal signal peptide is present in phylogenetically related levansucrases from the taxonomically distant families Burkholderiaceae and Acetobacteraceae. Within this clade, the levansucrase of *Gluconacetobacter diazotrophicus* is synthesized as a precursor protein with a 30–aa N–terminal signal peptide that is cleaved off during transport to the periplasm (Hernández et al., 1999a). This bacterium is equipped with the type–II secretory machinery for translocation of the mature and fully folded levansucrase across the outer membrane. The lack of a transport system to import sucrose prevents levan formation inside the cell (Arrieta et al., 1996 and 2004).

Most Gram−negative levansucrases are not proteolytically cleaved during secretion. Genera that secrete the enzyme by a signal−peptide independent pathway include *Zymomonas*, *Erwinia*, *Rahnella*, *Pseudomonas*, *Gluconobacter*, and *Halomonas* (Song et al., 1993; Geier and Geider, 1993; Song et al., 1998; Li and Ullrich, 2001; Jakob et al., 2012; Kirtel et al., 2018a). It remains unclear by which secretion mechanism(s) signal−peptide lacking levansucrases are transported to the cell periplasm and then to the extracellular medium.

The levansucrase reaction

Bacterial fructosyltransferases operate via a double−displacement (ping−pong) mechanism that involves the formation and subsequent hydrolysis of a covalent fructosyl−enzyme intermediate. The reaction occurs with an overall retention of the anomeric configuration of the fructosyl residue. Fructan elongation is catalyzed by distributive (non−processive) and/or processive mechanisms. In a distributive reaction, the product dissociates from the enzyme after every step of catalysis. In a processive reaction, the product remains bound to the enzyme, ready to be the acceptor of the next fructosyl transfer.

In a typical levansucrase reaction with sucrose as a single substrate, the glucose moiety is released upon cleavage of the glycosidic bond. Subsequently, the fructosyl moiety is transferred from the covalent enzyme intermediate to either water or a second sucrose molecule. Due to its low regioselectivity, levansucrase can synthesize 1−kestose, 6−kestose, or neo−kestose in the first fructosylation cycle (Iizuka et al., 2002; Caputi et al., 2013; Raga-Carbajal et al., 2018). The three FOS are potential initiators of levan synthesis reactions. Processive levansucrases synthesize high−DP levan (average MW 10^6–10^7 daltons) by adding the fructosyl moieties through β(2→6) bonds to the growing end of the fructan chain without releasing the intermediate oligofructans (Ozimek et al., 2006). Levan often contains lateral branches with β-(2→1) linkages at the connection points (Figure 1). Other alternative acceptor substrates are mono− and disaccharides released during the reaction course.

Sucrose hydrolysis is an intrinsic activity of bacterial fructosyltransferases. Most relevant factors that modulate the hydrolase/transferase ratio are: substrate concentration, temperature, ionic strength, and the presence of water−miscible

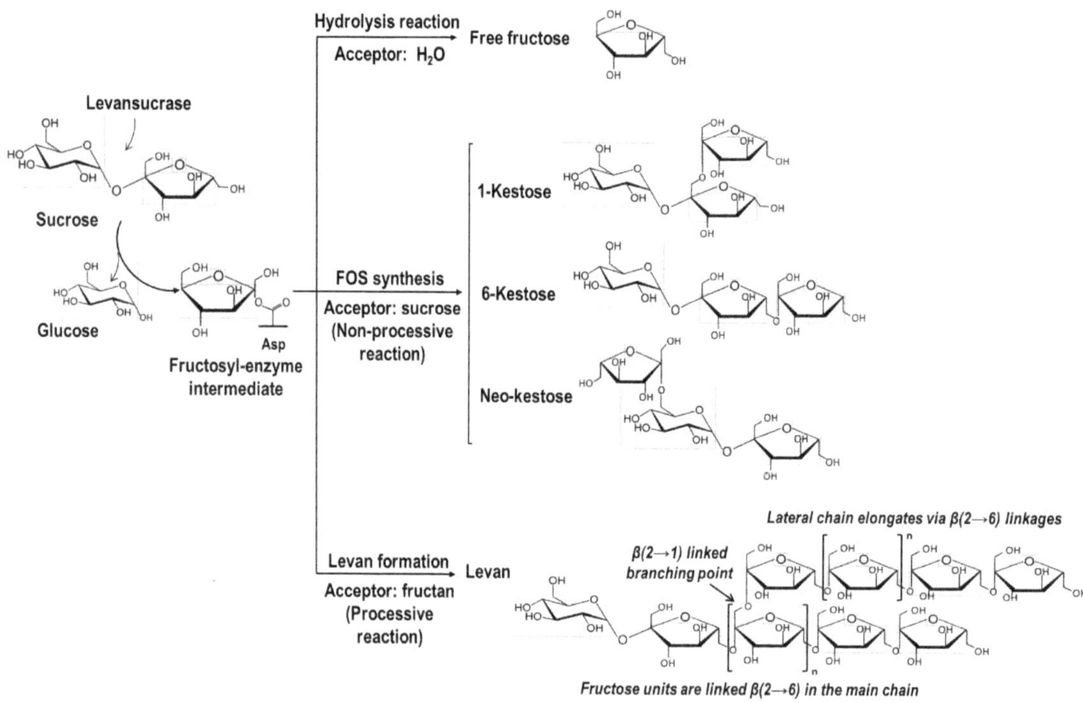

Figure 1. Schematic diagram of the main levansucrase activities on sucrose. Upon glycosidic bond cleavage, the glucose moiety is released; the fructosyl moiety is covalently linked to the nucleophile (Asp) and then transferred from the enzyme intermediate to the acceptor (water, sucrose or fructan) in a double–displacement (ping–pong) reaction. Sucrose hydrolysis is an intrinsic activity of levansucrases. The combined synthesis of 1–kestose, 6–kestose, and neo–kestose reflects the enzyme low regioselectivity in the fructosyl transfer to sucrose. Short–chain fructans are elongated in processive reactions conducting to rapid levan polymerization and eventual ramification. The monosaccharides released during the reaction course become alternative fructosyl acceptors (see text).

organic solvents. At low sucrose concentrations (< 200 mM) or high temperature (60ºC) hydrolysis is the predominant reaction (Támbara et al., 1999; Trujillo et al., 2001; Ozimek et al., 2006; Vigants et al., 2013; Visnapuu et al., 2015). The addition of NaCl or the presence of organic solvents in the reaction mixture generates water –restricted environments promoting transfructosylation over hydrolysis (Chambert and Petit-Glatron, 1989; Trujillo et al., 2004; Castillo and López-Munguía, 2004).

Besides the hydrolase/transferase ratio, levan yield depends on the enzyme affinity for the acceptor FOS and the fructosylation mechanism. Levansucrases from the Gram–positive bacteria *Bacillus subtilis, Streptococcus salivarius, Lactobacillus reuteri, Bacillus megaterium*, and *Bacillus licheniformis* rapidly polymerize the precursor FOS into high-DP levan (Chambert et al., 1974; Song and Jacques, 1999; Ozimek et al., 2006; Homann et al., 2007; He et al., 2018). Under similar reaction conditions, levansucrases from the Gram-negative bacteria *Gluconacetobacter diazotrophicus*, *Zymomonas mobilis, Rahnella aquatilis, Erwinia amylovora,* and *Pseudomonas syringae*) accumulate relative high levels of short-chain FOS (DP 3-5), with a consequent lower yield of levan (Hernández et al., 1995; Bekers et al., 2002; Kim et al., 2003; Caputi et al., 2013; Visnapuu et al., 2015).

The reaction conditions can also influence the enzyme fructosylation mechanism. *Bacillus subtilis* levansucrase, the most studied fructosyltransferase, synthesizes levan of low-molecular weight (LMW, average 7.2 kDa) and levan of high-molecular weight (HMW, average 2,300 kDa) using a distributive (non-processive) and a processive mechanism, respectively (Raga-Carbajal et al., 2016 and 2018). A low enzyme concentration (0.1 U/mL) favored the synthesis of HMW levan, while variations in sucrose concentration had negligible effect on promoting the processive elongation of the fructan chain (Porras-Domínguez et al., 2015; Raga-Carbajal et al., 2016). In the nineteen eighties, other authors reported that the presence of ethanol, polyethylene glycol or acetonitrile favored the synthesis of HMW levan, while an increase in the ionic force resulted in a shift from the bimodal type to a single LMW levan (Tanaka et al. 1980; Chambert and Petit-Glatron 1989). A bimodal molecular weight distribution was also observed in levan produced by levansucrase from *Bacillus licheniformis* RN-01. At high temperature (50°C), the enzyme synthesized HMW levan (612 kDa) as a major product, but when the temperature was decreased to 30°C or NaCl was added to the reaction, LMW levan (11 kDa) was mainly synthesized (Nakapong et al., 2013).

Bacterial levansucrases has the dual catalytic activity of a levan polymerase and a levan exo-hydrolase. In the absence of sucrose or another fructosyl donor, levansucrases cleave the β-(2→6) linkages of the levan chain in an exo-type manner releasing the non-reducing terminal fructose unit until the reaction stops at a branching point (Chambert and Gonzy-Treboul, 1976; Hernández et al., 1995; Méndez-Lorenzo et al., 2015). In this sense, the β-(2→1) fructosyl-fructose linkages appear to prevent branched levans from extensive hydrolysis by its own synthesizing enzyme. Most levansucrases are unable to hydrolyze β-(2→1) linked fructans.

Levansucrase does not need a metal cofactor for activity. The enzyme performs best in the pH range 5–7. In most cases, pH variations do not affect the ratio of hydrolase versus transferase activities (Hernández et al., 1995; Kim et al., 2003; Homann et al., 2007; Caputi et al., 2013; Visnapuu et al., 2015). Interestingly, levansucrase from *Zymomonas mobilis* catalyzed almost exclusively the synthesis of high–DP levan at pH values below 6 in its insoluble microfibril form, and behaved mostly as a hydrolase at pH values above 7 in its soluble dimer form (Goldman et al., 2008). The formation of microfibrils as a requirement for levan synthesis has not been reported for other levansucrases.

Substrate specificity

Sucrose [β-D-fructofuranosyl-(2↔1)-α-D-glucopyranoside] is the natural and preferred substrate of bacterial fructosyltransferases. Levansucrase can also catalyze the conversion of oligosaccharides containing a terminal sucrose with its fructose moiety being accessible, as raffinose [α-D-galactopyranosyl-(1→6)-α-D-glucopyranosyl-(1↔2)-β-D-fructofuranoside] and stachyose [α-D-galactopyranosyl-(1→6)-α-D-galactopyranosyl-(1→6)-α-D-glucopyranosyl-(1↔2)-β-D-fructofuranoside]. *Bacillus subtilis* levansucrase was found to poorly hydrolyze neo–kestose [β-D-fructofuranosyl (2→6)-α-D-glucopyranosyl-(1↔2)-β-D-fructofuranoside] (Raga-Carbajal et al., 2018). In general, 6–kestose [β-D-fructofuranosyl-(2→6)-β-D-fructofuranosyl-(2↔1)-α-D-glucopyranoside], 1–kestose [β-D-fructofuranosyl-(2→1)-β-D-fructofuranosyl-(2↔1)-α-D-glucopyranoside], and 1,1–nystose [β-D-fructofuranosyl-(2→1-(β-D-fructofuranosyl-(2→1)-β-D-fructofuranosyl-(2↔1)-α-D-glucopyranoside] are not donor substrates of levansucrases. As other β-fructofuranosidases, levansucrase does not react with melezitose [α-D-glucopyranosyl-(1→3)-β-D-fructofuranosyl-(2↔1)-α-D-glucopyranoside] because its fructose moiety is located between two monosaccharides.

The first fructosylation products 6–kestose, 1–kestose, and neo–kestose are the precursors for levan formation. Levansucrases from distinct origins show remarkable variations in the affinity for the initiator trisaccharide. Lineal levan chains are synthesized directly from 6–kestose by all levansucrases. Gram–positive enzymes successively elongate 1–kestose and 1,1–nystose through β-(2→6) bonds, when an efficient fructosyl donor (sucrose) is available (Hernández et al., 1995; Iizuka et al.,

2002; Ozimek et al., 2006; Raga-Carbajal et al., 2018). Contrasting, β-(2→1) linked FOS are poor fructosyl acceptors of Gram−negative levansucrases (Hernández et al., 1995; Yanase et al., 2002; Caputi et al., 2013). High concentrations of neo −kestose (20 mM) and 6−neo−nystose (5 mM) were needed to observe acceptor reactions by *Bacillus subtilis* levansucrase (Raga-Carbajal et al., 2018).

Levansucrases show a wide spectrum of non−conventional fructosyl−acceptor substrates, including mono−, di− and trisaccharides, as well as aromatic and aliphatic alcohols (Tanaka et al., 1981; Seibel et al., 2006; Beine et al., 2008; Han et al., 2009; Mena-Arizmendi et al., 2011; Visnapuu et al., 2011; Ruiz-Aceituno et al., 2017; Kirtel et al., 2018a; Xu et al., 2018). The enzyme ability to transfer the fructosyl moiety of sucrose to galactose, mannose, xylose, arabinose, maltose, isomaltose, lactose, cellobiose, maltotriose, and raffinose offers alternatives for the production of hetero−oligosaccharides with different applications. For instance, the fructosylation of lactose, the major carbohydrate in cow's milk, yields lactosucrose [β-D-galactopyranosyl-(1→4)-α-D-glucopyranosyl-(1↔2)-β-D-fructofuranoside], a low-calorie sugar of interesting prebiotic properties (Mu et al., 2013).

Kinetic properties

Levansucrases exhibit Michaelis−Menten type kinetics for both the hydrolase and transferase activities. Early studies using ^{14}C−labelled substrates confirmed the ping −pong mechanism and allowed determining the apparent rate constants for the different reaction steps in levansucrases of *Bacillus subtilis* (Chambert and Gonzy-Treboul, 1976), *Gluconacetobacter diazotrophicus* (Hernández et al., 1995), and *Streptococcus salivarius* (Song and Jacques, 1999). The transfer of the fructosyl unit from the covalent fructosyl−enzyme intermediate to the acceptor molecule is the rate −limiting step in the levansucrase reaction. The fructosylation of sucrose or the growing fructan occurs more rapidly than sucrose hydrolysis. The fructosyl transfer to glucose, when present at high levels (50−200 mM), partially inhibits other activities by reducing the steady−state concentration of the fructosyl−enzyme intermediate.

Inulosucrases show distinct catalytic behaviors for the hydrolase and transferase activities. Inulosucrase from *Lactobacillus reuteri* displays Michaelis−Menten type kinetics for the fructosyl transfer to water, which is the main reaction at sucrose concentrations lower than 200 mM (van Hijum et al., 2003; Ozimek et al., 2006). The rate of transfructosylation and the overall activity, measured as glucose release, gradually increased as the initial sucrose loading was raised up to the highest assayed concentration of 1.7 M). The apparent non-saturating performance revealed for the transferase activity of inulosucrase resembles that of plant 1-SST enzymes (Van den Ende et al., 1996; Koops and Jonker 1996; Ávila-Fernández et al., 2007).

Levansucrases and inulosucrases show comparable values of turnover rate (k_{cat}), affinity for the donor substrate (K_m), and catalytic efficiency (k_{cat}/K_m). Frasch et al., (2017) reported an exceptionally low k_{cat} (3.5 s^{-1}) for *Streptomyces viridochromogenes* inulosucrase, although the enzyme K_m for sucrose is within the range of values (4–50 mM) informed for *L. reuteri* inulosucrase (van Hijum et al., 2003) and most of the kinetically characterized levansucrases (Olvera et al., 2007, Caputi et al., 2013 and references therein). Unusually high K_m values of 122 and 160 mM were reported for the levansucrases of *Zymomonas mobilis* (Yanase et al., 2002) and *Pseudomonas syringae* pv. phaseolicola (Hettwer et al., 1995), respectively. In general, Gram-positive levansucrases have lower K_m and higher k_{cat} values than the enzymes from
Gram-negative bacteria.

Structure-function relationship

Bacterial fructosyltransferases (levansucrase and inulosucrase) belong to family 68 of glycoside hydrolases (GH), and along with family GH32 (invertases, sucrose-6P hydrolases, fructanases, and eukaryotic fructosyltransferases) comprise clan GH-J in the Carbohydrate-Active Enzyme (CAZy) classification. Mechanistic and folding similarities within clan GH-J members are broadly documented (Pons et al., 2004; Lammens et al., 2009; Yuan et al., 2012). Three highly conserved carboxylic residues, known as the catalytic triad, are responsible for catalysis by functioning as the nucleophile (Asp), the acid/base catalyst (Glu), and the transition-state stabilizer (Asp). The determined tertiary structures of GH32 and GH68 proteins share a catalytic 5-bladed β-propeller domain enclosing the substrate-binding central cavity. GH32 members have an additional C-terminal β-sheet domain of unclear function (Alberto et al., 2004), which is absent in family GH68.

The GH68 enzymes structurally characterized so far are levansucrases from *Bacillus subtilis* (Meng and Fütterer, 2003 and 2008), *Gluconacetobacter diazotrophicus* (Martinez-Fleites et al., 2005), *Bacillus megaterium* (Strube et al., 2011), and *Erwinia amylovora* (Wuerges et al., 2015); inulosucrase from *Lactobacillus johnsonii* (Pijning et al., 2011), and β-fructofuranosidase from *Microbacterium saccharophilum* K-1 (formerly known as *Arthrobacter* sp. K-1) (Tonozuka et al., 2012). These enzymes

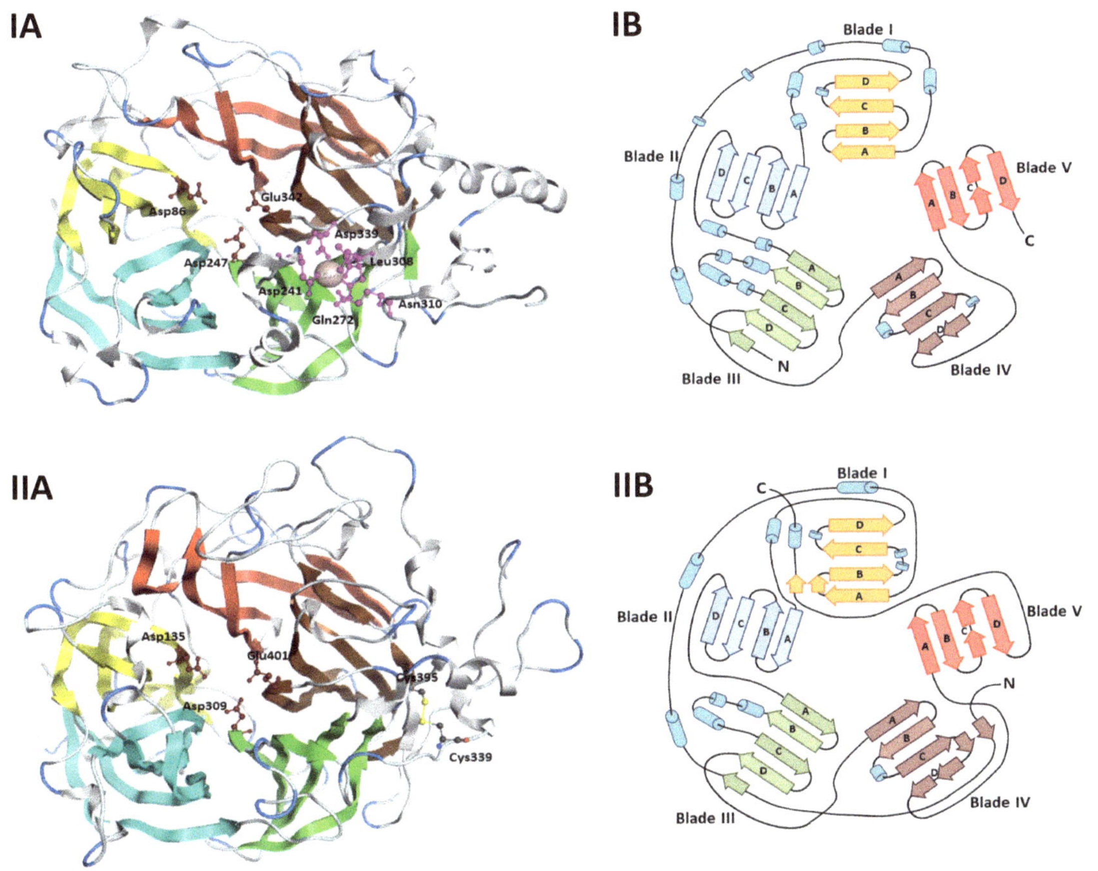

Figure 2. Crystal structures of representative levansucrases from *Bacillus subtilis* (I) and *Gluconacetobacter diazotrophicus* (II). (A) Ribbon diagram. (B) Topology scheme. The enzymes display a five–bladed β–propeller fold. The blades 1 to 5 are colored in yellow, blue, green, brown and red, respectively. The catalytic triad (Asp/Asp/Glu) at the bottom of the active site pocket of each enzyme is shown in ball–and–stick representation. The fold–stabilizing disulfide bond in *Gd* LS (PDB: 1W18) and the functionally/topologically complementary calcium–binding site in *Bs* LS (PDB: 1OYG) are mapped in the corresponding protein.

display a single domain 5–bladed β–propeller fold with the catalytic triad being perfectly superimposable at the bottom of a unique active site pocket (Figure 2). Available crystal structures of enzyme–ligand complexes as well as docking simulations reveal that the fructosyl donor is accommodated deep inside the active site in a similar position for all GH68 enzymes. Main structural differences reside at the surface of the catalytic cavity, in motifs presumably responsible for binding and correct orientation of the fructosyl acceptor.

Most fructosyltransferases from Gram–positive species require calcium ions for activity. The determined 3D structures of *B. subtilis* levansucrase and *L. johnsonii* inulosucrase contain a bound Ca^{2+} at an equivalent position, stabilizing a region of relevant connections between active site residues (Meng and Fütterer, 2003; Pijning et al., 2011). Mutation of a conserved Asp residue in the calcium–binding site of both levansucrase and inulosucrase of *Lactobacillus reuteri* resulted, in the two cases, in a strong reduction of activity and thermostability (Ozimek et al., 2005). Interestingly, inulosucrases from *Streptomyces viridochromogenes* and *Salipaludibacillus agaradhaerens* resemble Gram–negative levansucrases in being fully active in the absence of Ca^{2+} (Frasch et al., 2017; Kralj et al., 2018).

Gram–negative levansucrases lack a predictable calcium–binding site. Instead, the enzyme of *G. diazotrophicus* contains a disulfide bridge (Cys^{339}–Cys^{395}) that links the extended loop between β–strands IIIB and IIIC (Martinez-Fleites et al., 2005). The replacement of either Cys^{339} or Cys^{395} by serine did not alter the K_m for sucrose but reduced the k_{cat} approx. 60–fold. The disulfide bridge of *G. diazotrophicus* levansucrase and the structurally equivalent Ca^{2+}–binding site of Gram–positive fructosyltransferases play a similar fold–stabilizing role. Such a structural variation has no effect on the substrate and product specificities. The four cysteine residues of *E. amylovora* levansucrase are not interacting in the enzyme crystal structure (Wuerges et al., 2015). The absence of a predictable disulfide bridge is a conserved trait in other Gram–negative levansucrases lacking an N–terminal signal peptide (see section 2.3).

Levansucrases and inulosucrases display an invariant mechanism for the formation of the fructosyl–enzyme intermediate and the fructosyl transfer to water. Ortiz-Soto et al. (2017) succeeded to increase the transferase/hydrolase rate of *B. megaterium* levansucrase by rationally replacing active site residues that bind water and/or coordinate the position of the catalytic acid/base and the nucleophile. The structural determinants for the type of the fructosyl–fructose linkage and the elongation mechanism remain unknown.

Structure–function studies of levansucrases have provided some insights into oligosaccharide versus polysaccharide synthesis. Site–directed mutagenesis of amino acids at the surface and varied positions of the funnel–like channel leading towards the active site of the enzyme from three *Bacillus* species switched the reaction from polymer to FOS formation by changing from a processive to a distributive mechanism (Chambert and Petit-Glatron 1991; Homann et al., 2007; Strube et al., 2011; He et al., 2018). Levan formation has also been disrupted without inhibiting FOS synthesis in Gram–negative levansucrases through point mutations of positionally equivalent, but divergent, amino acids in structural motifs involved in substrate binding (Yanase et al., 2002; Mardo et al. 2014). As a general view, the loss of the polymer–synthesizing capability in the engineered enzymes drastically decreased the ratio of the transferase versus hydrolase activities. Site–directed mutagenesis remains unexplored as a tool to enhance levan production in a natural FOS–yielding levansucrase.

Sugar–binding subsites

A model for the organization of the sugar–binding subsites in bacterial fructosyltransferases was proposed by Ozimek et al. (2006). This model uses the general nomenclature previously defined for glycosyl hydrolases (Davies et al., 1997), and it is supported by available structural data of enzyme–ligand complexes. The scheme reflects the ping–pong mechanism of catalysis, and it serves to explain how variations in the product spectra of levansucrases from different sources (see section 2.4) are dependent upon the affinity of the fructosyl–enzyme intermediate for the acceptor substrate, and the FOS elongation mechanism.

Figure 3 shows a schematic drawing of the sugar–binding subsites in the active site of levansucrase, and the sequential steps of the three main reactions (sucrose hydrolysis, FOS synthesis, and fructan elongation). The presence of free trisaccharides (1–kestose, 6–kestose, and neo–kestose) in the reaction mixture of Gram–positive and Gram–negative levansucrases indicates that the first

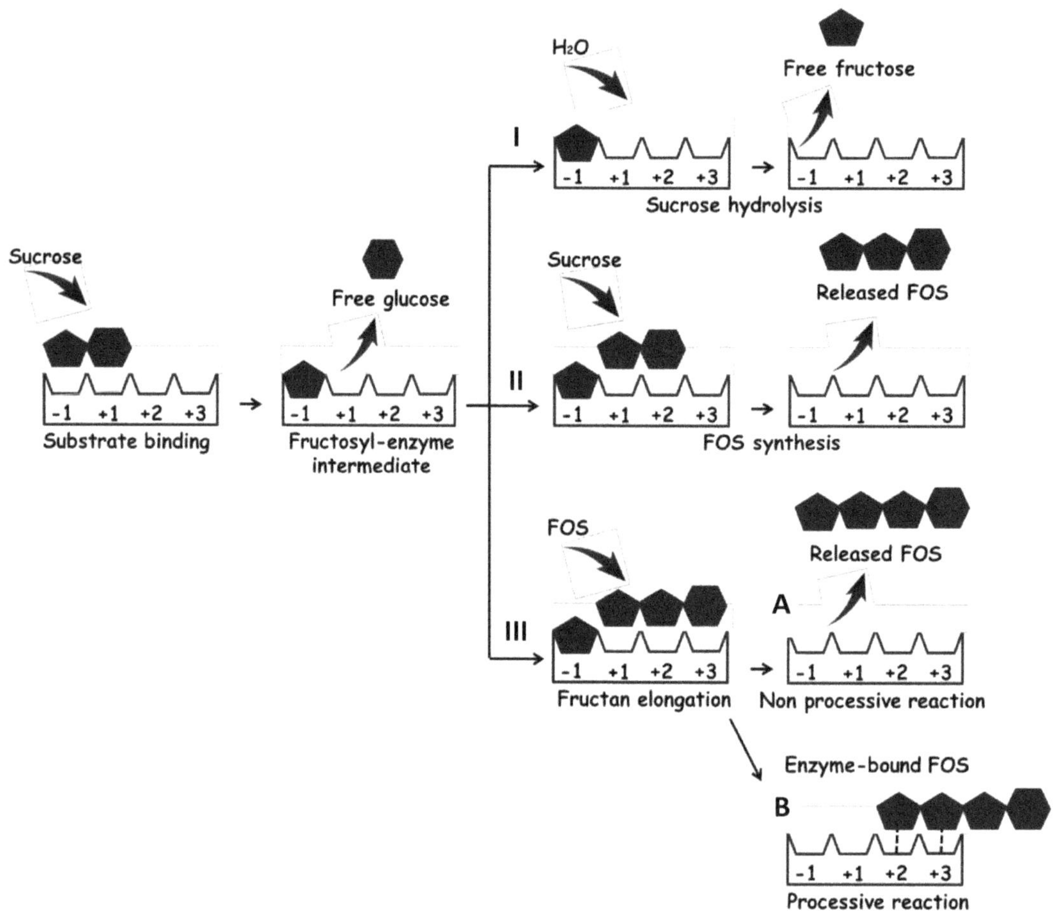

Figure 3. Schematic drawing of the sugar–binding subsites in the active site of levansucrases. The enzyme subsites are labelled from -1 to +n, with the numbers increasing positively from the bottom towards the surface of the unique substrate–entering channel of the active site cavity. Upon complex formation with the donor substrate (sucrose), the fructosyl and glucosyl moieties are accommodated at the -1 and +1 subsites, respectively. Cleavage of the glycosidic bond activates the covalent interaction between the fructosyl unit and the nucleophile (Asp) in subsite -1, and the release of free glucose through the funnel–like channel. The empty +1 subsite will be then occupied by the fructosyl acceptor. Different acceptors may alternatively enter the active site and react with the fructosyl–enzyme intermediate. The reaction with water results in hydrolysis and the release of free fructose (**I**). When a sucrose molecule acts as the acceptor, its fructosyl and glucosyl moieties are bound to the +1 and +2 subsites, respectively. The synthesized trisaccharide (1–kestose, 6–kestose, or neo–kestose) is released, regenerating the unmodified enzyme form (**II**). In next fructosylation steps, the FOS chain can be slightly further elongated via distributive (non–processive) reactions (**IIIA**) or it is rapidly polymerized in a processive manner (**IIIB**) (see text). The vertical dashed lines indicate that the fructan product remains bound to the enzyme after the fructosyl transfer.

fructosylation product dissociates from the enzyme (Hernández et al., 1995; Iizuka et al., 2002; Caputi et al., 2013; Raga-Carbajal et al., 2018). In further fructosylation steps, FOS are partially elongated in a distributive (non-processive) manner or extensively polymerized by sequential processive reactions using β-(2→6) linkages. At a certain size (DP ≥ 4), the intermediate FOS molecules remain bound to the enzyme, and directly serve as acceptors for rapid levan formation.

Inulosucrases were defined as non-processive enzymes (Ozimek et al., 2006). It has been postulated that subsites +2 and +3 of inulosucrase have a low affinity for the growing FOS chains, and the product is separated from the enzyme once the fructosyl unit had been transferred. The release of the extended acceptor after virtually each fructosyl transfer results in a broad range of inulooligosaccharides (DP 3–25) with a relatively small amount of polymeric inulin material (Ozimek et al., 2006, Anwar et al., 2008; Anwar et al., 2010; Peña-Cardeña et al., 2015; Frasch et al., 2017; Kralj et al., 2018). However, high-DP inulin (MW 10^3–10^5 daltons) is most likely synthesized through a processive mechanism (van Hijum et al., 2002; Anwar et al., 2008; Peña-Cardeña et al., 2015).

Evolutionary aspects

Family GH68 comprises prokaryotic fructosyltransferases (levansucrase and inulosucrase) and β-fructofuranosidases sequenced so far from about 400 species, comprising five bacterial phyla (mainly Protobacteria, Actinobacteria and Firmicutes) and five archaeal families. The vast majority of the GH68 entries in the CAZy database correspond to genome sequencing projects. Only bacterial members have been functionally or structurally characterized.

The biological function of a bacterial fructosyltransferase is influenced by the presence or not of other sucrose-transforming enzymes in the host species. Levansucrase is the enzyme responsible for both sucrose utilization and fructan synthesis in the endophytic species *Gluconacetobacter diazotrophicus* (Hernández et al., 2000). Contrarily, the contribution of levansucrase to sucrose metabolism is negligible in the soil resident *Bacillus subtilis*, which possesses a more efficient sucrose-utilizing pathway constituted by PTS-dependent permease specific for sucrose, and an intracellular sucrase (Lepesant et al., 1972).

In fructan-producing plants, distinct fructosyltransferases (1-SST, 6-SST/6-SFT, 1-FFT, and 6G-FFT) and fructan exo-hydrolases (1-FEH and 6-FEH) evolved from invertases by few amino acid substitutions in the catalytic β-propeller domain (Ritsema et al., 2006; Schroeven et al., 2008; Lammens et al., 2012; Versluys et al., 2018). In bacteria, it is clear the close evolutionary connection among GH32 enzymes (invertase, sucrose-6P hydrolase, levanase, and inulinase). GH32 and GH68 proteins share the β-propeller architecture necessary for catalysis, but their low sequence identities do not allow postulating that levansucrases and inulosucrases developed by mutational changes of invertases. Differently to the plants, invertases are not ubiquitous in bacteria.

Horizontal gene transfer has undoubtedly contributed to the widespread presence of fructosyltransferases in bacteria. The horizontal acquisition of levansucrase or inulosucrase genes from taxonomically distant species sharing a common habitat can be inferred by comparing the topologies of the GH68 protein tree against the corresponding 16S rRNA "reference" tree. We hypothesize that the biological advantage of utilizing exogenous sucrose both as a direct energy source for cell growth and as the precursor of a multifunctional polysaccharide has been a positive selective force driving retention of horizontally acquired GH68 genes. Archaea harboring putative GH68 enzymes are all halophiles (Kirtel et al., 2018b). The close phylogenetic relatedness of all archaeal GH68 sequences available so far reflects no evidence of horizontal gene acquisition from a bacterial donor.

Levan synthesis versus degradation

Levan synthesis is a specialized enzymatic process that fulfills diverse biological roles in nature. Levan, with other exopolysaccharides, contributes to the formation of bacterial biofilms in different habitats. High DP fructans enable adhesion of the cells to surfaces in dental plaque biofilms produced by *Actinomyces viscosus* and species of the genus *Streptococcus* (Rozen et al., 2001). Levan helps to agglutinate the soil bacterium *Bacillus subtilis* in a favorable environment, conferring protection against water deprivation and other stresses (Dogsa et al., 2013). Levan formation is a fitness factor for the *in planta* life of the pathogens *Erwinia amylovora* (Koczan et al., 2009) and *Pseudomonas syringae* (Mehmood et al., 2015), and the nitrogen-fixing endophyte *Gluconacetobacter diazotrophicus* (Hernández et al., 2000; Velázquez-Hernández et al., 2011). As a secondary but no less important role, levan in biofilms constitutes an extracellular carbon reservoir to be used as an energy source under starvation conditions (Toksoy Öner et al., 2016).

Bacteria can degrade native levan by combining the intrinsic β-(2→6) exo-type hydrolytic activity of levansucrase (see section 2.4) with the more general and efficient action of genuine exo-levanases. The expression of exo-levanase genes in levan-producing species is strictly controlled at the transcriptional level by repression mechanisms that sensor the energetic status of the cell. Figure 4 illustrates the finely regulated expression of the fructan metabolic operon in the endophytic bacterium *G. diazotrophicus*. Levan is synthesized by the constitutively

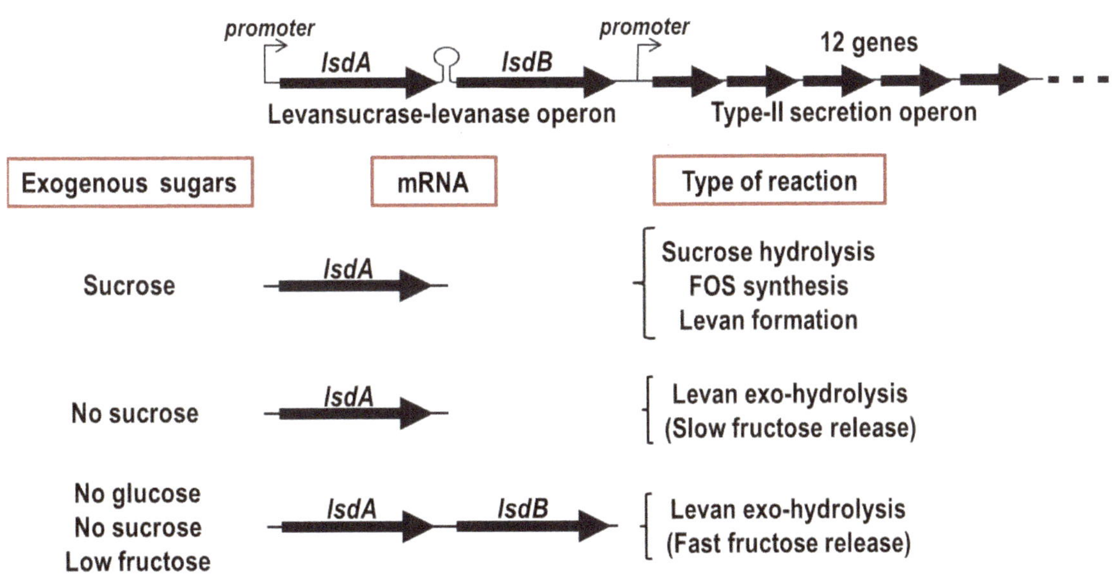

Figure 4. Schematic representation of the coordinated synthesis and secretion of the fructan metabolic enzymes in the endophyte *Gluconacetobacter diazotrophicus*. Two constitutive promoters enable transcription of the levansucrase-levanase (*lsdA-lsdB*) operon and the downstream 12-component operon encoding the type-II secretory machinery responsible for the extracellular release of both LsdA and LsdB (Arrieta et al., 2004). The codon usage of all members in the two operons mirrors that of the species. The *lsdA* gene is transcribed in the presence or absence of sucrose. Depending on the cell energetic status, the stem-loop structure at the intergenic region of the *lsdA-lsdB* operon functions as a terminator provoking the premature halt of the RNA polymerase. The synthesis of bicistronic *lsdAB* messengers is induced at low fructose concentrations (0.44 to 33 mM) and repressed by glucose (Menéndez et al., 2009). Catabolite repression of *lsdB* transcription ensures that levan is degraded only in the absence of sucrose or other preferred (rapidly metabolisable) carbon source.

expressed levansucrase (LsdA) as long as exogenous sucrose is available. When the favorite donor substrate is absent, the enzyme initiates the breakdown of levan through the non-reducing ends of the main chain and/or branches. The released fructose, once imported into the cell, activates the expression of a genuine exo-levanase (LsdB) to temporarily accelerate levan degradation, but only if preferred carbon sources are not accessible. LsdB transcription is repressed by glucose, a product which is continuously released during the levansucrase reaction on sucrose.

Fructose is known to regulate the expression of exo-fructanases in fructan-producing species by acting as an inducer at low concentrations and as a corepressor at high concentrations (Martin-Verstraete et al., 1995; Bergeron and Burne, 2001; Zeng et al., 2006; Menéndez et al, 2009). Fructose activation of exo-fructanase transcription is inhibited by glucose, indicating that the process is governed by catabolic repression. This dual control of exo-levanase expression is finely coordinated to safeguard the prevalence of levan synthesis over degradation whenever exogenous sucrose is available as an energetic substrate.

Concluding remarks

Fructan synthesis is widely spread in bacteria. Levan is the predominant bacterial fructose polymer and fulfills diverse biological roles in nature. Levansucrase catalyzes both sucrose hydrolysis and transfructosylation. Depending on the enzyme origin and the reaction conditions, there may be remarkable differences in the ratio of hydrolase versus transferase activities, as well as in the yield and size distribution of the fructosylated products. Most levansucrases from Gram-positive bacteria rapidly polymerize the precursor FOS molecules, while the enzyme from Gram-negative bacteria accumulates a significant fraction of the synthesized short-chain FOS, with the consequent lower yield of levan. Neither the distributive (non-processive) reaction of FOS synthesis nor the more specialized processive mechanism leading to the rapid polymerization of the bound levan intermediates, prevent the use of water as a fructosyl acceptor.

Evolution and researchers are working separately to solve the handicap of sucrose hydrolysis in bacterial fructosyltransferases. From the biological point of view, it might be advantageous for the host bacterium allocating the fructosyl moiety of exogenous sucrose to exclusively build a biofilm-forming exopolysaccharide, while using the released glucose to directly supply energy to the cell. From the biotech point of view, the fructosyl transfer to water represents a major waste of substrate in the production or either FOS or levan.

Declaration of Interest Statement

The authors of this article declare that they have no conflicts of interest of any kind.

References

Alberto, F., Bignon, C., Sulzenbacher, G., Henrissat, B., and Czjzek, M. (2004). The three-dimensional structure of invertase (beta-fructosidase) from *Thermotoga maritima* reveals a bimodular arrangement and an evolutionary relationship between retaining and inverting glycosidases. J. Biol. Chem. 279, 18903-18910. https://dx.doi.org/10.1074/jbc.M313911200

Anwar, M.A., Kralj, S., van der Maarel, M.J.E.C., and Dijkhuizen, L. (2008). The probiotic *Lactobacillus johnsonii* NCC 533 produces high-molecular-mass inulin from sucrose by using an inulosucrase enzyme. Appl. Environ. Microbiol. 74, 3426–3433. https://dx.doi.org/10.1128/AEM.00377-08

Anwar, M.A., Kralj, S., Villar Piqué, A., Leemhuis, H., van der Maarel, M.J.E.C., and Dijkhuizen, L. (2010). Inulin and levan synthesis by probiotic *Lactobacillus gasseri* strains: characterization of three novel fructansucrase enzymes and their fructan products. Microbiology 156, 1264-1274. https://dx.doi.org/10.1099/mic.0.036616-0

Arrieta, J.G., Hernández, L. Coego, A., Suárez, V., Balmori, E., Menéndez, C., Petit-Glatron, M.F., Chambert, R., and Selman-Housein, G. (1996). Molecular characterization of the levansucrase gene from the endophytic sugarcane bacterium *Acetobacter diazotrophicus* SRT4. Microbiology 142, 1077-1085. https://dx.doi.org/10.1099/13500872-142-5-1077

Arrieta, J.G., Sotolongo, M., Menéndez, C., Alfonso, D., Trujillo, L.E., Soto, M., Ramírez, R., and Hernández, L. (2004). A type II protein secretory pathway required for levansucrase secretion by *Gluconacetobacter diazotrophicus*. J. Bacteriol. 186, 5031-5039. https://dx.doi.org/.1128/JB.186.15.5031-5039.2004

Ateş, Ö., Arga, K.Y., and Toksoy Öner, E. (2013). The stimulatory effect of mannitol on levan biosynthesis: lessons from metabolic systems analysis of *Halomonas smyrnensis* AAD6T. Biotechnol. Progr. 29, 1386-1397. https://dx.doi.org/10.1002/btpr.1823

Ávila-Fernández, A., Olvera-Carranza, C., Rudiño-Piñera, E., Cassab, G.I., Nieto-Sotelo, J., and López-Munguía, A. (2007). Molecular characterization of sucrose:sucrose 1-fructosyltransferase (1-SST) from *Agave tequilana* Weber var. azul. Plant Science 173, 478–486. https://dx.doi.org/10.1016/j.plantsci.2007.07.009

Beine, R., Moraru, R., Nimtz, M., Na'amnieh, S., Pawlowski, A., Buchholz, K., and Seibel, J. (2008). Synthesis of novel fructooligosaccharides by substrate and enzyme engineering. J. Biotechnol. 138, 33-41. https://dx.doi.org/10.1016/j.jbiotec.2008.07.1998

Bekers, M., Laukevics, J., Upite, D., Kaminska, E., Vigants, A., Viesturs, U., Pankova, L., and Danilevics, A. (2002). Fructooligosaccharide and levan producing activity of *Zymomonas mobilis* extracellular levansucrase. Process Biochem. 38, 701-706. https://dx.doi.org/10.1016/S0032-9592(02)00189-9

Bergeron, L.J., and Burne, R.A. (2001). Roles of fructosyltransferase and levanase-sucrase of *Actinomyces naeslundii* in fructan and sucrose metabolism. Infect. Immun. 69, 5395-5402. https://dx.doi.org/10.1128/IAI.69.9.5395-5402.2001

Bergeron, L.J., Morou-Bermudez, E., and Burne, R.A. (2000). Characterization of the fructosyltransferase gene of *Actinomyces naeslundii* WVU45. J. Bacteriol. 182, 3649-3654. https://dx.doi.org/10.1128/JB.182.13.3649-3654.2000

Caputi, L., Nepogodiev, S.A., Malnoy, M., Rejzek, M., Field, R.A., and Benini, S. (2013). Biomolecular characterization of the levansucrase of *Erwinia amylovora*, a promising biocatalyst for the synthesis of fructooligosaccharides. J. Agric. Food Chem. 61, 12265-12273. https://dx.doi.org/10.1021/jf4023178

Castillo, E., and López-Munguía, A. (2004). Synthesis of levan in water-miscible organic solvents. J. Biotechnol. 114, 209-217. https://dx.doi.org/10.1016/j.jbiotec.2004.06.003

Chambert, R., and Gonzy-Treboul, G. (1976). Levansucrase of *Bacillus subtilis*: Kinetic and thermodynamic aspects of transfructosylation processes. European J. Biochem. 62, 55-64. https://dx.doi.org/10.1111/j.1432-1033.1976.tb10097.x

Chambert, R. and Petit-Glatron, M.F. (1988). Secretion mechanism of *Bacillus subtilis* levansucrase: characterization of the second step. J Gen Microbiol. 134, 1205-1214. https://dx.doi.org/10.1099/00221287-134-5-1205

Chambert, R. and Petit-Glatron, M.F. (1989). Study of the effect of organic solvents on the synthesis of levan and the hydrolysis of sucrose by *Bacillus subtilis* levansucrase. Carbohyd. Res. 191, 117-123. https://dx.doi.org/10.1016/0008-6215(89)85051-7

Chambert, R., and Petit-Glatron, M.F. (1991). Polymerase and hydrolase activities of *Bacillus subtilis* levansucrase can be separately modulated by site-directed mutagenesis. Biochem J. 279, 35–41. https://doi.org/10.1042/bj2790035

Chambert, R., Tréboul, G., and Dedonder, R. (1974). Kinetic studies of levansucrase of *Bacillus subtilis*. Eur. J. Biochem. 41, 285-300. https://dx.doi.org/10.1111/j.1432-1033.1974.tb03269.x

Cote, G.L. (1988). Production of a constitutive, extracellular levansucrase from *Erwinia herbicola* NRRL B-1678. Biotechnol. Lett. 10, 879–882. https://dx.doi.org/10.1007/BF01026999l

Crutz, A.M., Steinmetz, M., Aymerich, S., Richter, R., and Le Coq, D. (1990). Induction of levansucrase in *Bacillus subtilis*: an antitermination mechanism negatively controlled by the phosphotransferase system. J. Bacteriol. 172, 1043-1050. https://doi.org/10.1128/jb.172.2.1043-1050.1990

Daguer, J.P., Geissmann, T., Petit-Glatron, M.F., and Chambert, R. (2004). Autogenous modulation of the *Bacillus subtilis sacB–levB–yveA* levansucrase operon by the *levB* transcript. Microbiology 150, 3669–3679 https://dx.doi.org/10.1099/mic.0.27366-0

Davies, G. J., Wilson, K. S., and Henrissat, B. (1997). Nomenclature for sugar-binding subsites in glycosyl hydrolases. Biochem. J. 321, 557–559. https://doi.org/10.1042/bj3210557

Declerck, N., Minh, N.L., Yang, Y., Bloch, V., Kochoyan, M., and Aymerich, S. (2002). RNA recognition by transcriptional antiterminators of the BglG/SacY family: mapping of SacY RNA binding site. J. Mol. Biol. 319, 1035–1048. https://dx.doi.org/10.1016/S0022-2836(02)00373-X

Dogsa, I., Brloznik, M., Stopar, D., and Mandic-Mulec, I. (2013). Exopolymer diversity and the role of levan in *Bacillus subtilis* biofilms. PLoS ONE 8, e62044. https://doi.org/10.1371/journal.pone.0062044

Frasch, H.J., Leeuwen, S.S.V., and Dijkhuizen, L. (2017). Molecular and biochemical characteristics of the inulosucrase HugO from *Streptomyces viridochromogenes* DSM40736 (Tü494). Microbiology 163, 1030-1041. https://dx.doi.org/10.1099/mic.0.000493

Geier, G., and Geider, K. (1993). Characterization and influence on virulence of the levansucrase gene from the fireblight pathogen *Erwinia amylovora*. Physiol. Mol. Plant Pathol. 42, 387-404. https://dx.doi.org/10.1006/pmpp.1993.1029

Goldman, D., Lavid, N., Schwartz, A., Shoham, G., Danino, D., and Shoham, Y. (2008). Two active forms of *Zymomonas mobilis* levansucrase. An ordered microfibril structure of the enzyme promotes levan polymerization. J. Biol. Chem. 283, 32209-32217. https://dx.doi.org/10.1074/jbc.M805985200

Han, W.C., Byun, S.H., Kim, M.H., Sohn, E.H., Lim, J.D., Um, B.H., Kim, C.H., Kang, S.A., and Jang, K.H. (2009). Production of lactosucrose from sucrose and lactose by a levansucrase from *Zymomonas mobilis*. J. Microbiol. Biotechnol. 19, 1153-1160. https://doi.org/10.4014/jmb.0901.045

He, C., Yang, Y., Zhao, R., Qu, J., Jin, L., Lu, L., Xu, L., and Xiao, M. (2018) Rational designed mutagenesis of levansucrase from *Bacillus licheniformis* 8-37-0-1 for product specificity study. Appl. Microbiol. Biotechnol. 102, 3217-3228. https://dx.doi.org/10.1007/s00253-018-8854-3

Hernández, L., Arrieta, J., Betancourt, L., Falcón, V., Madrazo, J., Coego, A., and Menéndez, C. (1999a). Levansucrase from *Acetobacter diazotrophicus* SRT4 is secreted via periplasm by a signal-peptide-dependent pathway. Curr. Microbiol. 39, 146-152. https://dx.doi.org/10.1007/s002849900436

Hernández, L., Arrieta, J., Menéndez, C., Vazquez, R., Coego, A., Suárez, V., Selman, G., Petit-Glatron, M.F., and Chambert, R. (1995). Isolation and enzymic properties of levansucrase secreted by *Acetobacter diazotrophicus* SRT4, a bacterium associated with sugar cane. Biochem. J. 309, 113-118. https://dx.doi.org/10.1042/bj3090113

Hernández, L., Ramírez, R., Hormaza, J.V., Madrazo, J., and Arrieta, J. (1999b). Increased levansucrase production by a genetically modified *Acetobacter diazotrophicus* strain in shaking batch cultures. Lett. Appl. Microbiol. 28, 41-44. https://dx.doi.org/10.1046/j.1365-2672.1999.00470.x

Hernández, L., Sotolongo, M., Rosabal, Y., Menéndez, C., Ramírez, R., Caballero-Mellado, J., and Arrieta, J. (2000). Structural levansucrase gene (lsdA) constitutes a functional locus conserved in the species *Gluconacetobacter diazotrophicus*. Arch. Microbiol. 174, 120-124. https://dx.doi.org/10.1007/s002030000173

Hettwer, U., Gross, M., and Rudolph, K. (1995). Purification and properties of an extracellular levansucrase from *Pseudomonas syringae* pv. phaseolicola. J. Bacteriol. 177, 2834-2839. https://dx.doi.org/10.1128/jb.177.10.2834-2839.1995

Homann, A., Biedendieck, R., Götze, S., Jahn, D., and Seibel, J. (2007). Insights into polymer versus oligosaccharide synthesis: mutagenesis and mechanistic studies of a novel levansucrase from *Bacillus megaterium*. Biochem. J. 407, 189−198. https://dx.doi.org/10.1042/BJ20070600

Idogawa, N., Amamoto, R., Murata, K., and Kawai, S. (2014). Phosphate enhances levan production in the endophytic bacterium *Gluconacetobacter diazotrophicus* Pal5. Bioengineered 5, 173-179. https://dx.doi.org/10.4161/bioe.28792

Iizuka, M., Mima, T., Ammar, Y. Ben, Ito, K., and Minamiura, N. (2002). Mechanism of Synthesis of levan by Bacillus *natto* Levansucrase. J. Appl. Glycosci. 49, 229–237. DOI is not available.

Inthanavong, L., Tian, F., Khodadadi, M., and Karboune, S. (2013). Properties of *Geobacillus stearothermophilus* levansucrase as potential biocatalyst for the synthesis of levan and fructooligosaccharides. Biotechnol. Prog. 29, 1405-1415. https://dx.doi.org/10.1002/btpr.1788

Jakob, F., Meißner, D., and Vogel, R.F. (2012). Comparison of novel GH 68 levansucrases of levan-overproducing *Gluconobacter* species. Acetic Acid Bacteria 1, 6-14. https://dx.doi.org/org/10.4081/aab.2012.e2

Kim, H.J., Park, H.E., Kim, M.J., Lee, H.G., Yang, J.Y., and Cha, J.H. (2003). Enzymatic characterization of a recombinant levansucrase from *Rahnella aquatilis* ATCC 15552. J. Microbiol. Biotechnol. 13, 230-235. (Available from URL: http://xn--hy1b43dy9r4rl8nb.com/paper/view.html?no=2052199).

Kirtel, O., Menéndez, C., Versluys, M., van den Ende, W., Hernández, L., and Toksoy Öner, E. (2018a) Levansucrase from *Halomonas smyrnensis* AAD6T: first halophilic GH-J clan enzyme recombinantly expressed, purified, and characterized. Appl. Microbiol. Biotechnol. 102, 9207-9220. https://dx.doi.org/10.1007/s00253-018-9311-z

Kirtel, O., Versluys, M., van den Ende, W., and Toksoy Öner, E. (2018b) Fructans of the saline world. Biotechnol. Adv. 36, 1524-1539. https://dx.doi.org/10.1016/j.biotechadv.2018.06.009

Koczan, J.M., McGrath, M.J., Zhao, Y., and Sundin, G.W. (2009). Contribution of *Erwinia amylovora* exopolysaccharides amylovoran and levan to biofilm formation: implications in pathogenicity. Phytopathology 99, 1237-1244. https://dx.doi.org/10.1094/PHYTO-99-11-1237

Koops, A.J. and Jonker, H.H. (1996). Purification and characterization of the enzymes of fructan biosynthesis in tubers of *Helianthus tuberosus* Colombia. Plant Physiol. 110, 1167–1175. https://dx.doi.org/10.1104/pp.110.4.1167

Kralj, S., Leeflang, C., Sierra, E.I., Kempiński, B., Alkan, V., and Kolkman, M. (2018). Synthesis of fructooligosaccharides (FosA) and inulin (InuO) by GH68 fructosyltransferases from *Bacillus agaradhaerens* strain WDG185. Carbohydr. Polym. 179:350-359. https://dx.doi.org/10.1016/j.carbpol.2017.09.069

Lammens, W., Le Roy, K., Schroeven, L., van Laere, A., Rabijns, A., and van den Ende, W. (2009). Structural insights into glycoside hydrolase family 32 and 68 enzymes: functional implications. J. Exp. Bot. 60, 727-740. https://dx.doi.org/10.1093/jxb/ern333

Lammens, W., Le Roy, K., Yuan, S., Vergauwen, R., Rabijns, A., van Laere, A., Streikov, S.A., and van den Ende, W. (2012). Crystal structure of 6-SST/6-SFT from *Pachysandra terminalis*, a plant fructan biosynthesizing enzyme in complex with its acceptor substrate 6-kestose. Plant J. 70, 205-219. https://dx.doi.org/10.1111/j.1365-313X.2011.04858.x

Lepesant, J.A., Kunst, F., Lepesant-Kejzlarova, J., and Dedonder, R. (1972). Chromosomal location of mutations affecting sucrose metabolism in *Bacillus subtilis* Marburg. Mol. Gen. Genet. 118, 135–160. DOI is not available.

Li, H., and Ullrich, M.S. (2001). Characterization and mutational analysis of three allelic lsc genes encoding levansucrase in *Pseudomonas syringae*. J. Bacteriol. 183, 3282-3292. https://dx.doi.org/10.1128/JB.183.11.3282-3292.2001

Malang, S.K., Maina, N.H., Schwab, C., Tenkanen, M., and Lacroix, C. (2015). Characterization of exopolysaccharide and ropy capsular polysaccharide formation by *Weissella*. Food Microbiology, 46, 418–427. https://dx.doi.org/10.1016/j.fm.2014.08.022

Malik, A., Hapsari, M.T., Ohtsu, I., Ishikawa, S., and Takagi, H. (2015). Cloning and heterologous expression of the ftfCNC-2(1) gene from *Weissella confusa* MBFCNC-2(1) as an extracellular active fructansucrase in *Bacillus subtilis*. J. Biosci. Bioeng. 119, 515-520. https://dx.doi.org/10.1016/j.jbiosc.2014.09.024

Mardo, K., Visnapuu, T., Vija, H., Elmi, T., and Alamäe, T. (2014). Mutational analysis of conserved regions harboring catalytic triad residues of the levansucrase protein encoded by the lsc-3 gene (lsc3) of *Pseudomonas syringae* pv. tomato DC3000. Biotechnol. Appl. Biochem. 61, 11-22. https://dx.doi.org/10.1002/bab.1129

Martinez-Fleites, C., Ortiz-Lombardia, M., Pons, T., Tarbouriech, N., Taylor, E.J., Arrieta, J.G., Hernández, L., and Davies, G.J. (2005). Crystal structure of levansucrase from the Gram-negative bacterium *Gluconacetobacter diazotrophicus*. Biochem. J. 390, 19-27. https://dx.doi.org/10.1042/BJ20050324

Martin-Verstraete, I., Stulke, J., Klier, A., and Rapoport, G. (1995). Two different mechanisms mediate catabolite repression of the *Bacillus subtilis* levanase operon. J. Bacteriol. 177, 6919–6927. https://dx.doi.org/10.1128/jb.177.23.6919-6927.1995

Mehmood, A., Abdallah, K., Khandekar, S., Zhurina, D., Srivastava, A., Al-Karablieh, N., Alfaro-Espinoza, G., Pletzer, D. and Ullrich, M.S. (2015). Expression of extra-cellular levansucrase in *Pseudomonas syringae* is controlled by the in planta fitness-promoting metabolic repressor HexR. BMC Microbiol. 15, 1-11. https://dx.doi.org/10.1186/s12866-015-0349-0

Mena-Arizmendi, A., Alderete, J., Águila, S., Marty, A., Miranda-Molina, A., López-Munguía, A., and Castillo, E. (2011). Enzymatic fructosylation of aromatic and aliphatic alcohols by *Bacillus subtilis* levansucrase: reactivity of acceptors. J. Mol. Catal. B Enzym. 70, 41-48. https://dx.doi.org/10.1016/j.molcatb.2011.02.002

Méndez-Lorenzo, L., Porras-Domínguez, J.R., Raga-Carbajal, E., Olvera, C., Rodríguez-Alegría, M.E., Carrillo-Nava, E., Costas, M., and López-Munguía, A. (2015). Intrinsic levanase activity of *Bacillus subtilis* 168 levansucrase (SacB). PLoS ONE 10, e0143394. https://dx.doi.org/10.1371/journal.pone.0143394

Menéndez, C., Banguela, A., Caballero-Mellado, J., and Hernández, L. (2009). Transcriptional regulation and signal-peptide-dependent secretion of exolevanase (LsdB) in the endophyte *Gluconacetobacter diazotrophicus*. Appl. Environ. Microbiol. 75, 1782-1785. https://dx.doi.org/10.1128/AEM.01887-08

Meng, G., and Fütterer, K. (2003). Structural framework of fructosyl transfer in Bacillus subtilis levansucrase. Nat. Struct. Biol. 10, 935-941. https://dx.doi.org/10.1038/nsb974

Meng, G., and Fütterer, K. (2008). Donor substrate recognition in the raffinose-bound E342A mutant of fructosyltransferase *Bacillus subtilis* levansucrase. BMC Struct. Biol. 8, 16-27. https://dx.doi.org/10.1186/1472-6807-8-16

Mu, W., Chen, Q., Wang, X., Zhang, T., and Jiang, B. (2013). Current studies on physiological functions and biological production of lactosucrose. Appl. Microbiol. Biotechnol. 97, 7073-7080. https://dx.doi.org/10.1007/s00253-013-5079-3

Nakapong, S., Pichyangkura, R., Ito, K., Iizuka, M., and Pongsawasdi, P. (2013). High expression level of levansucrase from *Bacillus licheniformis* RN-01 and synthesis of levan nanoparticles. Int. J. Biol. Macromolec. 54, 30-36. https://dx.doi.org/org/10.1016/j.ijbiomac.2012.11.017

Olivares-Illana, V., Wacher-Rodarte, C., Le Borgne, S. and López-Munguía, A. (2002). Characterization of a cell-associated inulosucrase from a novel source: a

Leuconostoc citreum strain isolated from Pozol, a fermented corn beverage from Mayan origin. J. Ind. Microbiol. Biotechnol. 28, 112–117. https://doi.org/10.1038/sj/jim/7000224

Olvera, C., Centeno-Leija, S., and López-Munguía, A. (2007). Structural and functional features of fructansucrases present in *Leuconostoc mesenteroides* ATCC 8293. Anton. Leeuw. 92, 11-20. https://dx.doi.org/10.1007/s10482-006-9128-0

Ortiz-Soto, M.E., Possiel, C., Görl. J., Vogel, A., Schmiedel, R., and Seibel, J. (2017). Impaired coordination of nucleophile and increased hydrophobicity in the +1 subsite shift levansucrase activity towards transfructosylation. Glycobiology 27, 755-765. https://dx.doi.org/10.1093/glycob/cwx050

Ozimek, L.K., Euverink, G.J., van der Maarel, M.J., and Dijkhuizen, L. (2005). Mutational analysis of the role of calcium ions in the *Lactobacillus reuteri* strain 121 fructosyltransferase (levansucrase and inulosucrase) enzymes. FEBS Lett. 579, 1124-1128. https://dx.doi.org/10.1016/j.febslet.2004.11.113

Ozimek, L.K., Kralj, S., van der Maarel, M.J.E.C., and Dijkhuizen, L. (2006). The levansucrase and inulosucrase enzymes of *Lactobacillus reuteri* 121 catalyse processive and nonprocessive transglycosylation reactions. Microbiology 152, 1187-1196. https://dx.doi.org/10.1099/mic.0.28484-0

Peña-Cardeña, A., Rodríguez-Alegría, M.E., Olvera C., and Munguía, A. L. (2015). Synthesis of fructooligosaccharides by *IslA4*, a truncated inulosucrase from *Leuconostoc citreum.* BMC Biotechnol. 15, 2-9. https://dx.doi.org/10.1186/s12896-015-0116-1

Pereira, Y., Petit-Glatron, M. F., and Chambert, R. (2001). yveB, encoding endolevanase LevB, is part of the *sacB–yveB–yveA* levansucrase tricistronic operon in *Bacillus subtilis.* Microbiology, 147, 3413–3419. https://dx.doi.org/10.1099/00221287-147-12-3413

Petit-Glatron, M.F., Benyahia, F., and Chambert, R. (1987). Secretion of *Bacillus subtilis* levansucrase: a possible two-step mechanism. Eur. J. Biochem. 163, 379-387. https://doi.org/10.1111/j.1432-1033.1987.tb10810.x

Pijning, T., Anwar, M.A., Boger, M., Dobruchowska, J.M., Leemhuis, H., Kralj, S., Dijkhuizen, L., and Dijkstra, B.W. (2011). Crystal structure of inulosucrase from *Lactobacillus*: insights into the substrate specificity and product specificity of GH68 fructansucrases. J. Mol. Biol. 412, 80-93. https://dx.doi.org/10.1016/j.jmb.2011.07.031

Pons, T., Naumoff, D.G., Martínez-Fleites, C., and Hernández, L. (2004). Three acidic residues are at the active site of a β-propeller architecture in glycoside hydrolase families 32, 43, 62, and 68. Proteins 54, 424-432. https://dx.doi.org/10.1002/prot.10604

Porras-Domínguez, J.R., Á. Ávila-Fernández, Á., Miranda-Molina, A., Rodríguez-Alegría, M.E., and López-Munguía, A. (2015). *Bacillus subtilis* 168 levansucrase (SacB) activity affects average levan molecular weight. Carbohyd. Polym. 132, 338-344. https://dx.doi.org/10.1016/j.carbpol.2015.06.056

Raga-Carbajal, E., Carrillo-Nava, E., Costas, M., Porras-Dominguez, J., López-Munguía, A., and Olvera, C. (2016). Size product modulation by enzyme concentration reveals two distinct levan elongation mechanisms in *Bacillus subtilis* levansucrase. Glycobiology 26, 377-385. https://dx.doi.org/10.1093/glycob/cwv112

Raga-Carbajal, E., López-Munguía, A., Alvarez, L., and Olvera, C. (2018). Understanding the transfer reaction network behind the non-processive synthesis of low molecular weight levan catalyzed by *Bacillus subtilis* levansucrase. Sci Rep. 8, 15035-15045. https://dx.doi.org/10.1038/s41598-018-32872-7

Rathsam, C., and Jacques, N.A. (1998). Role of C-terminal domains in surface attachment of the fructosyltransferase of *Streptococcus salivarius* ATCC 25975. J. Bacteriol. 180, 6400-6403. Available from URL: http://www.ncbi.nlm.nih.gov/pmc/articles/PMC107731/).

Rathsam, C., Giffard, P.M., and Jacques, N.A. (1993). The cell-bound fructosyltransferase of Streptococcus salivarius: the carboxyl terminus specifies attachment in a *Streptococcus gordonii* model system. J. Bacteriol. 175, 4520-4527. https://dx.doi.org/10.1128/jb.175.14.4520-4527.1993

Ritsema, T., Hernández, L., Verhaar, A., Altenbach, D., Boller, T., Wiemken, A., and Smeekens, S. (2006). Developing fructan-synthesizing capability in a plant invertase via mutations in the sucrose-binding box. Plant J. 48, 228-237. https://dx.doi.org/10.1111/j.1365-313X.2006.02862.x

Rosell, K.G., and Birkhed, D. (1974). An inulin-like fructan produced by *Streptococcus mutans* strain JC2. Acta Chem. Scand. B28, 589. https://dx.doi.org/10.3891/acta.chem.scand.28b-0589

Rozen, R., Bachrach, G., Bronshteyn, M., Gedalia, I., and Steinberg, D. (2001). The role of fructans on dental biofilm formation by *Streptococcus sobrinus*, *Streptococcus mutans*, *Streptococcus gordonii* and *Actinomyces viscosus*. FEMS Microbiol. Lett. 195, 205-10. https://dx.doi.org/10.1016/S0378-1097(01)00009-X

Ruiz-Aceituno, L., Sanz, M.L., de Las Rivas, B., Muñoz, R., Kolida, S., Jimeno, M.L., and Moreno, F.J. (2017). Enzymatic synthesis and structural characterization of theanderose through transfructosylation reaction catalyzed by levansucrase from *Bacillus subtilis* CECT 39. J. Agric. Food Chem. 65, 10505-10513. https://dx.doi.org/10.1021/acs.jafc.7b03092

Schroeven, L., Lammens, W., Van Laere, A., and van den Ende, W. (2008). Transforming wheat vacuolar invertase into a high affinity sucrose:sucrose 1–fructosyltransferase. New Phytologist 180, 822-831. https://dx.doi.org/10.1111/j.1469-8137.2008.02603.x.

Seibel, J., Moraru, R., Götze, S., Buchholz, K., Na'amnieh, S., Pawlowski, A., and Hecht, H.J. (2006). Synthesis of sucrose analogues and the mechanism of action of *Bacillus subtilis* fructosyltransferase (levansucrase). Carbohyd. Res. 341, 2335-2349. https://dx.doi.org/10.1016/j.carres.2006.07.001

Senthilkumar, V., Rajendhran, J., Busby, S.J., and Gunasekaran, P. (2009). Characterization of multiple promoters and transcript stability in the *sacB-sacC* gene cluster in *Zymomonas mobilis*. Arch. Microbiol. 191, 529-541. https://dx.doi.org/10.1007/s00203-009-0479-6

Song, D.D., and Jacques, N.A. (1999). Purification and enzymic properties of the fructosyltransferase of *Streptococcus salivarius* ATCC 25975. Biochem. J. 341, 285-291. https://dx.doi.org/10.1042/bj3410285

Song, K-B., Joo, H-K., and Rhee, S-K. (1993). Nucleotide sequence of levansucrase gene (levU) of *Zymomonas mobilis* ZM1 (ATCC10988). Biochim. Biophys. Acta 1173, 320-324. https://dx.doi.org/10.1016/0167-4781(93)90130-6

Song, K.B., Seo, J.W., Kim, M.G., and Rhee, S.K. (1998). Levansucrase of *Rahnella aquatilis* ATCC33071: gene cloning, expression, and levan formation. Ann. N. Y. Acad. Sci. 864, 506-511. https://dx.doi.org/10.1111/j.1749-6632.1998.tb10369.x

Steinmetz, M., Le Coq, D., Aymerich, S., Gonzy-Treboul, G., and Gay, P. (1985). The DNA sequence of the gene for the secreted *Bacillus subtilis* enzyme levansucrase and its genetic control sites. Mol. Gen. Genet. 200, 220-228. https://dx.doi.org/10.1007/BF00425427

Strube, C.P., Homann, A., Gamer, M., Jahn, D., Seibel, J., and Heinz, D.W. (2011). Polysaccharide synthesis of the levansucrase SacB from *Bacillus megaterium* is controlled by distinct surface motifs. J. Biol. Chem. 286, 17593-17600. https://dx.doi.org/10.1074/jbc.M110.203166

Támbara, Y., Hormaza, J.V., Perez, C., Leon, A., Arrieta, J., and Hernández, L. (1999). Structural analysis and optimised production of fructo-oligosaccharides by levansucrase from *Acetobacter diazotrophicus* SRT4. Biotechnol. Lett. 21, 117-121. https://dx.doi.org/10.1023/A:1005474017009

Tanaka, T., Oi, S., and Yamamoto, T. (1980). The molecular structure of low and high molecular weight synthesized by levansucrase. J. Biochem. 87, 297-303. https://doi.org/10.1093/oxfordjournals.jbchem.a132737

Tanaka, T., Yamamoto, S., Oi, S., and Yamamoto, T. (1981). Structures of heterooligosaccharides synthesized by levansucrase. J. Biochem. 90, 521-526. https://doi.org/10.1093/oxfordjournals.jbchem.a133500

Tkachenko, A.A., Loitsyanskaya, M.S. and Yaremenko, T.G. (1980) Role of components of growth media in regulating the synthesis of levansucrase by *Gluconobacter oxydans*. Mikrobiologiya 49, 412–416. DOI is not available.

Toksoy Öner, E., Hernández, L., and Combie, J. (2016). Review of levan polysaccharide: From a century of past experiences to future prospects. Biotechnology Advances, 34, 827–844. https://dx.doi.org/10.1016/j.biotechadv.2016.05.002

Tonozuka, T., Tamaki, A., Yokoi, G., Miyazaki, T., Ichikawa, M., Nishikawa, A., Ohta, Y., Hidaka, Y., Katayama, K., Hatada, Y., Ito, T., and Fujita, K. (2012). Crystal structure of a lactosucrose-producing enzyme, *Arthrobacter* sp. K-1 β-fructofuranosidase. Enzyme Microb. Technol. 51, 359-365. https://dx.doi.org/10.1016/j.enzmictec.2012.08.004

Trujillo, L.E., Arrieta, J.G., Dafhnis, F., Garcia, J., Valdes, J., Tambara, Y., Perez, M., and Hernández, L. (2001). Fructo-oligosaccharides production by the *Gluconacetobacter diazotrophicus* levansucrase expressed in the methylotrophic yeast *Pichia pastoris*. Enzyme Microb. Technol. 28, 139-144. https://dx.doi.org/10.1016/S0141-0229(00)00290-8

Trujillo, L.E., Gómez, R., Banguela, A., Soto, M., Arrieta, J.G., and Hernández, L. (2004). Catalytical properties of N-glycosylated *Gluconacetobacter diazotrophicus* levansucrase produced in yeast. Elect. J. Biotechnol. 7, 115-123. https://dx.doi.org/10.2225/vol7-issue2-fulltext-4

van den Ende, W., van Wonterghem, D., Dewil, E., Verhaert, P., De Loof, A., and van Laere, A. (1996). Purification and characterization of 1-SST, the key enzyme initiating fructan biosynthesis in young chicory roots (*Cichorium intybus*).

Physiologia Plantarum 98, 455–466. https://dx.doi.org/10.1111/j.1399-3054.1996.tb05699.x

van Hijum, S.A.F.T., van der Maarel, M.J.E.C., and Dijkhuizen, L. (2003). Kinetic properties of an inulosucrase from *Lactobacillus reuteri* 121. FEBS Lett. 534, 207-210. https://dx.doi.org/10.1016/S0014-5793(02)03841-3

van Hijum, S.A.F.T., Szalowska, E., van der Maarel, M.J.E.C., and Dijkhuizen, L. (2004). Biochemical and molecular characterization of a levansucrase from *Lactobacillus reuteri*. Microbiology 150, 621-630. https://dx.doi.org/10.1099/mic.0.26671-0

van Hijum, S.A., van Geel-Schutten, G.H., Rahaoui, H., van der Maarel, M.J., and Dijkhuizen, L. (2002). Characterization of a novel fructosyltransferase from Lactobacillus reuteri that synthesizes high molecular-weight inulin and inulin oligosaccharides. Appl. Environ. Microbiol. 68, 4390–4398. https://dx.doi.org/10.1128/AEM.68.9.4390-4398.2002

Velázquez-Hernández, M.L., Baizabal-Aguirre, V.M., Cruz-Vázquez, F., Trejo-Contreras, M.J., Fuentes-Ramírez, L.E., Bravo-Patiño, A., Cajero-Juárez, M., Chávez-Moctezuma, M.P. and Valdez-Alarcón, J.J. (2011). *Gluconacetobacter diazotrophicus* levansucrase is involved in tolerance to NaCl, sucrose and desiccation, and in biofilm formation. Arch Microbiol. 93,137-49. https://dx.doi.org/10.1007/s00203-010-0651-z

Versluys, M., Kirtel, O., Toksoy Öner, E., and van den Ende, W. (2018). The fructan syndrome: evolutionary aspects and common themes among plants and microbes. Plant Cell Environ. 41, 16–38. https://dx.doi.org/10.1111/pce.13070

Vigants, A., Upite, D., Scherbaka, R., Lukjanenko, J., and Ionina, R. (2013). An influence of ethanol and temperature on products formation by different preparations of *Zymomonas mobilis* extracellular levansucrase. Folia Microbiol. 58, 75-80. https://dx.doi.org/10.1007/s12223-012-0185-7

Visnapuu, T., Mardo, K., and Alamäe, T. (2015). Levansucrases of a *Pseudomonas syringae* pathovar as catalysts for the synthesis of potentially prebiotic oligo- and polysaccharides. New Biotechnol. 32, 597-605. https://dx.doi.org/10.1016/j.nbt.2015.01.009

Visnapuu, T., Mardo, K., Mosoarca, C., Zamfir, A.D., Vigants, A., and Alamäe, T. (2011). Levansucrases from *Pseudomonas syringae* pv. tomato and *P. chlororaphis* subsp. aurantiaca: substrate specificity, polymerizing properties and usage of different acceptors for fructosylation. J. Biotechnol. 155, 338-349. https://dx.doi.org/10.1016/j.jbiotec.2011.07.026

Wada, T., Ohguchi, M., and Iwai, Y. (2003) A novel enzyme of *Bacillus* sp. 217C-11 that produces inulin from sucrose. Biosci. Biotechnol. Biochem. 67, 1327–1334. https://doi.org/10.1271/bbb.67.1327

Waldherr, F.W., Meissner, D., and Vogel, R.F. (2008). Genetic and functional characterization of *Lactobacillus panis* levansucrase. Arch. Microbiol. 190, 497-505. https://dx.doi.org/10.1007/s00203-008-0404-4

Wuerges, J., Caputi, L., Cianci, M., Boivin, S., Meijers, R., and Benini, S. (2015). The crystal structure of *Erwinia amylovora* levansucrase provides a snapshot of the products of sucrose hydrolysis trapped into the active site. J. Struct. Biol. 191, 290-298. https://dx.doi.org/10.1016/j.jsb.2015.07.010

Xu, W., Liu, Q., Yu, S., Zhang, T., and Mu, W. (2018) Synthesis of lactosucrose using a recombinant levansucrase from *Brenneria goodwinii*. Appl. Biochem. Biotechnol. 186, 292-305. https://dx.doi.org/10.1007/s12010-018-2743-1

Yanase, H., Maeda, M., Hagiwara, E., Yagi, H., Taniguchi, K., and Okamoto, K. (2002). Identification of functionally important amino acid residues in *Zymomonas mobilis* levansucrase. J. Biochem. (Tokyo) 132: 565–572. https://doi.org/10.1093/oxfordjournals.jbchem.a003258

Yuan, S., Le Roy, K., Venken, T., Lammens, W., van den Ende, W., and De Maeyer, M. (2012) pKa Modulation of the Acid/Base Catalyst within GH32 and GH68: A Role in Substrate/Inhibitor Specificity? PLoS ONE 7(5): e37453. https://dx.doi.org/10.1371/journal.pone.0037453

Zeng, L., Wen, Z.T., and Burne, R.A. (2006). A novel signal transduction system and feedback loop regulate fructan hydrolase gene expression in *Streptococcus mutans*. Mol. Microbiol. 62:187-200. https://dx.doi.org/10.1111/j.1365-2958.2006.05359.x

Exopolysaccharides from Genus *Weissella* and their Functional Applications

7

Rwivoo Baruah and Arun Goyal*

Carbohydrate Enzyme Biotechnology Laboratory, Department of Biosciences and Bioengineering, Indian Institute of Technology Guwahati, Guwahati 781039, Assam, India
*Corresponding author: arungoyl@iitg.ac.in

DOI: https://doi.org/10.21775/9781912530267.07

Abstract

The microbial exopolysaccharides that have a wide variety of applications are predominantly produced by Lactic acid bacteria (LAB). In recent times the focus on *Weissella*, a specific genus of LAB, has grown. In the genus *Weissella* two species, *cibaria* and *confusa* are the major producer of exopolysaccharides (EPS). These species chiefly produce the exopolysaccharide "dextran" a biopolymer of D-glucose. The physicochemical properties of dextran from *Weissella* makes it uniquely suitable for several food applications. Besides dextran the genus *Weissella* has been also reported to produce heteropolysaccharides containing alternating monomers such as mannose and galactose and also capsular polysaccharides (EPS covalently bound to the cells). Variety of applications using exopolysaccharide from *Weissella* such as improving the texture of bread in baking, its anti-oxidative and immunomodulatory properties are known. The ability to produce high amount of exopolysaccharides by members of genus Weissella have led to the development of novel food products such as quinoa-based yogurt made using dextran producing *Weissella cibaria*. This chapter describes the different types of EPS produced by genus *Weissella* and their characteristics with the functional properties.

Introduction

The genus *Weissella* belong to the phylum Firmicutes, class Bacilli, order Lactobacillales and family Leuconostocaceae (Collins et al., 1993). Bacteria belonging to the genus *Weissella* are Gram-positive, catalase-negative, non-endospore forming, facultative anaerobic cells having coccoid or rod-shaped morphology (Collins et al., 1993; Björkroth et al., 2009, 2014). The genus *Weissella* was named after the German microbiologist Norbert Weiss, known for his many contributions in the field of lactic acid bacteria research (Collins et al., 1993). They are difficult to separate from members of the genera *Leuconostoc* or the heterofermentative lactobacilli on the basis of phenotypic characteristics alone. Currently, the identification of *Weissella spp* are done using molecular biology techniques such as 16S rRNA gene sequencing, amplified ribosomal DNA restriction analysis (ARDRA) or ribotyping etc. Since the original description of the genus by Collins et al., (1993), various new species of *Weissella* have been described, so that currently the genus comprises 19 validated species (Figure 1). The *Weissella species* are grouped in five phylogenetic branches based on 16S phylogeny. *W. soli, W. diestrammenae, W. koreensis, W. kandleri* and *W. oryzae* are the members of the first branch. *W. cibaria* and *W. confusa* are the members of second and *W. thailandensis, W. hellenica* and *W. paramesenteroides* occurring in third branch. *W. ceti, W. halotolerans, W. viridescens, W. minor* and *W. uvarum* are associated with the fourth branch and *W. beninensis, W. fabalis, W. fabaria* and *W. ghanensis* with the fifth branch (Figure 1).

W. cibaria and *W. confusa* belong to the same branch and share numerous habitats. Both species have been found in fermented foods, animal, clinical samples and human faeces. These two species are difficult to distinguish since they share 99.6% of similarity in their 16S rRNA gene sequences. *W. cibaria* and *W. confusa* also share another similar trait which is the ability to produce extracellular exopolysaccharide dextran (biopolymer of D-glucose), fructan (biopolymer of fructose), heteropolysaccharides and non-digestible oligosaccharides (NDO) (Fusco et al., 2015). Some *Weissella* spp are known to possess antimicrobial activity by producing bacteriocins e.g. Weissellicin A showing antilisterial activity (Papagianni and Papamichael., 2012).

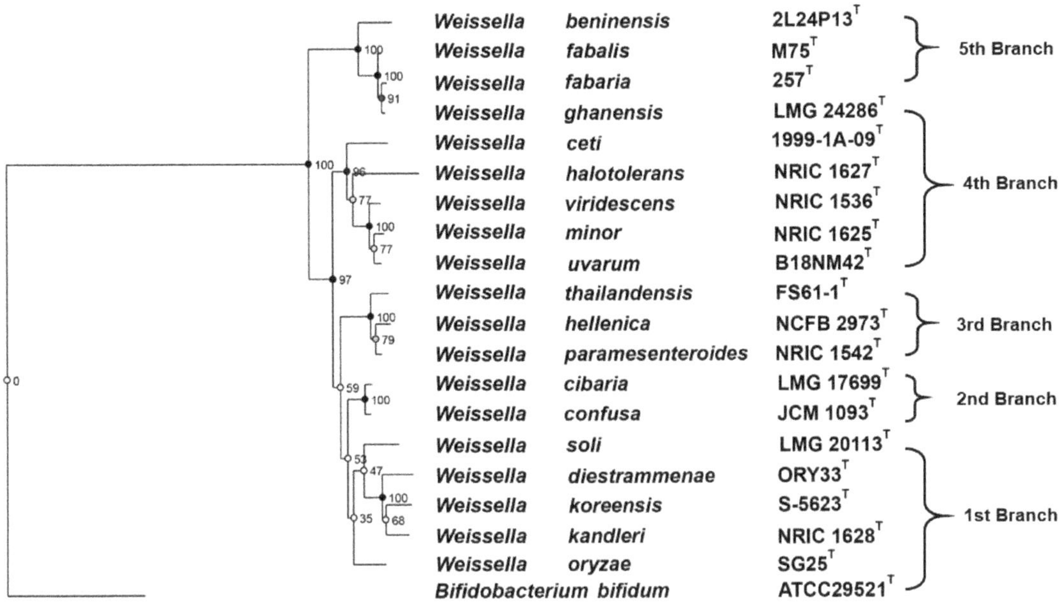

Figure 1. Neighbor-joining phylogenetic tree based on 16S rRNA sequences of *Weissella* species type strains. The 16S rRNA sequence of *Bifidobacterium bifidum* was used as an out group sequence. Bootstrap values (%) derived from 1000 replicates are given at branch points. Bar indicates % sequence divergence. Adapted from Fusco *et al.,* (2015).

Types of exopolysaccharide

Exopolysaccharides (EPS) are classified into two groups on the basis of their composition, these are Homopolysaccharides (HoPS) and Heteropolysaccharides (HePS). HoPS are biopolymers composed of a single monosaccharide residue e.g. HoPS composed of D-glucose is called glucan and those composed of fructose are called fructan. HoPs such as glucan is synthesized by LAB like *Weissella sp.* by its extracellular enzyme called glucansucrase. Glucansucrase hydrolyses sucrose and utilizes the free glucose moieties for biopolymer formation, releasing the fructose moieties free. HePS are biopolymers that are composed of more than one monosaccharide residue such as D-glucose, D-rhamnose, D-galactose and rarely with groups such as acetyl or phosphates. HePS are formed intracellularly with the help of various enzymes responsible for the polymerization of monosaccharide-nucleotide precursors e.g. UDP-glucose and UDP-galactose. The formation of HePS form precursors are controlled by glucosyltransferase gene expression. This gene is located in an operon commonly called EPS biosynthesis cluster which is responsible for regulation, chain-length determination, polymerization and export of HePS.

The vast majority of HoPS produced by the members in the genus *Weissella* are *W. confusa* and *W. cibaria*. The HoPs they produce is dextran, which is type of a glucan that comprises a liner chain of α-(1-6) linked D-glucose units with various branched linkages of D-glucose at α-(1,2), α-(1-3) and α-(1-4) positions (Capek et al., 2011). The presence of varying percentage of branched linkage in dextran contribute to different physico-chemical properties. Dextrans from *Weissella* have shown lower percentage (2.4-3.4%) of branched α-(1-3) linkages (Maina et al., 2008, Bounaix et al., 2009; Ahmed et al., 2012; Shukla et al., 2014; Baruah et al 2017). Dextran from *Weissella* tend to have higher molecular weights compared to dextrans produced from other lactic acid bacteria. Dextran produced from *L. mesenteroides* NRRL B-512F has a molecular weight ranging from 6.2 to 7.1 $x10^6$ Da (Lynch et al., 2014) whereas dextran produced from genus *Weissella* has a molecular weight ranging from 1.4 $x10^6$ to 7.2 x 10^8 Da (Fessard, and Remize et al., 2017). *Weissella* have been also reported to produce HoPS i.e. fructan which is a biopolymer composed of fructose monomer and depending on the linkages could be levan (β-2-6 linked) or inulin (β-2-1 linked) (Malang et al., 2015). *Weissella confusa* KR780676 isolated from an idli batter was reported for the production of linear galactan containing α-(1,6)-linked galactose units (Devi et al., 2016).

There are very fewer reports on HePS than on HoPS produced by *Weissella*. Malang et al., 2015 reported 3 *Weissella confusa* strains producing HePS. More recently *Weissella cibaria* GA44 has been reported to produce HePS composed of glucose and rhamnose sugar units with a molecular weight of 2.8×10^5 Da (Adesulu-Dahunsi et al., 2018). *Weissella confusa* produced HePS composed of galactose, mannose, glucose, fructose, rhamnose, arabinose, xylose and ribose sugars with galactose being the highest in percentage (Adebayo-Tayo et al., 2018).

Production and characterization of exopolysaccharide from genus *Weissella*

Dextran is the major EPS produced by genus *Weissella* and its yield is in the range, 10 mg/L to 9000 mg/L having molecular mass between 10^4 and 10^8 Da. Generally,

MRS medium (DeMan et al., 1960) containing glucose, modified MRS medium (Goyal, and Katiyar., 1994) containing sucrose or medium given by Tsuchiya et al. (1952) or an optimized medium can be used for the production of dextran in case of *Weissella* (Shukla, and Goyal., 2012). The dextran produced remains dissolved in the fermentation medium, firstly the bacterial cells are separated by centrifugation and subsequent organic solvent precipitation (Chilled ethanol) is used to obtain crude dextran. Dextran produced by *Weissella cibaria* RBA12 was obtained by ethanol precipitation of the cell free supernatant from the fermentation media (Baruah et al., 2017). Another way is to use purified enzyme dextransucrase to catalyse the polymerization reaction using pure sucrose to obtain highly purified dextran. Dextran from *Weissella confusa* Cab3 was obtained by the in vitro enzymatic reaction of dextransucrase with pure sucrose and subsequent ethanol precipitation (Shukla et al., 2014). The crude dextran obtained from the enzyme reaction is free from any other impurities present in the fermentation media and is subsequently precipitated. Dextran or any other EPS can be purified by Size Exclusion Chromatography (SEC) and Ion Exchange Chromatography. DEAE-Sepharose Fast-Flow chromatographic columns have many advantages for the separation of EPS according to their individual molecular masses and help in obtaining higher resolution with sharp elution peak (Wei et al., 2018). EPS from *Lactobacillus plantarum* LP6 was purified using DEAE-Sepharose Fast-Flow chromatographic column to obtain a fraction of two neutral EPSs and one fraction of acidic EPS (Li et al., 2013). The EPS obtained after purification is lyophilized to obtain it in a dry powdered form, which could be used for further characterization or any other purpose.

It is paramount to structurally characterize any newly isolated EPS produced within genus *Weissella* owing to vast structure variations. The first step in the structure characterization is the monosaccharide analysis that can be achieved by acid hydrolysis, periodate oxidation, Smith degradation and methylation (Wei et al., 2018). Some of these techniques are also coupled with High Performance Liquid Chromatography (HPLC) or Anion exchange chromatography (AEC) to obtain information on monosaccharide composition (Baruah et al., 2017). Chemical nature, functional groups and linkage analysis of the EPS are analysed by FT-IR and NMR spectroscopy. FT-IR primarily helps in revealing the important functional groups such as C–H, C=O and –OH or any other impurities that may be present with the purified

EPS. NMR analysis reveals the glycosidic bond configurations and the percentage of the linear or branched linkages. The study of physicochemical properties of EPS is also essential for the proper understanding for its applications. These physicochemical analysis includes light scattering such as multi angle laser light scattering (SEC-MALLS) and dynamic light scattering (DLS) which reveals characteristics of polymorphism and molecular size. Atomic force microscopy (AFM) and Electron microscopy are also used to understand the topology of EPS (Wei et al., 2018). Thermogravimetric analysis (TGA) can be used to calculate the degradation temperature of the EPS, that may lead to its application in the food industry (Tingirikari et al., 2014a). The study of the rheological properties of EPS is an important physicochemical property that determines the effectiveness of the EPS as a food hydrocolloid (Zhang et al., 2018).

Role of exopolysaccharide from *Weissella* in Food systems

Dextrans from *Weissella* have been widely studied in bread making for their capability in improving the softness of fresh bread (Di Cagno et al., 2006). The shelf-life of bread was also improved by the addition of high molecular weight dextran through sourdough in milk bread and in mixed rye breads (Lacaze et al., 2007). Katina et al., 2009 reported technological impacts of dextran-containing *W. confusa* sourdoughs. Their study revealed that a dextran producing *W. confusa* sourdough after baking leads to improved specific volume of wheat bread as compared to bread made using *L. mesenteroides* sourdoughs. Katina et al., 2009 also reported the development of a new assay method for the quantification of *Weissella* dextran, which utilizes a cocktail of dextran hydrolysing enzymes which specifically break down the dextran in sourdough to glucose monomer, which is then quantified and used for the calculation of dextran content produced in sourdough. *Weissella cibaria* MG1 and *Lactobacillus reuteri* VIP were compared for the effect of *in situ* EPS production and organic acids on the rheology of dough and bread quality (Sandra et al., 2012). *Weissella cibaria* MG1 produced far less amount of acetate as compared to *Lactobacillus reuteri* VIP, while giving similar amount of EPS production. Sourdough enriched by *Weissella cibaria* MG1 dextran (10% and 20%) provided mildly acidic conditions with improved bread quality. *Weissella cibaria* MG1 was evaluated for the production of EPS dextran in sourdough prepared from several gluten-free flours such as buckwheat, oat, quinoa and teff (Wolter et al., 2014). *Weissella cibaria* MG1 produced the highest EPS in buckwheat (4.2 g EPS/kg) followed by quinoa sourdoughs (3.2 g EPS/kg) but failed to grow in oats sourdough (Table 1).

Two strains of *Weissella confusa* were compared for *in situ* dextran production in wheat and rye bran systems (Kajala et al., 2016). Both the strains of *Weissella confusa* were reported to produce higher quantities of dextran in rye bran than in wheat bran. This difference was reported to be due to low acidification rate induced by slow fermentation in rye bran and the presence of higher amount of starch in wheat bran. The starch in wheat bran during fermentation gets converted to maltose by the action of endogenous amylases and interferes with the dextran biosynthesis and in turn isomalto-oligosaccharides are formed (Kajala et al., 2016). Similar results were reported for *Weissella cibaria* RBA12, where whole wheat flour, wheat bran and rye bran were used for *in situ* dextran production (Baruah et al., 2017). Rye bran showed the highest dextran of 3.3% d.w. Whole wheat flour yielded the lowest dextran production due the highest amount of starch present out of all the three matrices used. The presence of free maltose also helped in the formation of isomalto-oligosacchrides compromising the synthesis of dextran (Baruah et al., 2017). Dough of faba bean flour was fermented by *Weissella cibaria* Sj 1b and *Leuconostoc. pseudomesenteroides* DSM 20193 with the addition of sucrose to evaluate the *in situ* production of EPS (Xu et al., 2017). The dextrans from both the strains were also used to study the interactions between exopolysaccharide and Faba bean protein isolate (FPI) by Maillard reaction used in food related systems. *Weissella cibaria* Sj dextran displayed higher gel-strengthening ability than the *Leuconostoc. pseudomesenteroides* DSM 20193 dextran (Xu et al., 2018).

Weissella confusa E3403 and *Leuconostoc pseudomesenteroides* DSM 20193 were used for the study of *in situ* production of dextran in faba bean flour sourdough (Wang et al., 2018). Wheat bread was supplemented with 43% fermented faba bean flour sourdough containing EPS and then the rheological and textural properties of the final bread were evaluated. The addition of dextran from *Weissella confusa* E3403 increased the specific volume (by ~21%) and reduced crumb hardness (by ~12%) of the final bread improving the quality of wheat bread enriched with faba bean flour (Wang et al., 2018). *Weissella sp.* strains have also seen been used in Cheddar cheese manufacturing where dextran producing *Weissella cibaria* MG1 was used as adjunct culture (Lynch et al., 2014). *Weissella cibaria* MG1 helped in increasing the moisture retention of cheese without significantly affecting its proteolysis. *Weissella confusa* E-90392 was used for fermenting carrot medium, as a vegetable model to study its rheological, chemical and sensory properties (Juvonen et al., 2015). The carrot medium fermented with *Weissella confusa* E-90392 producing low branched dextran resulting in thick texture along with pleasant odour and flavour. The results showed the potential of dextran from *Weissella confusa* E-90392 as hydrocolloid additive of natural origin in carrot containing foods and ingredients replacing the synthetic additives (Juvonen et al., 2015). A novel beverage fermented with *Weissella cibaria* MG1 was developed using aqueous extracts of wholemeal quinoa flour (Zannini et al., 2018). The fermented wholemeal quinoa milk had significant growth of *Weissella cibaria* MG1 with EPS production of 40 mg/l, which was responsible for increased water holding capacity and viscosity. The interactions between EPS and protein resulted in formation of network like microstructures which in turn helped in improving the texture of the fermented beverage (Zannini et al., 2018). Dextrans from various species of genus *Weissella* in food systems are listed in Table 1.

Table 1. Dextran from Weissella genus in food systems.

Species	Source	Application	Reference
Weissella confusa VTT E-90392	Carrot mash	Improved texture in sourdough bread	Kajala et al., 2016
		Improved texture in fermented carrot medium.	Juvonen et al., 2015
Weissella cibaria MG1	Sourdough	Improved rheology and bread quality in sourdough bread.	Sandra et al., 2012
		Increase moisture retention in cheese.	Lynch et al., 2014
		Improved the texture of the fermented quinoa beverage.	Zannini et al., 2018
Weissella confusa Cab3	Fermented Cabbage	Production of *in situ* dextran in sourdough	Kajala et al., 2016
Weissella cibaria RBA12	Pummelo	Production of *in situ* dextran and isomaltooligosaccharides in sourdough	Baruah et al., 2017
Weissella cibaria Sj	Onion Powder	Production of *in situ* dextran in faba bean flour	Xu et al., 2017
Weissella confusa E3403	Faba Bean Flour	Improved rheology and bread quality in faba bean flour sourdough	Wang et al., 2018

Functional properties of exopolysaccharide from *Weissella*

Prebiotics are the compounds which stimulate the growth of beneficial probiotic microorganisms in the lower GI tract. EPS such as dextran which consists of α-(1, 6) glycosidic backbones are less susceptible to attack by human and animal digestive enzymes. As a result, upon reaching the colon they become fermentable carbohydrate for the resident microbiota. These EPS are fermented into short chain fatty acids (SCFA) which confer several beneficial effects in the colon (Salazar et al., 2008). Dextran produced by dextransucrase from *Weissella cibaria* JAG8 was evaluated for this prebiotic potential (Tingirikari et al., 2014a). Dextran from *Weissella cibaria* JAG8 showed resistance against artificial gastric juice and α-amylase. The dextran also stimulated the growth of probiotic bacteria and showed comparable result with commercial prebiotic inulin (Tingirikari et al., 2014a). Similar results were also seen in the case of *Weissella cibaria* RBA12 where its dextran showed minimal hydrolysis in the presence of artificial gastric juice, α-amylase and intestinal fluids compared to prebiotic inulin. Dextran from *Weissella cibaria* RBA12 also enhanced growth of probiotic *Bifidobacterium* and *Lactobacillus* spp., and controlled growth of non-probiotic enteric bacteria (Baruah et al., 2017) (Table 2).

Table 2. EPS from *Weissella sp.* with functional properties.

Species	Source	Type of EPS	Use	Reference
Weissella cibaria JAG8	Apple peel	Dextran	Prebiotic in nature.	Tingirikari et al., 2014a
			Food hydrocolloid.	Tingirikari et al., 2014b
			Used as coating for magnetic nanoparticles (MNPs)	Tingirikari et al., 2017
Weissella cibaria RBA12	Pummelo	Dextran	Prebiotic in nature	Baruah et al., 2017
Weissella cibaria CMS1	Human saliva	Dextran	Inhibits formation of biofilm and growth of *Streptococcus mutans*	Rosca et al., 2018
Weissella confusa KR780676	Idli Batter	Galactan	Food hydrocolloid.	Devi et al., 2016
Weissella confusa QS813	Sourdough	Dextran	Reduces syneresis rate of wheat starch	Tang et al., 2018

Weissella cibaria	Fermented gooseberries	Dextran	Reduces syneresis rate of wheat starch	Vasanthakumari et al., 2015
Weissella confusa AJ53	Fermented *Uttapam* batter	Dextran	Antioxidative in nature	Dubey and Jeevaratnam, 2015
Weissella cibaria GA44	*Gari* (a Nigerian traditional fermented cassava mash)	HePS (Glucose and rhamnose)	Antioxidative in nature	Adesulu-Dahunsi et al., 2018

Dextran produced by *Weissella cibaria* CMS1 isolated from human saliva was reported to inhibit the formation of biofilm and the growth of *Streptococcus mutans* found in human oral microbiota that causes dental caries (Rosca et al., 2018). A plaque index reduction of approximately 20.7% was reported after the subjects had mouth rinsed with solution containing *Weissella cibaria* CMS1. Similar results were also seen in the case of dextran from *Weissella confusa,* where at a concentration of 3mg/ml, it has been reported to break up 70% of biofilm produced by *Candida albicans* SC5314 strain (Rosca et al., 2018).

Dextran produced by dextransucrase from *Weissella cibaria* JAG8 showed high thermal stability, significant water holding capacity and flocculation activity (Tingirikari et al., 2014b). They also reported that when dextransucrase was incorporated in skimmed milk in the presence of sucrose, the milk started getting viscous and solidifying with increase in sucrose concentration. This showed that the high amount of dextran produced was responsible for this thickening property. These results along with its porous structure as observed in the Scanning electron micrograph makes this dextran an excellent candidate for use as a food hydrocolloid (Tingirikari et al., 2014a). A naturally produced food hydrocolloid should have good thickening, gelling, stabilizing, emulsifying and water-holding ability. Galactan produced from *Weissella confusa* KR780676 isolated from an idli batter showed high oil absorption capacity (5.1 mL/g) with good emulsifying activity (69.4%). The flocculating activity, 65.3–81.7% was observed with 0.2 mg/mL galactan in a wide range of pH. These properties make galactan from *Weissella confusa* KR780676, a potential thickening and gelling agent as well as a good emulsifier (Devi et al., 2016).

Dextran can also be used as a food additive as reported in the case of *Weissella confusa* QS813 isolated from traditional sourdough (Tang et al., 2018). The dextran (1%, w/v) from *Weissella confusa* QS813 when added to wheat starch, a significant reduction in the syneresis rate was observed. The study of the microstructure of the wheat starch gel showed denser and uniform structure in the presence of dextran as compared to that without dextran (Tang et al., 2018). Similar report on the reduction in syneresis rate has also been reported in the case of dextran from *Weissella cibaria* isolated from fermented gooseberries (Vasanthakumari et al., 2015). A reduction in the syneresis rate by 75% was observed in wheat starch preparations in the presence of 1% dextran. EPS from *Weissella* having potent antioxidative properties have also been reported. An unbranched, linear dextran from *Weissella confusa* AJ53 isolated from fermented Uttapam batter supplemented with *Piper betle* L. leaves was studied for its anti-oxidative activity (Dubey and Jeevaratnam, 2015). It showed activities for scavenging of hydroxyl, DPPH, and nitric oxide radicals and metal ion chelating and poor reducing abilities (Dubey and Jeevaratnam., 2015).

A HePS composed of glucose and rhamnose sugar produced by *Weissella cibaria* GA44 showed good scavenging effects on superoxide anion radical and hydroxyl radical (Adesulu-Dahunsi et al., 2018). The native dextran produced by dextransucrase from *Weissella cibaria* JAG8 was used as a coating material for magnetic nanoparticles (MNPs) (Tingirikari et al., 2017). The *in vitro* analysis of the effect of the dextran coated MNPs studied on human colon cancer (HT-29) cell lines showed that the dextran coated: MNPs in the ratio of 2:1 displayed good biocompatibility over the other ratios and uncoated MNPs (Tingirikari et al., 2017). Owing to the biocompatible nature of the dextran, it is used in coating of magnetic nanoparticles (MNPs). Dextran coated MNPs have extensive biomedical applications, such as a contrast agent in magnetic resonance imaging (MRI) and gene therapy (Uthaman et al., 2015), hyperthermia treatment and detoxification of biological fluids (Kim et al., 2008), cell separation and tissue repairing (Ito and Kamihira., 2011: Gamucci et al., 2014), drug delivery (Mody et al., 2014) and biosensors (Ciobanu et al., 2012). Various types of EPS from *Weissella sp.* with functional properties are listed in Table 2.

Conclusion

Even though the EPS produced by the genus *Weissella* possesses numerous functional properties, the use of *Weissella sp.* as a starter culture is still much debated. *Weissella sp.* even after 24 years of being described, do not possess the GRAS (Generally Regarded As Safe) status, which limits its use as a starter culture. Therefore, the use of EPS from *Weissella*, either purified from fermentation or produced *in vitro* by using enzymes can still be employed to produce novel products and reveal more of its beneficial properties. More research inputs are required in order to have cost effective production of EPS from *Weissella sp.* This can be achieved by using low cost substrates, optimising process parameters and medium composition, thereby enhancing the EPS productivity and yields. Discovering news types of EPS could lead to developing new applications. Exploring the molecular aspects of EPS production would throw light on the mechanism involved, yield and diverse structural forms. These advances will lead to the development of novel value added food products made through EPS from genus *Weissella*.

References

Adebayo-Tayo, B., Ishola, R., and Oyewunmi, T. (2018). Characterization, antioxidant and immunomodulatory potential on exopolysaccharide produced by wild type and mutant *Weissella confusa* strains. Biotechnol. Rep. 19. https://doi.org/10.1016/j.btre.2018.e00271

Adesulu-Dahunsi, A.T., Sanni, A.I., and Jeyaram, K. (2018). Production, characterization and *In vitro* antioxidant activities of exopolysaccharide from *Weissella cibaria* GA44. *LWT*- Food Sci. Technol. 87, 432-442. https://doi.org/10.1016/j.lwt.2017.09.013

Ahmed, R.Z., Siddiqui, K., Arman, M., and Ahmed, N. (2012). Characterization of high molecular weight dextran produced by *Weissella cibaria* CMGDEX3. Carbohydr. Polym. 90(1), 441-446. https://doi.org/10.1016/j.carbpol.2012.05.063

Baruah, R., Maina, N.H., Katina, K., Juvonen, R., and Goyal, A. (2017). Functional food applications of dextran from *Weissella cibaria* RBA12 from pummelo (*Citrus maxima*). Int. J. Food Microbiol. 242, 124-131. https://doi.org/10.1016/j.ijfoodmicro.2016.11.012

Björkroth, K.J., Schillinger, U., Geisen, R., Weiss, N., Hoste, B., Holzapfel, W.H., Korkeala, H.J. and Vandamme, P. (2002). Taxonomic study of *Weissella confusa* and description of *Weissella cibaria* sp. nov., detected in food and clinical samples. Int. J. Syst. Evol. Microbiol. 52(1), 141-148. doi: 10.1099/00207713-52-1-141

Björkroth, J., Dicks, L.M., and Endo, A. (2014). The genus *Weissella. Lactic Acid Bacteria: Biodiversity and Taxonomy. John Wiley and Sons.* pp 417-428. https://doi.org/10.1002/9781118655252.ch25

Bounaix, M.S., Gabriel, V., Morel, S., Robert, H., Rabier, P., Remaud-Siméon, M., Gabriel, B. and Fontagné-Faucher, C. (2009). Biodiversity of exopolysaccharides produced from sucrose by sourdough lactic acid bacteria. J. Agric. Food. Chem. 57(22), 10889-10897. DOI: 10.1021/jf902068t

Capek, P., Hlavoňová, E., Matulová, M., Mislovicova, D., Růžička, J., Koutný, M., and Keprdova, L. (2011). Isolation and characterization of an extracellular glucan produced by *Leuconostoc garlicum* PR. Carbohydr. Polym. 83(1), 88-93. https://doi.org/10.1016/j.carbpol.2010.07.024

Ciobanu, C.S., Iconaru, S.L., Gyorgy, E., Radu, M., Costache, M., Dinischiotu, A., Coustumer, P.L., Lafdi, K., and Predoi, D. (2012) Chem Cent J. 6: 17. doi: 10.1186/1752-153X-6-17

Collins, M.D., Samelis, J., Metaxopoulos, J. and Wallbanks, S. (1993). Taxonomic studies on some *Leuconostoc*-like organisms from fermented sausages: description of a new genus *Weissella* for the *Leuconostoc paramesenteroides* group of species. J. Appl. Bacteriol. 75(6), 595-603. https://doi.org/10.1111/j.1365-2672.1993.tb01600.x

De Man, J.C., Rogosa, D., and Sharpe, M.E. (1960). A medium for the cultivation of lactobacilli. J. Appl. Bacteriol. 23(1), 130-135. https://doi.org/10.1111/j.1365-2672.1960.tb00188.x

Devi, P.B., Kavitake, D., and Shetty, P.H. (2016). Physico-chemical characterization of galactan exopolysaccharide produced by *Weissella confusa* KR780676. Int. J. Biol. Macromol. 93, 822-828. https://doi.org/10.1016/j.ijbiomac.2016.09.054

Di Cagno, R., De Angelis, M., Limitone, A., Minervini, F., Carnevali, P., Corsetti, A., Gaenzle, M., Ciati, R. and Gobbetti, M. (2006). Glucan and fructan production by sourdough *Weissella cibaria* and *Lactobacillus plantarum*. J. Agric. Food. Chem. 54(26), 9873-9881. **DOI:** 10.1021/jf061393+

Dubey, A.K. and Jeevaratnam, K. (2015). Structural characterization and functional evaluation of an exopolysaccharide produced by *Weissella confusa* AJ53, an isolate from fermented Uttapam batter supplemented with Piper betle L. leaves. Food Sci. Biotechnol. 24(6), 2117-2124. https://doi.org/10.1007/s10068-015-0281-y

Fessard, A., and Remize, F. (2017). Why Are Weissella spp. Not Used as Commercial Starter Cultures for Food Fermentation? Ferment. *3*(3), 38. doi: 10.3390/fermentation3030038

Fusco, V., Quero, G.M., Cho, G.S., Kabisch, J., Meske, D., Neve, H., Bockelmann, W. and Franz, C.M. (2015). The genus Weissella: taxonomy, ecology and biotechnological potential. Front. Microbiol. 6, 155. https://doi.org/10.3389/fmicb.2015.00155

Gamucci O, Bertero A, Gagliardi M. and Bardi G (2014) Biomedical nanoparticles: overview of their surface immune-compatibility. Coat. 4, 139-159. https://doi.org/10.3390/coatings4010139

Goyal, A., and Katiyar, S.S. (1994). Fractionation of *Leuconostoc mesenteroides* NRRL B-512F dextran sucrase by polyethylene glycol: A simple and effective method purification. J. Microbiol. Methods. 20(3), 225-231. https://doi.org/10.1016/0167-7012(94)90007-8

Ito A. and Kamihira M (2011) Tissue engineering using magnetite nanoparticles. Prog. Mol. Biol. Transl. Sci. 104, 355-395. https://doi.org/10.1016/B978-0-12-416020-0.00009-7

Juvonen, R., Honkapää, K., Maina, N.H., Shi, Q., Viljanen, K., Maaheimo, H., Virkki, L., Tenkanen, M. and Lantto, R. (2015). The impact of fermentation with exopolysaccharide producing lactic acid bacteria on rheological, chemical and sensory properties of pureed carrots (*Daucus carota L.*). Int. J. Food Microbiol. 207, 109-118. https://doi.org/10.1016/j.ijfoodmicro.2015.04.031

Kajala, I., Mäkelä, J., Coda, R., Shukla, S., Shi, Q., Maina, N.H., Juvonen, R., Ekholm, P., Goyal, A., Tenkanen, M. and Katina, K. (2016). Rye bran as fermentation matrix boosts in situ dextran production by *Weissella confusa* compared to wheat bran. Appl. Microbiol. Biotechnol. 100(8), 3499-3510. https://doi.org/10.1007/s00253-015-7189-6

Kang, M.S., Chung, J., Kim, S.M., Yang, K.H., and Oh, J.S. (2006). Effect of *Weissella cibaria* isolates on the formation of *Streptococcus mutans* biofilm. Caries Res. 40(5), 418-425. https://doi.org/10.1159/000094288

Katina, K., Maina, N.H., Juvonen, R., Flander, L., Johansson, L., Virkki, L., Tenkanen, M. and Laitila, A. (2009). In situ production and analysis of *Weissella confusa* dextran in wheat sourdough. Food Microbiol. 26(7), 734-743. https://doi.org/10.1016/j.fm.2009.07.008

Kim, D.H., Nikles, D.E., Johnson, D.T. and Brazel, C.S. (2008) Heat generation of aqueously dispersed CoFe2O4 nanoparticles as heating agents for magnetically activated drug delivery and hyperthermia. J. Magn Magn Mater. 320, 2390-2396. https://doi.org/10.1016/j.jmmm.2008.05.023

Lacaze, G., Wick, M., and Cappelle, S. (2007). Emerging fermentation technologies: development of novel sourdoughs. Food Microbiol. 24(2), 155-160. https://doi.org/10.1016/j.fm.2006.07.015

Li, J.Y., Jin, M.M., Meng, J., Gao, S.M., and Lu, R.R. (2013). Exopolysaccharide from *Lactobacillus planterum* LP6: Antioxidation and the effect on oxidative stress. Carbohydr. Polym. 98(1), 1147-1152. https://doi.org/10.1016/j.carbpol.2013.07.027

Lynch, K.M., McSweeney, P.L., Arendt, E.K., Uniacke-Lowe, T., Galle, S., and Coffey, A. (2014). Isolation and characterisation of exopolysaccharide-producing *Weissella* and *Lactobacillus* and their application as adjunct cultures in Cheddar cheese. Int. Dairy J. 34(1), 125-134. https://doi.org/10.1016/j.idairyj.2013.07.013

Maina, N.H., Tenkanen, M., Maaheimo, H., Juvonen, R., and Virkki, L. (2008). NMR spectroscopic analysis of exopolysaccharides produced by *Leuconostoc citreum* and *Weissella confusa*. Carbohydr. Res. 343(9), 1446-1455. https://doi.org/10.1016/j.carres.2008.04.012

Malang, S.K., Maina, N.H., Schwab, C., Tenkanen, M., and Lacroix, C. (2015). Characterization of exopolysaccharide and ropy capsular polysaccharide formation by *Weissella*. Food Microbiol. 46, 418-427. https://doi.org/10.1016/j.fm.2014.08.022

Mody, V.V., Cox, A., Shah, S., Singh, A., Bevins, W. and Parihar, H. (2014). Magnetic nanoparticle drug delivery systems for targeting tumor. Appl. Nanosci. 4, 385-392. https://doi.org/10.1007/s13204-013-0216-y

Papagianni, M., and Papamichael, E.M. (2012). Production of the antimicrobial protein weisselin A by *Weissella paramesenteroides* DX in batch fermentations: the type of carbohydrate used as the C-source in the substrate affects the association of production with growth. Appl. Biochem. Biotechnol. 168(5), 1212-1222. https://doi.org/10.1007/s12010-012-9851-0

Rosca, I., Petrovici, A.R., Peptanariu, D., Nicolescu, A., Dodi, G., Avadanei, M., Ivanov, I.C., Bostanaru, A.C., Mares, M. and Ciolacu, D. (2018). Biosynthesis of dextran by *Weissella confusa* and its In vitro functional characteristics. Int. J. Biol. Macromol. 107, 1765-1772. https://doi.org/10.1016/j.ijbiomac.2017.10.048

Salazar, N., Gueimonde, M., Hernández-Barranco, A.M., Ruas-Madiedo, P., and Clara, G. (2008). Exopolysaccharides produced by intestinal *Bifidobacterium* strains act as fermentable substrates for human intestinal bacteria. Appl. Environ. Microbiol. 74(15), 4737-4745. **DOI:** 10.1128/AEM.00325-08

Sandra, G., Schwab, C., Dal Bello, F., Coffey, A., Gänzle, M., and Arendt, E. (2012). Comparison of the impact of dextran and reuteran on the quality of wheat sourdough bread. J. Cereal Sci. 56(3), 531-537. https://doi.org/10.1016/j.jcs.2012.07.001

Shukla, S., and Goyal, A. (2011). 16S rRNA-based identification of a glucan-hyperproducing *Weissella confusa*. Enzyme Res. 2011. http://dx.doi.org/10.4061/2011/250842

Shukla, S., Shi, Q., Maina, N.H., Juvonen, M., Tenkanen, M., and Goyal, A. (2014). *Weissella confusa* Cab3 dextransucrase: Properties and in vitro synthesis of dextran and glucooligosaccharides. Carbohydr. Polym. *101*, 554-564. https://doi.org/10.1016/j.carbpol.2013.09.087

Tang, X., Liu, N., Huang, W., Cheng, X., Wang, F., Zhang, B., Chen, J., Jiang, H., Omedi, J.O. and Li, Z. (2018). Syneresis rate, water distribution, and microstructure of wheat starch gel during freeze–thaw process: Role of a high molecular weight dextran produced by *Weissella confusa* QS 813 from traditional sourdough. Cereal Chem. 95(1), 117-129. https://doi.org/10.1094/CCHEM-08-17-0174-R

Tingirikari, J.M.R., Kothari, D., and Goyal, A. (2014a). Superior prebiotic and physicochemical properties of novel dextran from *Weissella cibaria* JAG8 for potential food applications. Food Funct. 5(9), 2324-2330. https://doi.org/10.1039/C4FO00319E

Tingirikari, J.M.R., Kothari, D., Shukla, R., and Goyal, A. (2014b). Structural and biocompatibility properties of dextran from *Weissella cibaria* JAG8 as food additive. Int. J. Food Sci. Nutr. 65(6), 686-691. https://doi.org/10.3109/09637486.2014.917147

Tingirikari, J.M.R., Rani, A., and Goyal, A. (2017). Characterization of super paramagnetic nanoparticles coated with a biocompatible polymer produced by dextransucrase from *Weissella cibaria* JAG8. J. Polym. Environ. 25(3), 569-577. https://doi.org/10.1007/s10924-016-0836-x

Tsuchiya, H.M., Koepsell, H.J., Corman, J., Bryant, G., Bogard, M.O., Feger, V.H., and Jackson, R.W. (1952). The effect of certain cultural factors on production of dextransucrase by *Leuconostoc mesenteroides*. J. Bacteriol. 64(4), 521.

Uthaman, S., Lee, S.J., Cherukula, K., Cho, C.S. and Park, I.K. (2015) Polysaccharide-coated magnetic nanoparticles for imaging and gene therapy. Biomed Res Int. 2015: 959175. http://dx.doi.org/10.1155/2015/959175

Vasanthakumari, D.S., Harikumar, S., Beena, D.J., Pandey, A., and Nampoothiri, K.M. (2015). Physicochemical characterization of an exopolysaccharide produced by a newly isolated *Weissella cibaria*. Appl. Biochem. Biotechnol. 176(2), 440-453. https://doi.org/10.1007/s12010-015-1586-2

Wang, Y., Sorvali, P., Laitila, A., Maina, N.H., Coda, R., and Katina, K. (2018). Dextran produced in situ as a tool to improve the quality of wheat-faba bean composite bread. Food Hydrocolloids. 84, 396-405. https://doi.org/10.1016/j.foodhyd.2018.05.042

Wei, D.I., ZHANG, Y.C., Hua-Xi, Y.I., Xue, H.A.N., Shu-Mei, W.A.N.G. and ZHANG, L.W. (2018). Research Methods for Structural Analysis of Lactic Acid Bacteria Induced Exopolysaccharides. Chin. J. Anal. Chem. 46(6), 875-882. https://doi.org/10.1016/S1872-2040(18)61091-6

Wolter, A., Hager, A.S., Zannini, E., Galle, S., Gänzle, M.G., Waters, D.M., and Arendt, E K. (2014). Evaluation of exopolysaccharide producing *Weissella cibaria* MG1 strain for the production of sourdough from various flours. Food Microbiol. 37, 44-5 https://doi.org/10.1016/j.fm.2013.06.009

Xu, Y., Pitkänen, L., Maina, N.H., Coda, R., Katina, K., and Tenkanen, M. (2018). Interactions between fava bean protein and dextrans produced by *Leuconostoc pseudomesenteroides* DSM 20193 and *Weissella cibaria* Sj 1b. Carbohydr. Polym. 190, 315-323. https://doi.org/10.1016/j.carbpol.2018.02.082

Xu, Y., Wang, Y., Coda, R., Säde, E., Tuomainen, P., Tenkanen, M., and Katina, K. (2017). In situ synthesis of exopolysaccharides by *Leuconostoc* spp. and *Weissella* spp. and their rheological impacts in fava bean flour. Int. J. Food Microbiol. 248, 63-71. https://doi.org/10.1016/j.ijfoodmicro.2017.02.012

Zannini, E., Jeske, S., Lynch, K.M., and Arendt, E.K. (2018). Development of novel quinoa-based yoghurt fermented with dextran producer *Weissella cibaria* MG1. Int. J. Food Microbiol. 268, 19-26. https://doi.org/10.1016/j.ijfoodmicro.2018.01.001

Zhang, Y., Guo, L., Xu, D., Li, D., Yang, N., Chen, F., Jin, Z. and Xu, X. (2018). Effects of dextran with different molecular weights on the quality of wheat sourdough breads. Food Chem. 256, 373-379. https://doi.org/10.1016/j.foodchem. 2018.02.146

Fructans as Natural, Bioactive Cosmeceutical Ingredients

8

Merve Erginer Hasköylü[1], Margarita Kambourova[2], Ebru Toksoy Öner[1*]

[1]IBSB Research Group Marmara University, Department of Bioengineering, İstanbul, Turkey
[2]The Stephan Angeloff Institute of Microbiology, Bulgarian Academy of Science, Sofia, Bulgaria
*Corresponding author: ebru.toksoy@marmara.edu.tr

DOI: https://doi.org/10.21775/9781912530267.08

Abstract
Fructans have novel characteristics and have many applications in cosmetic market as bioactive ingredient or rheology and formulation improver. Cosmeceuticals are not only used for aesthetic reasons but also for the treatment of ailments. Here in this chapter we focus on cosmeceuticals, applications, natural polysaccharides and their needs, fructans and their usage and activity in cosmetic formulations. Main attention is paid on fructan based skin care cosmeceuticals (moisturizing, antiaging, and whitening) based on new formulations with natural ingredients. The annual exponential growth of consumers that are fully aware of natural, and bioactive ingredients determines the special impact of this review for evaluation the future of the pharmaceutical industry.

Cosmeceuticals

Conserving and improving youth and beauty is important since ancient times and using cosmeceuticals to improve look became a routine in modern life and is increasing rapidly with the effect of social media. Consumers become self-aware about skin wellness, sun protection and aging day by day. Various products are produced due to high social demand (Feetham et al., 2018; Priyan Shanura

Fernando et al., 2018; Razak et al., 2018). Cosmeceuticals is a rapidly growing industry in worldwide and 25% of women are known to use 15 or more products on their skin daily. In 2016 in the United States alone, $607 million and $1.2 billion are spent for acne and sunscreen products correspondingly. In 2018, up to now, $428 billion worth of sale is recorded for cosmeceuticals worldwide. This amount is expected to reach $429.8 billion by 2022 (Feetham et al., 2018; Juhász et al., 2018).

Cosmetic refers to any substance that come in contact with external parts of human body and used for cleaning, beautifying, promoting attractiveness and improving appearance (Aranaz et al., 2018; Priyan Shanura Fernando et al., 2018). The term "Cosmetics" is first defined by Reed as a product that is designed scientifically for external application to human body to achieve desired aesthetic results with rigid chemical, physical and medical standards (Reed, 1962). Further, cosmeceuticals have determined as medicinal or drug like features and used for protection, whitening, anti-aging, tanning, deodorizing, hair and nail caring with their biologically active ingredients (Kligman and Kligman, 1993). The term "cosmeceutical" is the combination of "cosmetic" and "pharmaceutical" and differ from cosmetics with physiologically or biologically active pharmaceutical ingredients in it. They not only improve appearance in aesthetic ways but also help to treat skin ailments with medical like features (Agrawal et al., 2018; Aranaz et al., 2018). Cosmeceuticals are complex and functional systems, which are applied to body topically (epidermis, nails, lips, hair and external genital organs) for cleaning, appearance enhancing and protection of those body parts. They renovate skin, reduce scar formation, normalize cellular turnover, support skin's barrier function and can be used against aging and skin blemishes. This group not only contains high-quality cosmetics but also prescription drugs that are applied topically such as retinoid, minoxidil, sun creams and antiperspirants (Aranaz et al., 2018; Feetham et al., 2018) (Figure 1).

Skin Morphology and Aging Changes

As mentioned above many of external body parts are in focus for cosmeceutical product development but skin care products such as cleansers, anti-aging creams or moisturizers hold popularity. Researchers develop formulations depending on chemistry, biology, pharmacy and so on. Understanding anatomy and physiology of skin, mechanism of aging and interaction of skin with cosmetic ingredients give sight to improve formulations with desired effects.

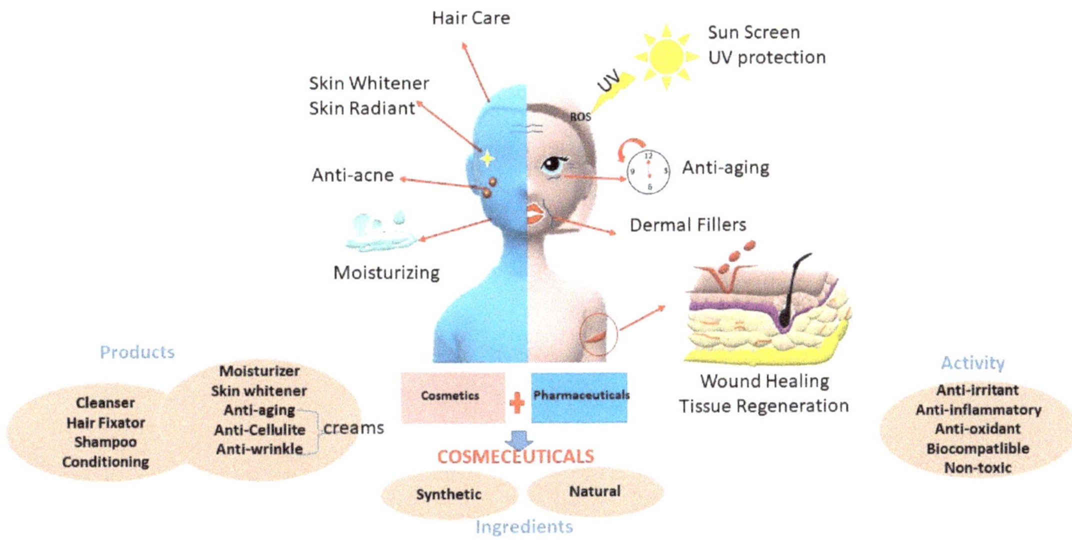

Figure 1. Cosmeceuticals and their usage.

Skin Anatomy

Skin protects inner part of body from external damage; maintain body hydration and electrolyte balance. Skin not only serves as a protective barrier but also has perceptual, neural, cosmetic and regulatory features and have self- renewal ability and form new skin layer in every 2-3 weeks. Skin is composed of three different layers called epidermis, dermis and hypodermis from outside to inside and has the largest surface area (approx. 18 m^2 for man and 16 m^2 for female). Epidermis has five different sublayers, namely, *stratum basale, stratum spinosum, stratum granulosum* and *stratum corneum*. Epidermis consist of squamous epithelium, pigment containing melanocyte cells, Langerhans cells, and pressure sensing Merkel cells and keratinocyte cells. *Stratum basale* (basement membrane) is the deepest layer and contains basal keratinocytes. Cells in this layer proliferate and differentiate to form keratinocytes in the upper layer (*stratum spinosum*) and secrete lipid and protein containing lamellar bodies, extracellularly, between *s. granulosum* and *s. corneum*. When cells reach outer layers, *corneum,* they become flat, tightly packed, water resistant and keratinized. *Stratum corneum* is the outermost layer and protects skin from infection, dehydration or any chemical and mechanical damages. Dermis is separated from epidermis with basement membrane and it is known to be thicker than epidermis and responsible for the mechanical properties of skin, sense of heat, touch and elasticity. It is composed of fibrous tissue and dermal matrix, and has three sublayers called; papillary layer, sub papillary layer and reticular layer. Fibroblasts, extra cellular matrix (ECM), fibronectin, proteoglycans, elastin, collagen, vessels and nerve endings are main components of dermis. Elastin and collagen (the 70% of dry weight of dermis) are structural proteins that are responsible for the shape, function and elasticity of animal body. They bind cells together (Bailey, 2001;Wong and Chang, 2009; Yildirimer et al., 2012; Bhushan, 2017; Vig et al., 2017; Aranaz et al., 2018). Fibroblasts and keratinocytes synthesize high molecular weight linear glycosaminoglycan called Hyaluronic acid (HA) which plays vital roles in skin hydration and barrier functions (Adrien et al., 2017; Patel, 2018). Hypodermis is a fat rich tissue which is the mediate layer between dermis and muscular fascia and responsible for heat generation, protect body from external physical pressure and preserve natural fat (Aranaz et al., 2018).

Many of the cosmetics target skin care including improved cellular proliferation, barrier function, induced collagen and HA production, reduced degradative enzyme activity of those components and delayed aging. In order to develop anti-aging formulations mechanism of aging should be understood clearly.

Mechanism of Aging

Aging is the result of both genetic and environmental damages that occur in all living organisms and skin is the first organ where signs of aging become visible. Since it is, a cumulative process that is triggered both inside and outside, once it has started it cannot be reversed because of the ticking biological clock inside us and environmental factors expedite this process. Euro monitor International reported that 22% of the global skin care market is covered by anti-aging products and 66 billion $ value recoded in 2007 which is expected to reach 216 billion $ by 2021. Aged population in United States is predicted to reach 20% of the population by 2030 (Agrawal et al., 2018). Desire of longevity, youthful and healthier body and skin are the demands of our era and age-related skin changes are major concerns in worldwide for both cosmetic and pharmaceutical industries. Thus, researchers seek alternative novel products to delay aging and its observable effects on skin. Process of aging is divided into two as intrinsic aging (physiological aging) and extrinsic aging (photo aging) depending on its origin (Wiegand et al., 2017; Otlatıcı et al., 2018). Intrinsic aging depend on genetic regulation, cellular metabolism, toxicity of products on repair and metabolism and hormonal changes (Mahawar et al., 2018; Otlatıcı et al., 2018). This type of aging starts by epidermal thinning, occurrence of fine wrinkles due to cell loss, flattening of dermal- epidermal junction, shortening of telomeres, increased fragility and elasticity loss (Baumann, 2007; Berardesca et al., 2018; Campa and Baron, 2018). Extrinsic aging recognized by environmental factors such as sun radiations: ultraviolet, visible light and infrared radiation, nicotine, smoking, air pollution, nutrition, life style (sedentary life, sleep deprivation), stress, chemicals and unconscious use of products that meet skin. Dryness of skin, formation of fine lines and deep wrinkles, dyspigmentation, dark spots, loss of elasticity, skin laxity, decrease in collagen, elastin and HA levels and reactive oxygen species (ROS) generation are results of extrinsic aging (Berardesca et al., 2018; Campa and Baron, 2018; Mahawar et al., 2018; McDaniel et al., 2018; Otlatıcı et al., 2018).

Aging effects both epidermis and dermis layers of skin (Table 1). Decrease in water content, total lipid content, amino acid composition, corneocyte desquamation on stratum corneum reduce natural moisturizing factor (NMF) and result in dry, rough

Table 1: Changes in Epidermis and Dermis with Aging

EPIDERMIS	DERMIS
Dermal-Epidermal Junction Thinning	Decreased thickness
Variation in size, shape and thickness	Decreased vascularity
Decreased number of melanocytes	Decrease in fibroblast and mast cell population
Inhibition of keratinocytes	Decreased collagen, HA and elastin levels
Decrease in epidermal cell growth	Decrease in ECM level, content and production
Decreased water binding capacity	

Derived from (Campa and Baron, 2018; Cole et al., 2018; Mahawar et al., 2018)

and crusted look. Cellular senescence (end of replicative lifespan, growth arrest at G1phase) in melanocytes is observed in epidermal layer. Since dermal layer is thicker and more complex than epidermis, several changes can be seen at the same time. Fibroblasts become senescent and elastin gene expression is known to decrease by aging. Collagen production decreases; present fibers become thicker, disorganized, production of collagen degrading enzymes increases and Type I – Type III collagen (produced by fibroblasts and support skin, nail and hair growth) ratio changes when skin ages. Oxidative stress related aging is another process and causes generation of ROS and free radicals and cellular senescence, lipid protein and DNA damage. Oxygen metabolism of mitochondria generates ROS and high levels are detrimental in body. Antioxidant system of our body tries to alter oxidative stress with enzymes such as superoxide dismutase (SOD), catalase, ceruloplasmin, and transferrin but cannot succeed 100%. UV radiation, especially UVA and UVB, initiates fibroblast and keratinocyte cells to synthesize matrix metalloproteinase (MMP) through ROS generation and initiate membrane growth factor activation of those cells. Receptor activation through protein kinase cascade signal transduction and activator protein 1 (AP-1) in nucleus than follow. Activation of AP-1 induces MMPs in epidermis and dermis and results in collagen and ECM degradation. UVA and UVB (help to synthesize vitamin D) initiate changes in DNA mutagenesis, P53 related DNA repair mechanism, pigmentation, production of inflammatory cytokines, and hormones. Elastin degrades and calcification starts which result in elasticity loss in dermis. HAs, in the ECM fill spaces, play shock-absorbing roles, and decrease by age. Chemical pollutants like ozone (O_3), carbon dioxide (CO_2), and nitrogen dioxide (NO_2) are higher in skin than in other organs. Although epidermis has higher antioxidant activity compared to other tissues, exposure to such pollutants affect antioxidant levels and oxidation markers of the outermost layer, induce cellular stress response cascades of the skin and damage skin defense system (Jenkins, 2002; Valacchi et al., 2015; Naidoo and Birch-Machin, 2017; Wiegand et al., 2017; Berardesca et al., 2018; Campa and Baron, 2018; Cole et al., 2018; Markiewicz and Idowu, 2018; Priyan Shanura Fernando et al., 2018).

Exopolysaccharides as Ingredients of Skin Care Products

Wound healing, pigmentation, aging, microbial infections, wrinkles and cancer are main skin concerns worldwide and cosmeceuticals are developed by researches working in these areas (Agrawal et al., 2018) Anti-oxidant, anti-wrinkle, anti-inflammatory, anti-acne, anti-tyrosine activity, biocompatibility and non-toxicity are desired key functionalities of natural polysaccharides that are used in skin care cosmetics' development (Priyan Shanura Fernando et al., 2018).

Since time immemorial, natural products are used for development of cosmetic formulations. Although various synthetic and chemical formulations are developed in the light of pharmacy, medicine and chemistry, natural ingredients take back their popularity with the demand of natural cosmeceuticals by conscious consumers. Cosmetic industry started to look for alternative natural ingredients for use in cosmetic products that mimic naturally occurring growth factors and cytokines (Adrien et al., 2017; Gazitaeva et al., 2017). safe and effective natural ingredients of New generation cosmeceuticals are expected to be developed that can solve various skin problems in alternative ways and protect skin against oxidative damage by their antioxidant activities (Agrawal et al., 2018; Razak et al., 2018). Functional polysaccharides are highly preferred in cosmetic formulation technologies because of their film forming, thickening, emulsifying, conditioning, gelling and suspending properties. These polysaccharides are used in skin moisturizers as hydrogel or hydrocolloid that mobilize water to the skin (Kanlayavattanakul and Lourith, 2015). Plant, marine organisms, animal, microbial derived polysaccharides are some of them and they are highly preferred with their bioactivity, biocompatibility and biodegradability features (Freitas et al., 2015; Adrien et al., 2017). Despite the fact that plant derived polysaccharides are one of the oldest sources for cosmetic applications, their chemical composition and stability vary between seasons and large terrestrial fields are needed for production of these slowly growing plants, which in turn restricts their usage. Marine organisms are also known to produce diverse biopolymers. Large scale or non-sustainable production of those ingredients in oceans is discussed to cause damage of the marine ecosystems. Luckily marine polysaccharides can be produced rapidly in large scale under controlled environments like fermenters but extraction, downstream processes and deep water sampling still count in limitation list (Agrawal et al., 2018). Bacterial polysaccharides are commonly used in cosmeceutical industry with their biocompatible, biodegradable, bioactive features and production easiness. Bacteria can synthesize polymers intracellularly or extracellularly but intracellular production requires detailed purification methods. Exopolysaccharides (EPS) are superior with their production and purification easiness under controlled conditions in large scale. Levan, bacterial cellulose, xanthan gum, fucapol, HA are some of the widely used bacterial EPS for cosmetic applications (Balkrishna et al., 2018). Even they are preferable with their production advantages mentioned above, risk of contamination is the main drawback of those polymers' usage (Kambourova et al., 2015) and stability of those microbes polymers under extreme conditions is another important problem for industrial application. Extremophilic microorganisms, which live in extreme conditions like high salinity, acidity etc. are in focus among microbes for biopolymer production with their adaptations to survive in extreme conditions. Those microorganisms produce novel polysaccharides that are stable under extreme conditions and have high value applications in various industries (Nicolaus et al., 2010; Öner et al., 2016). A variety of extremophilic microorganism are studied and used in applications but halophilic microorganisms are in focus because of their several advantages over others. Halophilic microorganisms belong to extremophiles and they balance external stress like high salinity and hyperosmotic conditions to maintain physiological activity and survive in those niches. Their biopolymers with novel characteristics are preferable in industries that require extreme temperature, salinity and pH processes. Non-sterile continuous production in high saline water make them good candidates for biopolymer production (Öner et al., 2016; Kindzierski et al., 2017; Chen and Jiang, 2018; Erkorkmaz et al., 2018; Radchenkova et al., 2018).

Inulin, levan, curdlan, xanthan, gellan and pullulan are some of the widely produced bacterial exopolysaccharides but fructans (inulin, levan) are the most common polysaccharides among others and known as extreme condition facilitators. Thus their usage in cosmetics is important not only to improve the stability and quality of products but also their bioactivity is an outstanding facility.

Fructans

Fructan Producers

Fructans are non-structural fructose based oligo-and polysaccharides synthesized by fructosyltrasnferases using sucrose and occur in bacteria, algae, flowering plants and mosses (Haskoylu et al. 2017; Kırtel et al., 2018; Van den Ende, 2018; Versluys et al., 2018). Fructans can be classified in two main structural groups, namely inulin with β-(1-2) linked fructose residues and β-(2-6) linked levan. There are more complex forms of fructans in plants like graminans, neo-fructans and agavins, which are branched fructans harboring both β-(1-2) and β-(2-6) linkages in their structure

(Di Bartolomeo et al., 2013; Van den Ende and El-Esawe, 2014). Flowering plants, liliales, poals, astrales, campanules, palemoniaceae, ericales, dipsacales, barley, Crops like chicory and artichoke (*Chicorium ıntybus, Cynara cardunculus*), rye grass (*Lolium perenne*), oat (*Avena sativa*) onion (*Allium cepa*), wheat (*Triticum aestivum*) are known as fructan containing plants. Many microorganisms including *Pseudomonas sp., Xanthamonas sp., Azotobacter chroococum, Streptococcus salivarius, Bascillus subtilis, Rothis dentocarciosa, Arthrobacter ureafascilis, Zymomonas mobilis, Lactobacillus sp., Leuconostoc sp., Streptococus sp., Weissella sp., Erwinia sp., Gluconacetobacter sp., Halomonas sp., Pseudomonas sp.* are also reported as fructan producers (Kim et al., 2005; Versylus et al., 2017). Some plants and microorganisms are known to produce fructan selectively when they are under stress. Aloe vera (*Aloe barbadensis miller*) is one of them and this succulent is reported to contain fructan under water stress conditions (Salinas et al., 2016). Plant and microbial fructans differ in their place of accumulation and degree of polymerizations (DP), plants produce shorter DP fructans accumulated intracellularly (vacuoles) while microbes produce much higher DP fructans extracellularly (Kırtel et al., 2018). These polysaccharides are not only storage compounds for energy but also protect plants and microbes from extreme climate conditions like drought and freezing (Pommering et al., 2018; Versylus et. al., 2018). Intestinal and extraintestinal pathogen resistance, symbiotic association with probiotic bacteria, blood parameter improver, allergy reducer and immune system modulator properties make fructans good candidates for prebiotic and cosmetic applications (Peshev and Van den Ende, 2014).

Inulin

Inulin is a β-(1-2) linked fructose polymer with terminal glucose and extracted from plants such as chicory (*Cichorium intybus L*) , atichoke (*Helianthus tuberosus*), oats agave cactus and microorganisms *Lactobacillus, Leuconostoc, Streptococcus, Weissella* (Haskoylu et al., 2017; Versylus et al., 2018). Inulin is commonly used in anti-oxidant, anti-allergic, skin prebiotic and other skin care products. Many patent applications and products are common in market. Inulin Lauryl carbamate for example is reported as surfactant in fast absorbing cosmetic products (Paik et al., 2018)

Fructooligosaccharides

Sucrose originated enzymatically synthesized or hydrolyzed high DP fructose oligomers are named as fructooligosaccharides (FOSs) with low DP and can be grouped as inulin type or levan type. FOSs are consumed as prebiotics for probiotic bacterial growth with intestinal immune response improver, mineral absorption enhancer, and food consumption and obesity reducer activities. Microorganisms, plants or enzymes of those organisms can synthesize FOSs. They are also obtained by hydrolysis of inulin by endo-inulinase (EC 3.2.1.7), synthesized from sucrose by *Aspergillus aculeatus* fructosyltransferase (EC 2.4.1.9), *Aspergillus niger* β-fructosyltansferase (EC 3.2.1.26), and can be found in chicory, Jerusalem Artichoke, asparagus and onion family members (Thamer and Penna, 2005; Wack and Blaschek, 2006; Ghazi et al., 2007; van den Heuvel et al., 2009; Singh and Singh, 2010; Venema, 2015; Haskoylu et al., 2017).

Levan

Levan is the most common β-(2-6) linked fructose homopolymer found in nature and produced by *Bacillus, Erwinia, Gluconacetobacter, Halomonas, Pseudomonas, Streptococcus salivarius, Zymomonas* bacteria commonly (Oner et al., 2016). This nontoxic, biocompatible, water soluble, film former and strongly adhesive polymer has diverse applications in fields such as pharmacy, chemistry, food, medical and cosmeceutical industries (Erginer et al., 2016; Öner et al., 2016). Bacterial levan production is more advantageous because extraction of levan from plants in high yield is very hard (Belghith et al., 2012). Extracellular production of levan by various microorganisms and easiness of downstream processes without cellular disruption of microorganism make it a strong candidate for high yield production on an industrial scale. Levan is known to be more soluble in water than inulin because of β (2-6) linkage however weak stability in aqueous formulations or hydrolysis to fructose under acidic conditions and risk of contamination during production limits it's usage in cosmetics (Balkrishna et al., 2018). Since halophilic microorganisms are advantageous over other microorganisms in low cost and high yield production as mentioned before with novel biopolymer characteristics, halophilic levan producers gained popularity. *Halomonas smyrnensis* AAD6T is known to be the first levan producing halophilic bacterium (Öner et al., 2016). Numerous studies with levan as adhesive multilayer and freestanding films (Costa et al., 2013; Gomes et al., 2018), antioxidant and anticancer (Sarilmiser and Oner, 2014), anticoagulant (Erginer et al.,

2016; Avsar et al., 2018), biocompatible peptide and protein drug micro carrier systems (Sezer et al., 2011; Sezer et al., 2017), bioactive laser deposited surfaces (Sima et al., 2011; Sima et al., 2012; Axente et al., 2014; Mihailescu et al., 2018), prebiotic activity and antidiabetic activity (Dahech et al., 2011), anti-irritant and cosmetic (Kang et al., 2009; Kim et al., 2005) applications are reported.

Fructan in Personal Care Products

Many personal care products like skin creams, hair care products and whiteners contain fructans with anti-irritant, anti-inflammatory, moisturizing, anti-aging, whitener, stability controller gelling agent activity, volume and rheology improver properties. Scientific research, patent applications and commercially available products are present. fructan containing cosmetic products or ingredients and their activity is discussed in detail below and summarized in Table 2.

Table 2: Activity of fructan containing cosmetic products and ingredients.

Brand	Supplier	Origin	Type of Fructan	Activity
APOLLUSKIN	Silab	*Taraxacum officinale*	-	Anti-irritant Free Radical Scavanger Skin Radiant
Belides™ (NP)	CLR Berlin	*Bellis perennis*	Inulin	Anti-oxidant Whitener Anti-microbial
Belides™ ORG	CLR Berlin	*Bellis perennis*	Inulin	Anti-oxidant Whitener

				Anti-microbial
Burdock	GREENTECH	*Arctium lappa*	inulin	Anti-Acne Varicose Wound Healing
Burdock Extract-NS	The Garden of Naturalsolution	*Arctium lappa*	inulin	Anti-allergic Anti-oxidant
Chicory Extract-NS	The Garden of Naturalsolution	*Cichorium ıntybus*	Inulin	Anti-oxidant Anti-inflammatory Anti-microbial
COBIOBALANCE	Cobiosa	*Polymnia sonchifolia*	FOS	Skin Prebitoic
Dermalab Elecampane extract	Dermalab	*Inula helenium*	Inulin	Anti-inflammatory Anti-microbial
Detoxium™	Lucas Meyer Cosmetics		Inulin	Anti-oxidant Cell Purifier
ECOSKIN®	Solabia	*Polymnia sonchifolia*	FOS	Skin Radiant
Elecampane Extract-NS	The Garden of Naturalsolution	*Inula helenium*	Inulin	Anti-oxidant Anti-microbial
Fructan	Presperse (Sumitomo Corporation)	-	-	Moisturizer Anti-acne Anti-irritant
FRUCTAN GUM	Sensient Cosmetic Technologies	*Cichorium intybus*	Inulin	Thickener Film Former
Fructan™	SK Bio	-	levan	Moisturizer Anti-inflammatory

Ginja	Campo Research	-	Inulin	Moisturizer Anti-inflammatory Sun Screen
Herb Ex™ Fructovan	Clariant	-	-	Moisturizing Anti-irritant
IBR-UrBioTect 1901	IBR	*Inula helenium*	Inulin	Anti-aging Anti-wrinkle
MultiMoist CLR™	CLR Berlin	-	FOS	Moisturizer Hair Conditioner
Nonomist	SK Bio	-	levan	Anti-acne Anti-inflammatory
PoLevan S	GREENTECH	Fruit Extract	levan	Moisturizer Anti-irritant Anti-inflammatory Anti-aging Anti-wrinkle Sun Care
PreBio™ Defense	Vantage Specialty Ingredients	-	Inulin	Moisturizer Skin Prebiotic
PROTEOLEA®	Rahn	-	Levan	Moisturizer Anti-Aging
Skin Pure-SD	The Garden of Naturalsolution	*Asparagus cochinchinensis*	Inulin	Anti-inflammatory Anti-microbial
SLIMEXIR®	Rahn	-	levan	Anti-dark cycle Slimming

				Silhouette refiner
SoyPol	SK Bioland	*Glycine max* *Bascillus Subtilis* Natto	levan	Moisturizer Whitener
Wu Qing	Campo Research	-	Inulin	Moisturizer Anti-inflammatory Sun Screen

Derived from https://www.specialchem.com/

As shown in Table 2, moisturizing, anti-aging, anti-inflammatory and anti-oxidant activities are common properties of fructan containing products and they are selling by several companies worldwide.

Skin Moisturizer

In order to maintain skin barrier function, hydration, and delay of aging and wrinkle formation, moisturization of skin is important. Barrier function of skin is disrupted by many factors including, diet, and life style, cosmetics we use such as cleansers, which cannot distinguish sebum and intercellular lipid and clean all of them from skin surface. Removing intercellular lipid damages barrier function and increase dehydration. Skin creams with hydration activity take role here. Moisturizers increase skin hydration by delaying water loss from skin surface which is known as Trans Epidermal Water Loss (TEWL), regular packing of lipid lamella, normalizing cellular turnover, improve optical appearance repairing function of skin and help to make skin surface smoother and softer. These are main objectives of skin cream development (Freitas et al., 2015; Draelos, 2018). A tropical fruit *Durio zibethinus* is known to have fructose branched polygalacturonan (Kanlayavattanakul and Lourith, 2015) and *in vivo* skin hydrating activity of polysaccharide gel from Durian analyzed with 18 volunteers for 8 week and younger volunteers (under 30) showed improved skin hydration over older ones (Futrakul et al., 2010). Levan has cell proliferating, skin irritation alleviating and skin-moisturizing activities thus is commonly preferred in cosmetic industries as skin whitener and moisturizer (Kim et al., 2003; Freitas et al., 2015; Öner et al., 2016). SK bioland sells fructan (levan) as a cosmetic ingredient under the name of Fructan^{TM}, Nanomist and SoyPol SP (Table 2). Fructan^{TM} is known to have skin moisturizing and anti-inflammatory activities and is shown in 3D artificial skin models by decrease of IL-1. Nanomist is used as anti-acne and anti-inflammatory agent. SoyPol produced from fermented soybean by *Bacillus subtillis* Natto, and reported to have skin whitening and moisturizing activity even better than water itself (www.skbioland.com). *Rahnella aquatilis*, *Zymomonas mobilis* and *Pseudomonas aurantiaca* levan is repoted to have cell proliferating, skin moisturizing and skin irritation ailment relieving activities as reported by Kim et al. (2004) with the use of *in vitro* artificial skin tests and human studies. *Zymomonas mobilis* levan (β(2-6) linked β(2-6) branched) with mid molecular weight and self-nanoparticle forming abilities is studied for cosmeceutical properties via analyzing cytotoxicity, cellular proliferation, anti-inflammation and moisturizing effects by Kim et al. (2005). Sodium lauryl sulfate (SLS) induced cells showed improved anti-inflammatory and proliferative activity and good cellular proliferation while no cytotoxicity is observed. Moisturizing effect of polymer is tested with 22-37 year-old female subjects via observing transepidermal water loss (TEWL), skin moisture content and similar moisturizing activity with HA was observed. Tests are supported by 3D artificial skin models for proliferative and anti-inflammatory activities and decreased IL- α levels (responsible in inflammatory response) are observed. Usage of levan as subcutaneous filler in dental and medical applications and healing wounds are reported (Kang et al., 2009). Levan produced from sucrose from levansucare recombinant *Erwinia amylovora* formed a water-soluble microgel. The high molecular weight levan is proposed as a strong candidate for cosmeceutical applications due to its high yield production, high conversion rate, high decomposition temperature, concentration dependent Newtonian fluid and shear thinning behavior and gel forming features (Peng et al., 2018).

Skin Whitener

Another commonly used skin care product is skin whiteners, which improve appearance of skin and normalize color inequalities. UV radiation initiated melanogenesis and results in darkening of skin caused by excess of dark brown melanin. Tyrosinase is a membrane located glycoprotein in melanocytes that plays role in melanogenesis with tyrosine related protein -1 and 2 (Trp 1, Trp 2) and pigmentation occurs through L-DOPA-L-dopaquinone conversion. Tyrosinase inhibitors are widely used to combat melanin associated dermatological diseases like hyperpigmentation, melisma, and solar lentigo. However, the poor water stability, light oxidation and skin irritation limits their usage in skin whitener cosmeceuticals. So whitening industry seek natural stable and bioactive alternatives to suppress the process of skin melanogenesis (Öner et al., 2016; Agrawal et al., 2018). Levan or inulin fructosyltransferase enzyme combined with ascorbic acid improves stability against oxidation with skin whitening and anti-oxidative activities. Kim et al. (2005 b) is reposted fructans as possible agent to improve stability of cosmeceutical ingredients. Skin whitening activity of levan via decrease of tyrosinase activity in hyperpigmented skin is reported by patent application of Masayo and Takayuki (2016).

Hair Care

Maintaining soft and good-looking hair, supporting its growth, preventing hair loss and damage are main reasons to use hair care cosmetics. Several factors are known to damage hair including heat stress, UV exposure, chemicals in dying, bleaching, washing hair and so on. Hair care products are expected to have low stickiness, to be preferably transparent and clear, and to have good film formation ability, high hair stability and easiness in removing from hair. Natural hair care ingredients are in focus not only because of their biological origin but also because of their rheological properties (Aranaz et al., 2018). Levan is used in hair care and fixative products together with hair holding gels, mousses, muds and anti-frizz products because of its film forming and hair holding activities. Any effect of levan on physiological properties of final product is known because of its low intrinsic viscosity in aqueous solutions and nonionic properties (Öner et al., 2016). Use of fructan (levan from Montana Polysaccharides, USA) based surfactant with cationic substituent in conditioning shampoo to give high conditioning activity at low concentrations is developed by Rhoda Inc. (Gunn et al., 2009). Increase in tensile strength of chemically damaged hair via bleaching by use of carboxymethylated levan to decline lifting of cuticle layers due to bleaching was studied by Kim et al. (2012). Carson (2015) reported use of levan in hair spray resin with water and water/ethanol solution spraying possibility with current pump spray technologies.

Hair loss is another aesthetic problem and current therapies include drugs, synthetic chemicals but have limitations. Use of natural formulations to support hair growth are important and gaining escalating popularity because of their less side effects, easiness in availability and low cost (Herman and Herman, 2017). Application of hydrophilic inulin and taranexamic acid nanoemulsions performed on hairless and hairy mouse for transport facilitating activity and sebum compatible environment with the hair follicle and efficient transport is observed (Wu et al., 2001). Allium species have high fructan content (Roberford, 2007) and topical application of onion (*Allium cepa*) juice is tested on 23 volunteers with hair loss syndrome (Alopecia areata) for 2 months ,twice daily, and increased hair regrowth is observed just after 2 weeks (Sharquie and Obaidi, 2002). In another study, garlic gel (*Allium sativum*) is applied together with corticosteroid (1h gel application and after corticosteriod treatment) for 3 months on 40 alopecia aerate patient by Hajheydari et al. (2007). Increased total and terminal hair observed when applied garlic gel.

Anti-Aging, Anti-inflammatory and Wound Healer Products

Antioxidants are known to neutralize free radicals through accepting or donating electrons to eliminate their dangerous effects in body. Topical use of natural substances that are made up from vitamins, minerals, active polysaccharides is a good strategy to support skins own antioxidant mechanism. Coenzyme Q-10, polyphenols, vitamins (A, B, C and E) are some of antioxidants currently used together with biological growth factors, kinetin, herbal ingredients, retinoid, sun filters and other bioactive natural polysaccharides in fighting against skin aging. Use of those components is also important for cosmetic ingredients to protect lipid peroxidation since odor and color of cosmetics are important parameters for consumers and many of them have limited shelf life (Valacchi et al., 2015; Mahawar et al., 2018; Otlatıcı et al., 2018;). Svendsen et al (1994) reported anti-aging and youth preserving effect of garlic via keeping human fibroblast cells at maximum proliferative activity. Levan (2 types of levan with different molecular weight) and sulfated derivative from *B. subtilis* is shown to have free radical scavenger activity when tested against diphenyl-2-pycryl-hydrazyl (DPPH) radicals (Abdel-Fattah et al., 2012). Liu et al. (2014) investigated antioxidant activity of water soluble polysaccharide, fructan, with D-fructose and D-glucose content, from *Arctium lappa* L. (burdock) root *in vitro* and *in vivo*. Hydroxyl radical, ABTS and H_2O_2 scavenging activities are measured *in vitro*, SOD, GSH-PX and CAT enzyme activities were analyzed *in vivo* with ICR mice and strong antioxidant activity, and increased enzyme activity were observed. Rahn AG sells fructan-containing cosmeceuticals like PROTEOLEA® and Slimexir® (Rahn 2012; Rahn 2013). Fructan containing Slimexir® (artichoke extract and levan) improves body shape, reduces eye bags resulting in younger looking eyes and silhouette refiner features (Rahn, 2013). White fructan (GF) (1-kestose type fructan) is extracted from garlic (*Allium sativum L.*) and protective effect on UVB induced human keratinocyte cells (HaCaT) line at an increased cell viability was shown (Chen et al., 2013). High antioxidant activity of *Acetobacter xylinum* levan (β-(2-1) branched) reaching 81% of that of the ascorbic acid was observed (Srikanth et al. (2015). As a result of this investigation, a strong candidature of the thermally stable and bioactive levan for pharmaceutical, medical and cosmetic applications was suggested. Blue light irradiated human keratinocyte (HaCaT) cells incubated with levan containing anti-oxidant cosmetic PROTEOLEA® showed ROS decrease (Hettwer et al., 2017). Antioxidant activity of inulin from Jerusalem artichoke (Beijin Wede Biological Technology) is studied *in vitro* and *in vivo* by Shang et al. (2018). DPHH, ABTS radical scavenging activity and ferric reducing power are observed to be lower than vitamin C while dietary supplementation of inulin improved egg production in hens and serum SOD, CAT, GSH-Px antioxidant enzyme activity and antioxidant capacity increase were observed *in vivo*.

Immune modulatory and anti-irritant activities of cosmetics are important because of controlling skin allergies via maintaining skins pathogen defense mechanism, toxin and inflammatory responses of skin problems like redness, sensitivity and allergies like atopic dermatitis etc. Immune modulatory effect of levan on peritoneal macrophage and T cells is shown in mice experiments (Robertson et al., 1977). Inulin from *Arctium lappa L.* (Compositae) (burdock) is known to be beneficial in skin diseases including eczema, ulcers and fasten wound healing (Azizov et al., 2012). Immune modulatory activity of onion (*Allium cepa*) inulin-type fructans (FOSs with threesaccharide to hexasaccharide range with β-D-Gly terminal end) were tested *in vitro* with primary murine peritoneal macrophage, peritoneal exudate cell (PECs) culture. Results showed increased cellular proliferation, nitric oxide, intracellular free radical production and phagocytic activity (Kumar et al., 2015). APOLLUSKIN® (Silab) containing fructan from *Taraxacum officinale* (dandelion) is reported to have skin radiant and anti-irritant activities in a patent application (Stout and Murphy, 2018).

Healing skin wounds fast without any mark on skin surface after trauma, diabetes related conditions or surgery is important because of aesthetic reasons. Providing microenvironment for cells, non-toxic, antimicrobial, antiallergic, anti- scar tissue former and nonsticky wound dressings are required for skin wound healing (Şahiner et al., 2016; Verma et al., 2017; Zhang et al., 2017) while maintaining moisture and hydration of wounded area is known to enhance the quality of the formed tissue (Castaño et al., 2018). Since skin, it the first defense system of our body, any wound occurred in this tissue effects life severely and needs to be treated with minimal scar tissue formation in short time. Thus, development of cosmeceuticals that contain biologically active wound healing supporting polysaccharides are in need. Fructans are strong candidates for components of wound healing creams with their hydrophilicity, high moisturizing activity and cellular proliferation supporter abilities. Collagen is known to support fibroblasts' cell-cell and cell-ECM interactions and allows new tissue formation together with growth factors, cytokines, integrin's and MMPs (Ramanathan et al., 2017; Ishida et al., 2018).

Effect of *Zymomonas mobilis* NCIB 11163 levan on matrix metalloproteinase (MMP) activity which play key role in tissue remodeling in burns or wounds is investigated by Sturzoiu et al. (2011) on skin burned or wounded Wistar rats and new tissue formation responsible for MMP (MMP-2, MMP-9) activation is observed. This study concluded that levan modulate MMP activity is closely related with antioxidant protection and cellular proliferation mechanisms. Prevention of median sternotomy wound in children and hypertrophic scar formation by an onion extract is observed Upadhyay (2016). This study concluded it has therapeutic effect on skin fibroblast cell line and keloid (scar tissue, overgrowth of skin cells on wound side) treatment. Carboxymethyl cellulose, pluronic acid and levan (Real Biotech) injectable dermal filler hydrogels are synthesized and tested with human dermal adult fibroblasts (HDF) and enhanced cell proliferation and collagen expression are reported by Choi et al (2018). Anti-wrinkle formation of this hydrogel tested with 1α25-dihydroxyvitamin D3 (1, $25(OH)_2VD_3$) mice model revealed a wrinkle decrease and using of levan hydrogel in cosmetic applications was suggested due to its concentration dependent resident stability, elastic modulus, biocompatibility, collagen production and cell proliferation inducer abilities. Aloe vera (*Aloe barbadensis miller*) succulent is known to have several application field including cosmetics and topical applications against dermatitis, erythema and photo toxicity (Boudreau and Beland, 2006; Shi et al., 2019). Tissue regeneration ability of aloe vera makes it a strong candidate for tissue engineering applications and cosmetics with its antifungal, antiseptic, antiviral, antibacterial, anti-inflammatory, anti-oxidant activities and wound healing, cell migration, proliferation, collagen synthesis and growth promoting abilities (Rahman et al., 2017). Acemannan and other components of aloe vera are studied but fructan content of stressed aloe vera (Salinas et al., 2016) and its bioactivity need to be investigated in detail.

Nutricosmetics

Since it is believed that beauty comes from inside, diet is in focus to reduce acnes, cellulite and many remedies are applied that contain raw vegetables, their extracts, tea and egg as skin masks. When people became conscious about nutrition and importance of intestinal health, several studies are performed to highlight their effect on metabolism. Skin is one of them that can be supported via prebiotic consumption or application topically. Bacteria living on skin can be grouped as resident or transient (contaminant) and those two kinds of bacteria form skin microbiome. *Propionibacteria* (*P. acnes, P. avidum, P. granulosum*), *Staphylococci* (*Staphlyococcus epidermidis*), *Micrococci, Corynebacteria*, *Acinetobacter* and Malasseiza yeasts are resident microbes while *Staphlyococcus aureus*, *Escherichia coli*, *Pseudomonas aeroguinosa* and *Bacillus* species are classified as contaminants. Those beneficial bacteria are known to produce vitamins (A, B1, B2, B5, B6, B7, B9, B12, E, K), acetate, butyrate, propionate, neurotransmitters, neuropeptides that regulate innate or adaptive immune system on skin (Krutmann, 2009; Simmering and Breves, 2010). Skin microflora, skin barrier function, epidermal keratinocytes, lifestyle and other exogenous factors play active role in skin sensitivity, chronic inflammatory diseases, infections and allergies. Keratinocytes effect skin microbiota through production of anti-microbial peptides that result in imbalanced microflora and several infections and ailments depending on species overgrowth occur. Treatment of such diseases or ailments without drugs and antibiotics to allow stimulation of beneficial bacteria are possible with the use of natural polysaccharides. The term "Nutricosmetics" is in focus in the light of probiotic, prebiotic and symbiotic approaches to those cases via either consuming or applying through surface with selectively allowing beneficial growth while effecting contaminants in antibiotic freeway. pH and water content of skin are known to play crucial roles on skin microbiota and *S. epidermidis* protect skin from infections and suppress growth of contaminant bacteria. Free water is an important parameter for microbiota and skin care moisturizers bind water to skin cells thus free water is reduced. Prebiotic containing moisturizers help to balance skin microbiota. Fructans are in the group of prebiotics that are used for skin hydration. Use of prebiotics and probiotics suppress overgrowth of those contaminants such as *P. acnes*, which cause skin acne. Gut microbiota affect skin via overall immunity enhancing activities (Krutmann, 2009; Priyan Shanura Fernando et al., 2018; Seite and Misery, 2018). Consumption of probiotics or prebiotics (nonliving, non-digestible but selectively fermented dietary fibers that support growth of probiotics) is known to support gut microbiota, immunity and skin health (Krutmann, 2009; Haskoylu et al., 2017). Selective activity of prebiotic oligofructans on skin microbiota is reported (Smiricky-Tjardes et al., 2003). Use of inulin is reported to support growth of *S.epidermidis* over *S. aures* and *E. coli* (Carolan et al., 2008). Moro et al (2006) investigated the effect of galacto and long chain fructooligosaccharide prebiotic mixture on formula fed infants (6 months old) with atopic dermatitis (AD) and improved bowel flora associated postnatal immunity priming and primary allergy prevention is observed. Gerasimov et al. (2010) investigated effect of symbiotic supplement on atopic dermatitis (AD) in 90 children aged 1-3. rice maltodextrin (DDS® Junior) containing a mixture of *L. acidophilus* DDs-1, *B. lactis* UABLA-12 and fructooligosaccharides was used for treatment for 8 weeks twice daily and results showed significant improvements in AD syndrome and a lymphocyte increase in peripheral blood. The superior effect and reduced AD symptoms were observed after a twice daily treatment for 10 weeks of 60 children aged between 2-14 with *Lactobacillus salivarius* plus FOS or FOS alone (Wu et al. (2012). A reduced skin wrinkle formation due to an increase of hyperkeratosis, skin SOD and CAT enzyme activity while MMP1 and MMP2 activity decreased after dietary supply of garlic on UV-induced skin of hairless mice for eight weeks (Kim (2016). Study concluded that dietary supplementation of garlic decreased UV induced photo aging through anti-oxidant activity increase and matrix metalloproteinase activity decrease in skin.

Conclusion

Protection of skin from outside agents such as UV radiation and toxic chemicals resulted in a reduced wrinkle formations, increased antioxidant activity and balancing its barrier function. Microbiome on it and hydration are main concerns of skin cosmetics/nutricosmetics development as a result of natural product aware consumers. Fructans are in focus as natural cosmeceutical ingredient with their novel features and levan stands out with its high hydrophilicity and biocompatibility but the limited aqueous and acidic stability limits its usage. Thus, new technologies and production methods need to be developed in the light of biotechnology to produce novel polysaccharides with desired features. New microbial fructan producers should be introduced to cosmetic industry to improve skin look in cellular based smart way by preventing target DNA damage and oxidative stress, and improving skin barrier function and microbiome. Studies that contain direct interaction between skin hydration and fructans when applied topically, their exact role in anti-inflammatory or anti-irritant response and interaction with skin microbiome need to be investigated in cellular and molecular level for next generation nutria-cosmeceutical development.

References

Abdel-Fattah, A.M., Gamal-Eldeen, A.M., Helmy, W.A., and Esawy, M.A. (2012). Antitumor and antioxidant activities of levan and its derivative from the isolate *Bacillus subtilis* NRC1aza. Carbohydr Polym 89, 314-322. https://doi.org/10.1016/j.carbpol.2012.02.041.

Adrien, A., Bonnet, A., Dufour, D., Baudouin, S., Maugard, T., and Bridiau, N. (2017). Pilot production of ulvans from *Ulva sp.* and their effects on hyaluronan and collagen production in cultured dermal fibroblasts. Carbohydrate polymers 157, 1306-1314. https://doi.org/10.1016/j.carbpol.2016.11.014.

Agrawal, S., Adholeya, A., Barrow, C.J., and Deshmukh, S.K. (2018). Marine fungi: An untapped bioresource for future cosmeceuticals. Phytochemistry Letters 23, 15-20. https://doi.org/10.1016/j.phytol.2017.11.003.

Aranaz, I., Acosta, N., Civera, C., Elorza, B., Mingo, J., Castro, C., Gandía, M.d.l.L., and Heras Caballero, A. (2018). Cosmetics and Cosmeceutical Applications of

Chitin, Chitosan and Their Derivatives. Polymers 10, 213. https://doi.org/10.3390/polym10020213.

Avsar, G., Agirbasli, D., Agirbasli, M.A., Gunduz, O., and Oner, E.T. (2018). Levan based fibrous scaffolds electrospun via co-axial and single-needle techniques for tissue engineering applications. Carbohydrate polymers 193, 316-325. https://doi.org/10.1016/j.carbpol.2018.03.075.

Axente, E., Sima, F., Elena Sima, L., Erginer, M., Eroglu, M.S., Serban, N., Ristoscu, C., Petrescu, S.M., Toksoy Oner, E., and Mihailescu, I.N. (2014). Combinatorial MAPLE gradient thin film assemblies signalling to human osteoblasts. Biofabrication 6, 035010. https://doi.org/10.1088/1758-5082/6/3/035010.

Azizov, U., Khadzhieva, U., Rakhimov, D., Mezhlumyan, L., and Salikhov, S. (2012). Chemical composition of dry extract of *Arctium lappa* roots. Chemistry of Natural Compounds 47, 1038-1039.

Bailey, A.J. (2001). Molecular mechanisms of ageing in connective tissues. Mech Ageing Dev 122, 735755. https://doi.org/10.1016/S0047-6374(01)00225-1.

Balkrishna, A., Agarwal, V., Kumar, G., and Gupta, A.K. (2018). Applications of Bacterial Polysaccharides with Special Reference to the Cosmetic Industry. In Microbial Bioprospecting for Sustainable Development (Springer), pp. 189-202.

Baumann, L. (2007). Skin ageing and its treatment. J Pathol 211, 241-25. https://doi.org/10.1002/path.2098.

Belghith, K.S., Dahech, I., Belghith, H., and Mejdoub, H. (2012). Microbial production of levansucrase for synthesis of fructooligosaccharides and levan. Int J Biol Macromol 50, 451-458. https://doi.org/10.1016/j.ijbiomac.2011.12.033.

Berardesca, E., Cameli, N., and Mariano, M. (2018). Aging of the Skin. In Gender and Dermatology (Springer), pp. 25-30.

Bhushan, B. (2017). Skin and Skin Cream. In Biophysics of Skin and Its Treatments (Springer), pp. 1124. https://doi.org/10.1007/978-3-319-45708-6-2.

Boudreau, M.D., and Beland, F.A. (2006). An evaluation of the biological and toxicological properties of *Aloe barbadensis (miller*), Aloe vera. J Environ Sci Health C Environ Carcinog Ecotoxicol Rev 24, 103154. https://doi.org/10.1080/10590500600614303.

Campa, M., and Baron, E. (2018). Anti-aging Effects of Select Botanicals: Scientific Evidence and Current Trends. Cosmetics 5, 54. https://doi.org/10.3390/cosmetics5030054.

Carolan, H., Watkins, S., and Bradshaw, D. (2008). The prebiotic concept-a novel aproach for skin health. Euro Cosmetics 7, 22-27.

Castano, O., Perez-Amodio, S., Navarro-Requena, C., Mateos-Timoneda, M.A., and Engel, E. (2018). Instructive microenvironments in skin wound healing: Biomaterials as signal releasing platforms. Advanced Drug Delivery Reviews 129, 95-117. https://doi.org/10.1016/j.addr.2018.03.012.

Chen, G.Q., and Jiang, X.R. (2018). Next generation industrial biotechnology based on extremophilic bacteria. Curr Opin Biotechnol 50, 94-100. https://doi.org/10.1016/j.copbio.2017.11.016.

Chen, J., Cheong, K., Song, Z., Shi, Y., and Huang, X. (2013). Structure and protective effect on UVB induced keratinocyte damage of fructan from white garlic. Carbohydr Polym 92, 200-205. https://doi.org/10.1016/j.carbpol.2012.09.068.

Choi, W.I., Hwang, Y., Sahu, A., Min, K., Sung, D., Tae, G., and Chang, J.H. (2018). An injectable and physical levan-based hydrogel as a dermal filler for soft tissue augmentation. Biomaterials science 6, 2627-2638. https://doi.org/10.1039/C8BM00524A.

Cole, M.A., Quan, T., Voorhees, J.J., and Fisher, G.J. (2018). Extracellular matrix regulation of fibroblast function: redefining our perspective on skin aging. Journal of cell communication and signaling, 1-9. https://doi.org/10.1007/s12079-018-0459-1.

Costa, R.R., Neto, A.I., Calgeris, I., Correia, C.R., Pinho, A.C., Fonseca, J., Öner, E.T., and Mano, J.F. (2013). Adhesive nanostructured multilayer films using a bacterial exopolysaccharide for biomedical applications. Journal of Materials Chemistry B 1, 2367-2374. https://doi.org/10.1039/C3TB20137F.

Dahech, I., Belghith, K.S., Hamden, K., Feki, A., Belghith, H., and Mejdoub, H. (2011). Oral administration of levan polysaccharide reduces the alloxan-induced oxidative stress in rats. Int J Biol Macromol 49, 942-947. https://doi.org/10.1016/j.ijbiomac.2011.08.011.

Di Bartolomeo, F., Startek, J.B., and Van den Ende, W. (2013). Prebiotics to fight diseases: reality or fiction? Phytother Res 27, 1457-1473. https://doi.org/10.1002/ptr.4901.

Draelos, Z.D. (2018). The science behind skin care: Moisturizers. J Cosmet Dermatol 17, 138-144. https://doi.org/10.1111/jocd.12490.

Erginer, M., Akcay, A., Coskunkan, B., Morova, T., Rende, D., Bucak, S., Baysal, N., Ozisik, R., Eroglu, M.S., and Agirbasli, M. (2016). Sulfated levan from Halomonas smyrnensis as a bioactive, heparinmimetic glycan for cardiac tissue engineering applications. Carbohydrate polymers 149, 289-296. https://doi.org/10.1016/j.carbpol.2016.04.092.

Erkorkmaz, B.A., Kirtel, O., Ates Duru, O., and Toksoy Oner, E. (2018). Development of a cost-effective production process for *Halomonas* levan. Bioprocess Biosyst Eng 41, 1247-1259. https://doi.org/10.1007/s00449018-1952-x.

Feetham, H.J., Jeong, H.S., McKesey, J., Wickless, H., and Jacobe, H. (2018). Skin care and cosmeceuticals: Attitudes and trends among trainees and educators. Journal of cosmetic dermatology 17, 220-226. https://doi.org/10.1111/jocd.12460.

Freitas, F., Alves, V.D., and Reis, M.A. (2015). Bacterial polysaccharides: production and applications in cosmetic industry. Polysaccharides: Bioactivity and Biotechnology, 2017-2043

Futrakul, B., Kanlayavattanakul, M., and Krisdaphong, P. (2010). Biophysic evaluation of polysaccharide gel from durian's fruit hulls for skin moisturizer. International journal of cosmetic science 32, 211-215. https://doi.org/10.1111/j.1468-2494.2009.00552.x.

Gazitaeva, Z.I., Drobintseva, A.O., Chung, Y., Polyakova, V.O., and Kvetnoy, I.M. (2017). Cosmeceutical product consisting of biomimetic peptides: antiaging effects in vivo and in vitro. Clin Cosmet Investig Dermatol 10, 11-16. https://doi.org/10.2147/CCID.S97573.

Gerasimov, S.V., Vasjuta, V.V., Myhovych, O.O., and Bondarchuk, L.I. (2010). Probiotic supplement reduces atopic dermatitis in preschool children: a randomized, double-blind, placebo-controlled, clinical trial. Am J Clin Dermatol 11, 351-361. https://doi.org/10.2165/11531420-000000000-00000.

Ghazi, I., Fernandez-Arrojo, L., Garcia-Arellano, H., Ferrer, M., Ballesteros, A., and Plou, F.J. (2007). Purification and kinetic characterization of a fructosyltransferase from *Aspergillus aculeatus*. Journal of Biotechnology 128, 204-211. https://doi.org/10.1016/j.jbiotec.2006.09.017

Gomes, T.D., Caridade, S.G., Sousa, M.P., Azevedo, S., Kandur, M.Y., Oner, E.T., Alves, N.M., and Mano, J.F. (2018). Adhesive free-standing multilayer films containing sulfated levan for biomedical applications. Acta Biomater 69, 183-195 DOI: 10.1016/j.actbio.2018.01.027.

Gunn, E., Gabbianelli, A., Crooks, R., and Shanmuganandamurthy, K. (2009). Derivatized polysaccharide polymer (Google Patents).

Hajheydari, Z., Jamshidi, M., Akbari, J., and Mohammadpour, R. (2007). Combination of topical garlic gel and betamethasone valerate cream in the treatment of localized alopecia areata: a double-blind randomized controlled study. Indian J Dermaetol Venereol Leprol 73, 29-32.

Haskoylu Erginer, M., Yasar Yildiz, S., Toksoy Oner, E. (2017). Fructan rich diet to improve gut microbiota in disease and health. In: Science within Food: Up-to-date Advances on Research and Educational Ideas. (Formatex Research Center).

Herman, A., and Herman, A.P. (2017). Topically used herbal products for the treatment of hair loss: preclinical and clinical studies. Archives of dermatological research 309, 595-610. https://doi.org/10.1007/s00403-017-1759-7.

Hettewer, S., Gyenge, E., and Obermayer, B. (2017). Blue Light Protecting Cosmetic Active Ingredients: A Case Report. J Dermat Cosmetol 1, 00023.

http://www.skbioland.com

https://www.specialchem.com.

Ishida, Y., Kuninaka, Y., Furukawa, F., Kimura, A., Nosaka, M., Fukami, M., Yamamoto, H., Kato, T., Shimada, E., Hata, S., et al. (2018). Immunohistochemical analysis on aquaporin-1 and aquaporin-3 in skin wounds from the aspects of wound age determination. Int J Legal Med 132, 237-242. https://doi.org/10.1007/s00414-017-1725-0.

Jenkins, G. (2002). Molecular mechanisms of skin ageing. Mech Ageing Dev 123, 801-810. https://doi.org/10.1016/S0047-6374(01)00425-0.

Juhász, M.L., Levin, M.K., and Marmur, E.S. (2018). The use of natural ingredients in innovative Korean cosmeceuticals. Journal of cosmetic dermatology 17, 305-312. https://doi.org/10.1111/jocd.12492.

Kambourova, M., Oner, E.T., and Poli, A. (2015). Exopolysaccharides from Prokaryotic Microorganisms—Promising Sources for White Biotechnology Processes. In Industrial Biorefineries & White Biotechnology (Elsevier), pp. 523-554. https://doi.org/10.1016/B978-0-444-63453-5.00017-3.

Kang, S.A., Jang, K.-H., Seo, J.-W., Kim, K.H., Kim, Y.H., Rairakhwada, D., Seo, M.Y., Lee, J.O., Ha, S., and Kim, C. (2009). Levan: applications and perspectives. Microbial production of biopolymers and polymer precursors Caister Academic Press, Norwich, 145-161.

Kanlayavattanakul, M., and Lourith, N. (2015). Biopolysaccharides for skin hydrating cosmetics. Polysaccharides: Bioactivity and Biotechnology, 1867-1892.

Kim, C.H., Kim, G.H., Ko, G.I., Lee, J.S., Lee, S.G., Park, G.S. (2004). Cosmetic composition containing levan extract having elasticity enhancing effect by cell proliferation and skin moisturization. Korean patent. KR100440357B1.

Kim, H.K. (2016). Garlic Supplementation Ameliorates UV-Induced Photoaging in Hairless Mice by Regulating Antioxidative Activity and MMPs Expression. Molecules 21, 70. https://doi.org/10.3390/molecules21010070.

Kim, K., Kim, K., Ryo, O., Lee, T., and Kim, T. (2003). Cosmetic composition containing levan having cell proliferation, skin-moisturizing and irritation-alleviating effects. Japanese Patent 2003277225.

Kim, K.H., Chung, C.B., Kim, Y.H., Kim, K.S., Han, C.S., and Kim, C.H. (2005). Cosmeceutical properties of levan produced by *Zymomonas mobilis*. Journal of cosmetic science 56, 395-406.

Kim, Y.H., Kim, J.G., Jang, J.W., Lee, S.Y., Heo, W.B., Kim, K.H. (2012). Study on the synthesis of modified levan and its properties on the hair. International Federation of Societies of Cosmetic Chemists, Johannesburg, South Africa, poster session 5, number 302.

Kindzierski, V., Raschke, S., Knabe, N., Siedler, F., Scheffer, B., Pfluger-Grau, K., Pfeiffer, F., Oesterhelt, D., Marin-Sanguino, A., and Kunte, H.J. (2017). Osmoregulation in the Halophilic Bacterium *Halomonas elongata*: A Case Study for Integrative Systems Biology. PLoS One 12, e0168818. https://doi.org/10.1371/journal.pone.0168818.

Kırtel, O., Versluys, M., Van den Ende, W., and Öner, E.T. (2018). Fructans of the saline world. Biotechnology Advances. 36(5), 1524-1539. https://doi.org/10.1016/j.biotechadv.2018.06.009.

Kligman, L. H., ve Kligman, A. M. (1986). The nature of photoaging: its prevention and repair. Photodermatology, 3(4), 215-227.

Krutmann, J. (2009). Pre- and probiotics for human skin. J Dermatol Sci 54, 1-5. https://doi.org/ 10.1016/j.jdermsci.2009.01.002.

Kumar, V.P., Prashanth, K.V., and Venkatesh, Y.P. (2015). Structural analyses and immunomodulatory properties of fructo-oligosaccharides from onion (*Allium cepa*). Carbohydr Polym 117, 115-122. https://doi.org/10.1016/j.carbpol.2014.09.039.

Liu, W., Wang, J., Zhang, Z., Xu, J., Xie, Z., Slavin, M., and Gao, X. (2014). In vitro and in vivo antioxidant activity of a fructan from the roots of *Arctium lappa* L. International journal of biological macromolecules 65, 446-453. https://doi.org/10.1016/j.ijbiomac.2014.01.062.

Mahawar, V., Patidar, K., and Joshi, N. (2018). Unmet Need and Challenges of Skin Aging by Herbal Anti-aging Cosmeceuticals: An Overview. Asian Journal Of Pharmaceutics 12, 410-418.

Markiewicz, E., and Idowu, O.C. (2018). Involvement of the nuclear structural proteins in aging-related responses of human skin to the environmental stress. Clin Cosmet Investig Dermatol 11, 297-307. https://doi.org/10.2147/CCID.S163792.

Masayo, F., Takayuki, T. 2006. Beautifully whitening agent, Japanese patent appl. JP2006052146A.

McDaniel, D., Farris, P., and Valacchi, G. (2018). Atmospheric skin aging—Contributors and inhibitors. Journal of cosmetic dermatology 17, 124-137. https://doi.org/10.1111/jocd.12518.

Mihailescu, N., Haskoylu, M.E., Ristoscu, C., Bostan, M.S., Sopronyi, M., Eroğlu, M.S., Chifiriuc, M.C., Mustaciosu, C.C., Axente, E., and Oner, E.T. (2018). Gradient multifunctional biopolymer thin film assemblies synthesized by Combinatorial MAPLE. Applied Surface Science, 466, 628-636. https://doi.org/10.1016/j.apsusc.2018.10.077.

Moro, G., Arslanoglu, S., Stahl, B., Jelinek, J., Wahn, U., and Boehm, G. (2006). A mixture of prebiotic oligosaccharides reduces the incidence of atopic dermatitis during the first six months of age. Arch Dis Child 91, 814-819. https://doi.org/10.1136/adc.2006.098251.

Naidoo, K., and Birch-Machin, M.A. (2017). Oxidative stress and ageing: The influence of environmental pollution, sunlight and diet on skin. Cosmetics 4, 4. https://doi.org/10.3390/cosmetics4010004.

Nicolaus, B., Kambourova, M., and Oner, E.T. (2010). Exopolysaccharides from extremophiles: from fundamentals to biotechnology. Environ Technol 31, 1145-1158. https://doi.org/10.1080/09593330903552094.

Öner, E.T., Hernández, L., and Combie, J. (2016). Review of levan polysaccharide: From a century of past experiences to future prospects. Biotechnology advances 34, 827-844. https://doi.org/ 10.1016/j.biotechadv.2016.05.002.

Otlatıcı, G., Yeğen, G., Güngör, S., and Aksu, B. (2018). Overview on nanotechnology based cosmeceuticals to prevent skin aging. Istanbul Journal of Pharmacy 48, 55-62. https://doi.org/ 10.5152/IstanbulJPharm.2018.424278.

Paik, B.R., Chang, H.K., Lee, M.R., NA, Y.J., Kwon, L.K., Kim, Y.S., and Han, S.H. (2018). Low viscous cosmetic composition using a natural emulsifying agent (Google Patents).

Pareek, S., Sagar, N.A., Sharma, S., and Kumar, V. (2018). Onion (*Allium cepa* L.). Fruit and Vegetable Phytochemicals: Chemistry and Human Health, 2nd ed; Yahia, EM, Ed, 1145-1161. https://doi.org/10.1002/9781119158042.ch58.

Patel, S. (2018). Role of injectable hyaluronic acid in skin rejuvenation: a literature review. Journal of Aesthetic Nursing 7, 76-80. https://doi.org/10.12968/joan.2018.7.2.76.

Peng, J., Xu, W., Ni, D., Zhang, W., Zhang, T., Guang, C., and Mu, W. (2018). Preparation of a novel water-soluble gel from *Erwinia amylovora* levan. Int J Biol Macromol. https://doi.org/10.1016/j.ijbiomac.2018.10.093.

Peshev, D., and Van den Ende, W. (2014). Fructans: prebiotics and immunomodulators. Journal of Functional Foods 8, 348-357. https://doi.org/10.1016/j.jff.2014.04.005.

Pommerrenig, B., Ludewig, F., Cvetkovic, J., Trentmann, O., Klemens, P.A.W., and Neuhaus, H.E. (2018). In Concert: Orchestrated Changes in Carbohydrate Homeostasis Are Critical for Plant Abiotic Stress Tolerance. Plant Cell Physiol 59, 1290-1299. https://doi.org/10.1093/pcp/pcy037.

Priyan Shanura Fernando, I., Kim, K.N., Kim, D., and Jeon, Y.J. (2018). Algal polysaccharides: potential bioactive substances for cosmeceutical applications. Crit Rev Biotechnol, 1-15. https://doi.org/10.1080/07388551.2018.1503995.

Radchenkova, N., Boyadzhieva, I., Atanasova, N., Poli, A., Finore, I., Di Donato, P., Nicolaus, B., Panchev, I., Kuncheva, M., and Kambourova, M. (2018). Extracellular polymer substance synthesized by a halophilic bacterium *Chromohalobacter canadensis* 28. Appl Microbiol Biotechnol 102, 4937-4949. https://doi.org/10.1007/s00253-018-8901-0.

Rahman, S., Carter, P., and Bhattarai, N. (2017). Aloe Vera for Tissue Engineering Applications. J Funct Biomater 8, 6. https://doi.org/10.3390/jfb8010006.

Rahn AG, Proteolea®. (2012) a new source of strength to rebalance the signs of time.

Rahn AG, Slimexir®. (2013) Cosmetics for the silhouette, an overview of the slimming market. Presentation.

Ramanathan, G., Singaravelu, S., Muthukumar, T., Thyagarajan, S., Perumal, P.T., and Sivagnanam, U.T. (2017). Design and characterization of 3D hybrid collagen matrixes as a dermal substitute in skin tissue engineering. Mat Sci Eng C-Mater 72, 359-370. https://doi.org/10.1016/j.msec.2016.11.095.

Razak, D.L.A., Jamaluddin, A., Rashid, N.Y.A., Mohd, N.H., Fadzil, N.A.S., and Manan, M.A. (2018). Comparative Evaluation of *Schizophyllum* commune Extracts as Potential Cosmeceutical BioIngredient.

Reed, E. (1962). The definition of cosmeceuticals". Journal of Cosmetic Science, Society of Cosmetics and Chemists, 13, 103–110.

Roberord, M. B. (2007). Inulin-type fructans: functional food ingredients. The Journal of nutrition, *137*(11), 2493-2502. https://doi.org/10.1016/j.tibtech.2012.08.004.

Robertson, T.A., Papadimitriou, J.M., Walters, M.N., and Wolman, M. (1977). Effects of exposure of murine peritoneal exudate and resident macrophages to high molecular levan: a morphological study. J Pathol 123, 157-164. https://doi.org/10.1002/path.1711230305.

Sahiner, N., Sagbas, S., Sahiner, M., Silan, C., Aktas, N., and Turk, M. (2016). Biocompatible and biodegradable poly(Tannic Acid) hydrogel with antimicrobial and antioxidant properties. Int J Biol Macromol 82, 150-159. https://doi.org/10.1016/j.ijbiomac.2015.10.057.

Salinas, C., Handford, M., Pauly, M., Dupree, P., and Cardemil, L. (2016). Structural Modifications of Fructans in *Aloe barbadensis Miller* (Aloe Vera) Grown under Water Stress. PLoS One 11, e0159819. https://doi.org/10.1371/journal.pone.0159819.

Sarilmiser, H.K., and Oner, E.T. (2014). Investigation of anti-cancer activity of linear and aldehydeactivated levan from *Halomonas smyrnensis* AAD6T. Biochemical engineering journal 92, 28-34. https://doi.org/10.1016/j.jbiosc.2014.09.019.

Seite, S., and Misery, L. (2018). Skin sensitivity and skin microbiota: Is there a link? Experimental dermatology. https://doi.org/10.1111/exd.13686.

Sezer, A.D., Kazak Sarilmiser, H., Rayaman, E., Cevikbas, A., Oner, E.T., and Akbuga, J. (2017). Development and characterization of vancomycin-loaded levan-based microparticular system for drug delivery. Pharm Dev Technol 22, 627-634. https://doi.org/10.3109/10837450.2015.1116564.

Sezer, A.D., Kazak, H., Öner, E.T., and Akbuğa, J. (2011). Levan-based nanocarrier system for peptide and protein drug delivery: optimization and influence of experimental parameters on the nanoparticle characteristics. Carbohydrate Polymers 84, 358-363. https://doi.org/10.1016/j.carbpol.2010.11.046.

Shang, H.M., Zhou, H.Z., Yang, J.Y., Li, R., Song, H., and Wu, H.X. (2018). In vitro and in vivo antioxidant activities of inulin. PLoS One 13, e019227. https://doi.org/10.1371/journal.pone.0192273.

Sharquie, K.E., and Al-Obaidi, H.K. (2002). Onion juice (*Allium cepa* L.), a new topical treatment for alopecia areata. J Dermatol 29, 343-346. https://doi.org/10.1111/j.1346-8138.2002.tb00277.x.

Shi, X.-D., Yin, J.-Y., Zhang, L.-J., Li, O.-Y., Huang, X.-J., and Nie, S.-P. (2019). Studies on polysaccharides from leaf skin of *Aloe barbadensis Miller*: Part II. Structural characteristics and molecular properties of two lower molecular weight fractions. Food Hydrocolloids 86, 50-61. https://doi.org/10.1016/j.foodhyd.2018.01.038.

Sima, F., Axente, E., Sima, L., Tuyel, U., Eroglu, M., Serban, N., Ristoscu, C., Petrescu, S., Toksoy Oner, E., and Mihailescu, I. (2012). Combinatorial matrix-assisted pulsed laser evaporation: single-step synthesis of biopolymer compositional gradient thin film assemblies. Applied Physics Letters 101, 233705. https://doi.org/10.1063/1.4769987.

Sima, F., Mutlu, E.C., Eroglu, M.S., Sima, L.E., Serban, N., Ristoscu, C., Petrescu, S.M., Oner, E.T., and Mihailescu, I.N. (2011). Levan nanostructured thin films by MAPLE assembling. Biomacromolecules 12, 2251-2256. https://doi.org/10.1021/bm200340b.

Simmering, R., and Breves, R. (2010). Prebiotic cosmetics. In Nutrition for Healthy Skin (Springer), pp. 137-147.

Singh, R.S., and Singh, R.P. (2010). Production of fructooligosaccharides from inulin by endoinulinases and their prebiotic potential. Food Technology and Biotechnology 48, 435.

Smiricky-Tjardes, M.R., Flickinger, E.A., Grieshop, C.M., Bauer, L.L., Murphy, M.R., and Fahey, G.C., Jr. (2003). In vitro fermentation characteristics of selected oligosaccharides by swine fecal microflora. J Anim Sci 81, 2505-2514. https://doi.org/10.2527/2003.81102505x.

Srikanth, R., Siddartha, G., Sundhar Reddy, C.H., Harish, B.S., Janaki Ramaiah, M., and Uppuluri, K.B. (2015). Antioxidant and anti-inflammatory levan produced from *Acetobacter xylinum* NCIM2526 and its statistical optimization. Carbohydr Polym 123, 8-16. https://doi.org/10.1016/j.carbpol.2014.12.079.

Stout, H.S., and Murphy, K.A. (2018). Skin rejuvenation and defense system (Google Patents).

Sturzoiu, C., Petrescu, M., Galateanu, B., Anton, M., Nica, C., Simionca, G.I., Dinischiotu, A., and Stoian, G. (2011). *Zymomonas mobilis* Levan is Involved in Metalloproteinases Activation in Healing of Wounded and Burned Tissues. Scientific Papers Animal Science and Biotechnologies 44, 453-458.

Svendsen, L., Rattan, S.I., and Clark, B.F. (1994). Testing garlic for possible anti-ageing effects on longterm growth characteristics, morphology and macromolecular synthesis of human fibroblasts in culture. J Ethnopharmacol 43, 125-133. https://doi.org/10.1016/0378-8741(94)90009-4.

Thamer, K.G., and Penna, A.L. (2005). Efeito do teor de soro, açúcar e de frutooligossacarídeos sobre a população de bactérias lácticas probióticas em bebidas fermentadas. Revista Brasileira de Ciências Farmacêuticas, São Paulo 41.

Upadhyay, R.K. (2016). Nutraceutical, pharmaceutical and therapeutic uses of Allium cepa: A review. International Journal of Green Pharmacy (IJGP) 10.

Valacchi, G., Sticozzi, C., Belmonte, G., Cervellati, F., Demaude, J., Chen, N., Krol, Y., and Oresajo, C. (2015). Vitamin C compound mixtures prevent ozone-induced oxidative damage in human keratinocytes as initial assessment of pollution protection. PLoS One 10, e0131097. https://doi.org/10.1371/journal.pone.0131097.

Valacchi, G., Sticozzi, C., Belmonte, G., Cervellati, F., Demaude, J., Chen, N., Krol, Y., and Oresajo, C. (2015). Vitamin C compound mixtures prevent ozone-induced oxidative damage in human keratinocytes as initial assessment of pollution protection. PLoS One 10, e0131097. https://doi.org/10.1371/journal.pone.0131097.

Van den Ende, W. (2018). Novel fructan exohydrolase: unique properties and applications for human health. J Exp Bot 69, 4227-4231. https://doi.org/10.1093/jxb/ery268.

Van den Ende, W., and El-Esawe, S.K. (2014). Sucrose signaling pathways leading to fructan and anthocyanin accumulation: a dual function in abiotic and biotic stress responses? Environmental and experimental botany 108, 4-13. https://doi.org/10.1016/j.envexpbot.2013.09.017.

van den Heuvel, E.G., Muijs, T., Brouns, F., and Hendriks, H.F. (2009). Short-chain fructooligosaccharides improve magnesium absorption in adolescent girls with a low calcium intake. Nutr Res 29, 229-237. https://doi.org/10.1016/j.nutres.2009.03.005.

Venema, K. (2015). Probiotics and prebiotics (Caister Academic Press).

Verma, J., Kanoujia, J., Parashar, P., Tripathi, C.B., and Saraf, S.A. (2017). Wound healing applications of sericin/chitosan-capped silver nanoparticles incorporated hydrogel. Drug delivery and translational research 7, 77-88. https://doi.org/10.1007/s13346-016-0322-y.

Versluys, M., Kirtel, O., Toksoy Öner, E., and Van den Ende, W. (2018). The fructan syndrome: Evolutionary aspects and common themes among plants and microbes. Plant, cell & environment 41, 16-38. https://doi.org/10.1111/pce.13070.

Vig, K., Chaudhari, A., Tripathi, S., Dixit, S., Sahu, R., Pillai, S., Dennis, V.A., and Singh, S.R. (2017). Advances in skin regeneration using tissue engineering. International journal of molecular sciences 18, 789. https://doi.org/10.3390/ijms18040789.

Wack, M., and Blaschek, W. (2006). Determination of the structure and degree of polymerisation of fructans from *Echinacea purpurea* roots. Carbohydrate research 341, 1147-1153. https://doi.org/10.1016/j.carres.2006.03.034.

Wiegand, C., Raschke, C., and Elsner, P. (2017). Skin aging: a brief summary of characteristic changes. Textbook of aging skin, 55-65.

Wong, D.J., Chang, H.Y. (2009). Skin tissue engineering. In: StemBook, (Harvard Stem Cell Institute) DOI: 10.3824/stembook.1.44.1.

Wu, H., Ramachandran, C., Weiner, N.D., and Roessler, B.J. (2001). Topical transport of hydrophilic compounds using water-in-oil nanoemulsions. International Journal of Pharmaceutics 220, 63-75. https://doi.org/10.1016/S0378-5173(01)00671-8.

Wu, K.G., Li, T.H., and Peng, H.J. (2012). *Lactobacillus salivarius* plus fructo-oligosaccharide is superior to fructo-oligosaccharide alone for treating children with moderate to severe atopic dermatitis: a double-blind, randomized, clinical trial of efficacy and safety. British journal of dermatology 166, 129136. https://doi.org/10.1111/j.1365-2133.2011.10596.x.

Yildirimer, L., Thanh, N.T., and Seifalian, A.M. (2012). Skin regeneration scaffolds: a multimodal bottom-up approach. Trends in biotechnology 30, 638-648. https://doi.org/10.1016/j.tibtech.2012.08.004.

Zhang, W., Chen, L., Chen, J., Wang, L., Gui, X., Ran, J., Xu, G., Zhao, H., Zeng, M., and Ji, J. (2017). Silk fibroin biomaterial shows safe and effective wound healing in animal models and a randomized controlled clinical trial. Advanced healthcare materials 6, 1700121. https://doi.org/10.1002/adhm.201700121.

Levan as a Bioactive Material for Medical Applications

9

Sinem Selvin Selvi[1], Elif Piranlioğlu[1], Tuğçe Doğruel[1], Edina Eminagic[1], Muhammed Yusuf Kandur[1], Ebru Toksoy Öner[1*]

[1]IBSB, Department of Bioengineering, Marmara University, Istanbul, Turkey
*Corresponding Author: ebru.toksoy@marmara.edu.tr

DOI: https://doi.org/10.21775/9781912530267.09

Abstract

Levan is a fructan homopolymer with β-(2, 6) glycosidic bonds between its fructose monomers and can be obtained from plants and numerous microorganisms. This extracellular polysaccharide distinguishes from other natural polysaccharides with its unique properties and hence, its wide range of possible applications have been studied thoroughly in the past decade resulting with interesting findings related to its wound healing, heparin-mimetic, anti-cancer, anti-inflammatory, antiviral, non-toxic, antihyperlipidemic, hypocholesterolemic, hyperglycaemic inhibitory and anti-AIDS properties. Besides medical properties, it is also digestible by human gastrointestinal (GI) tract and shows prebiotic activity allowing it to be used in functional foods. This chapter will mainly focus on levan polysaccharide and its medical applications such as bioactive carrier in drug delivery systems and bioactive surfaces produced by laser technologies.

Levan

Levan is one of the most interesting homopolysaccharides that can be obtained from numerous microorganisms and a few plant species. This β-(2, 6) linked fructose biopolymer can have some β-(2,1) linked branches along its structure. Levan obtained from plants can have a molecular weight that can range from 2,000 to 33,000 Da whereas bacterial levan can weigh as much as 100 million Da. Microorganisms produce levan as exopolysaccharides (EPS) where it contributes to the matrix and biofilm formation while plants synthesize it as a multifunctional carbohydrate (Kırtel et al., 2018). In plants such as *Poa secunda*, *Agropyron cristatum*, wheat and barley levan can be located within the leaf sheaths and/or stems where it provides carbohydrates for grain filling within the growing season. Levan has a unique nature with biocompatible, anti-microbial, water soluble, adhesive, heparin-mimetic properties for medicine, food and cosmetic industries and studies regarding its industrial application potentials have increased in recent years (Öner et al., 2016; Erkorkmaz et al., 2018; González-Garcinuño et al., 2018; Kwon et al., 2018).

Levan in biomedical and cosmetic applications

Anti-tumour, anti-bacterial, anti-carcinogenic, anti-irritant, anti-viral, fibrinolytic activity, anti-oxidant, hyperglycaemic inhibition, anti-inflammatory, prebiotic, biocompatibility, biodegradability, flexibility, renewability, natural strong adhesive properties are only some of the properties that make levan a beneficial biopolymer that can be used in medicine, cosmetic and food industry (Figure 1). It can also be used as emulsifier, stabilizer, thickener, encapsulating agent, feed supplement for aquaculture, osmo-regulation, cryoprotectant, a plasma substitute, drug activity prolongator. These properties of levan are usually attributed to its ability to form solutions both in water and oil and its compatibility with various salts.

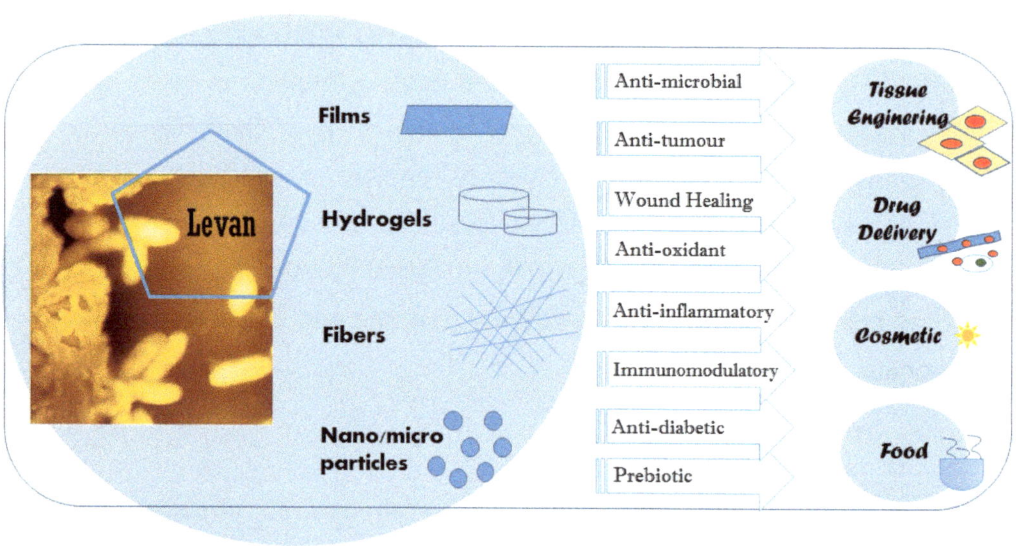

Figure 1. Significant properties of levan polysaccharide and its applications.

Nanotechnology is a significant field that affect many diverse areas including mechanics, electronics, optics, medicine, aerospace and material-biomedical engineering. The biological, physical and chemical properties of biomaterials have crucial roles in tissue engineering, drug delivery, diagnosis and wound dressing. Variety of structure that mimics extracellular matrix (ECM) formed by different bio-macromolecules and the way they are organized. An important feature of native ECM is the nanoscale dimensions of their physical structure. Recently, nanofiber scaffold systems are being researched substantially as scaffolds for tissue engineering. Nanofibrous biomimetic scaffolds are formed by biodegradable polymer nanofibers. The high surface area to volume ratio and their high porosity are two important properties of nanofibrous biomimetic scaffolds for aid to cell adhesion, proliferation, migration, and differentiation. Many approaches for the fabrication of nanofibers exist including electrospinning, self-assembly, synthesis of template-like scaffolds and thermal-induced phase separation and freeze-drying (Li et al., 2013; Park et al. 2017; Kenry and Lim 2017, Lee et al., 2018). However, nowadays electrospinning, self-assembly and thermal-induced phase separation are common techniques for the fabrication of biomimetic scaffolds.

With thermally induced phase separation (TIPS) technique highly porous scaffold can be produced. In this process, the polymer rich phase solidifies, and the polymer-poor phase is distracted, so producing polymer membranes with highly porous structures. Several natural and synthetic biodegradable polymers and composites such as PGA, PLLA, gelatine, alginate and chitosan have been produced using this procedure (Smith and Ma, 2004; Vasita and Katti 2006; Park et al. 2017). However, in the literature, there is no published research about levan nanofiber produced by TIPS.

Self-assembly is the autonomous organization of molecules into patterns or structures. Nano-scaled supramolecular structures and sometimes nanofibers can be produced from natural or synthetic macromolecules by self-assembly. However, there is no reported research about levan nanofibers *via* self-assembly as well.

Electrospinning is the most common technique used for the processing of polymeric biomaterials into nanofibers. This technique is simple, cost-effective and a broad range of polymers (e.g., polyolefin, polyamides, polyester, aramid, acrylic as well as biopolymers like proteins, DNA, polypeptides, or others like electric conducting and photonic polymer) are feasible for using electrospinning. Besides the nano-fibrous matrix, micro-fibrous structures can also be produced by electrospinning technique. This technique also has several advantages like controlling the thickness and composition of the nanofibers along with porosity of the nanofiber using a relatively simple experimental setup. In the electrospinning process, a syringe pump, a high voltage source, and a collector are needed. In electrospinning, high voltage between 5 and50 kV is applied to a capillary tube, which is held at the tip of the capillary via surface tension. A wide range of natural or synthetic polymers used in tissue engineering, drug delivery, wound dressing and other biomedical field have been fabricated into nanofiber via electrospinning (Ma et al., 2005; Vasita and Katti 2006; Park et al. 2017; Kenry and Lim 2017). In the literature, many diverse researches reported that examination of polysaccharide based on fibrous matrix fabricated by electrospinning and its applications.

Hydrogels are known as hydrated polymeric networks and comprised of chemically or physically cross-linked three-dimensional structures. They have high water content up to 90% and have ability to swell and de-swell in reversible ways. Due to their flexible, sensitive, water content and low foreign body response, they have been the main focus in medical studies (Jeong et al., 2017). Wichterle and Lim were the first scientists to use hydrogel as a material and apply cross-linked poly (2-hydroxyethyl methacrylate) in contact lens (Wichterle and Lim, 1960). First polymeric hydrogel structures were formed in 1980 by Lim and Sun and used for cell encapsulation (Lim and Sun, 1980). Nowadays, hydrogels are used in wound healing as a medical application for their various properties including 3D structures (El-Sherbiny et al., 2013; Bhardwaj et al., 2018).

Levan has been reported to biocompatible, biodegradable, anti-oxidant, anti-inflammatory characteristics and due to its effects, levan hydrogels have been developing for pharmaceutical, medical and cosmetic industry (Choi et al., 2018). Also, it shows notable characteristics in production of films being the most promising polysaccharide as biomaterial and in medical, pharmaceutical and nano-technological industries. One of the most important characteristics of levan polysaccharide is that it doesn't swell easily in water and it has low intrinsic viscosity. Films based on levan are shown good oxygen barrier properties. It has good heat stability and is relatively acid stable and levan can also absorb UV radiation, particularly in the UV-C range. (Sima et al., 2011; Öner et al., 2016; Gomes et al., 2018; González-Garcinuño et al., 2018). Significant applications of levan in last decade are given Table 1.

Table 1. Different applications of levan in last decade.

Bacterial species and strain	Application	Reference
Acetobacterxylinum NCIM2526	Anti-oxidant and Anti-inflammatory activities	Srikanth et al., 2015
Acinetobacter nectaris CECT 8127	Anti-tumor activity *via* drug delivery	Tabernero et al., 2017
Bacillus licheniformis	Anti-tumor activity	Dahech et al., 2012
	Anti-diabetic activity	Dahech et al., 2011
	Anti-oxidant activity	Dahech et al., 2013
Bacillus polymyxa. PTCC1020	Anti-bacterial activity	Taran et al., 2017
Bacillus spp	Anti-viral activity	Esawy et al., 2012
	Anti-microbial and prebiotic activities	Abdel et al., 2018
Bacillus subtilis M	Anti-tumour activity	Esawy et al., 2013
Bacillus subtilis NRC1aza	Anti-tumour and Anti-oxidant activities	Abdel-Fattah et al., 2012
	Anti-proliferative activity	Esawy et al., 2013
	Anti-proliferative activity	Queiroz et al., 2017
Halomonas smyrnensis AAD6T	Drug delivery and Antibiotic activity	Sezer et al., 2017
	Heparin mimetic activity and Anti-coagulant activity	Erginer et al., 2016
	Anti-cancer activity	Kazak Sarilmiser et al., 2014

Paenibacillus bovis sp. Nov BD3526	Immunomodular	Xu et al., 2016
Paenibacillus polymyxa EJS-3	Anti-oxidant and Anti-tumour activities	Liu et al., 2012
	Anti-oxidant activity	Liu et al., 2010
Pseudomonas syringae	Decrease toxicity	Bondarenko et al., 2015
Zymomonas mobilis	Wound healing	Sturzoiu et al., 2011

Levan and its anti-tumour activity

Despite the advanced technology and comprehensive researches, cancer is still a major cause of death worldwide. According to The World Health Organization, new cases of cancer will be increasing every year and it is very important to develop treatment for cancer that minimizes side effects. Various studies have suggested that polysaccharides can inhibit proliferation of tumour cells and tumour growth and levan is one of those polymers (Zong et al., 2012; Queiroz et al. 2017).

In 1997 and 2000, Calazans et al. in their studies have shown anti-tumour activity of levan produced by *Zymomonas mobilis* strains against sarcoma 180 in mice. Their results suggested that the anti-tumour activity of levan is associated to its molecular weight (MW) (Calazans et al. 1997; Glícia M.T Calazans et al. 2000). Another study examined the anti-tumour activity of levan produced by *Zymomonas mobilis, Microbacterium laevaniformans, Gluconoacetobacter xylinus* and *Rahnella aquatilis* against eight different tumour cell lines for *in vitro* assay and on sarcoma 180 for *in vivo* assay. The effectiveness of anti-tumour activities of levan showed a significant variation according to MW and its branch structure (Yoo et al. 2004; Yoon et al. 2004; Dahech et al. 2012). However, there is no comprehensive explanation to why the molecular weight and branching structure is important for anti-tumour activity.

Nowadays some studies are focused on different issues such that, the effects of modified levan on tumour growth or which method is used by levan for inducing that anti-tumour activity. Levan isolated from *B. subtilis* and its chemically modified forms were examined for the anti-tumour activity against five different tumour cell lines and the mechanistic role in the apoptosis pathway by Abdel-Fattah et al (2012). The result of this study showed that sulphated levan possessed a strong cytotoxic effect on hepatocellular carcinoma HepG2 cells. Moreover, treatment of HepG2 cells with sulphated *B. subtilis* levan induced apoptosis via mitochondrial pathway depending on DNA damaging and fragmentation (Abdel-Fattah et al. 2012). In another study it was revealed that acetylated levan (AL), phosphorylated levan (PL) and benzylated levan (BL) displayed higher anti-proliferative activity against human gastric cancer BGC-823 cells *in vitro* than native levan and its derivatives showed higher reducing power and scavenging activity against superoxide radical than native levan. It was suggested that the reason for this result could be the fact that acetyl, benzyl or phosphoryl groups in the modified levan increase electron-donating ability and affinity with the receptors on immune cells. The interaction of polysaccharides and immune cells trigger immunological responses, including anti-tumour activity (Liu et al. 2012; Kazak Sarilmiser et al., 2014). Also highlighted in this study, that the some physical characters such as water solubility and conformation are also important factors affecting the modified polysaccharide bioactivities (J. Liu et al. 2012). Investigation of anti-tumour activities of levan by *H. smyrnensis* AAD6 and its chemically modified form were tested against A549 (human lung adenocarcinoma), HepG2/C3A (human liver hepatocellular carcinoma), AGS (human gastric adenocarcinoma), and MCF-7 (human breast adenocarcinoma) cell lines. According to these results, aldehyde-activated levan exhibited higher anti-cancer activity than unmodified levan against all cancer cell lines while native levan showed the highest anti-cancer effect against AGS cells. Due to induction of apoptosis, decreased cancer cell viability were shown (Kazak Sarilmiser et al., 2014). In another study of *Halomonas smyrnensis* AAD6^T that investigated anti-proliferative and pro-apoptotic effects in MCF-7 breast cancer cells. The treatment of MCF-7 cells with the levan a time- and concentration-dependent anti-proliferative activity was observed. Gene expression of pro-apoptotic proteins p53 and p27 by levan increased apoptosis and oxidative stress (Queiroz et al. 2017).

In the literature, various studies have suggested that polysaccharides can show anti-tumour activity mediated through common three mechanisms: (1) direct cytotoxic effects on cancer cells, (2) activating different immune responses, (3) combination treatment with conventional anti-cancer drugs (Zong et al., 2012; Kazak Sarilmiser et al., 2014). In this line, one study showed that the immobilization of *Erwinia* L-asparaginase to levan from *Zymomonas mobilis* might have therapeutic potential for drug delivery system because of being enhanced the stabilization of anti-leukemic enzyme L-asparaginase (Vina et al., 2001).

In another work, Indocyanine green (ICG) was encapsulated in levan nanoparticles *via* self-assembly and the nanoparticles used for targeted breast cancer imaging. The nanoparticles accumulated in breast cancer *via* specific interaction between fructose moieties in levan and overexpressed glucose transporter 5 (Glut5), which is a transmembrane protein, in breast cancer cells. Glut5 is involved in transporting sugars and specifically interacts with fructose. Additionally, as other result of the study indicated that levan interacts with Glut5 more strongly than fructose (Kim et al., 2015). In the literature, it is known that cancer cells and tissues use more glucose and have overexpressed glucose transporters to obtain the energy they need (Liberti and Locasale 2016).

In consideration of these results of studies, it is safe to say levan is a promising and potential polysaccharide for developing polysaccharide-based anticancer therapeutics.

Levan and its anti-oxidant and anti-inflammatory activity

Oxidative stress and inflammation are common main reasons of all chronic diseases such as heart diseases, stroke, atherosclerosis, gastric ulcer, cancer and diabetes (Abdel-Fattah et al. 2012; Öner et al., 2016). Reactive oxygen species (ROS) and oxygen-derived free radicals are formed as a result of normal cellular metabolism or constitute exogenous factors and agents. ROS at high concentrations can contribute to a variety of pathological effects such as DNA damages, carcinogenesis and cellular degeneration. An imbalance between oxidant/ antioxidant molecules is named “oxidative stress” (Birben et al. 2012).

The recent studies on this field has focused on approaches that enhance the removal of ROS by using antioxidants. In this line the study which mentioned before, *Bacillus subtilis* levan and its sulphated derivative has shown to have potent free radical scavenging activity (Abdel-Fattah et al., 2012).

One study has examined that *A. xylinum* levan for its anti-oxidant and anti-inflammatory activities which showed anti-oxidant activity by increasing the activities of superoxide dismutase (SOD) and catalase (CAT) and have exhibited dose-dependent anti-inflammatory activity (Srikanth et al., 2015). Another research investigated antioxidant activities of levan produced by *Paenibacillus polymyxa* EJS-3. By result, both crude and purified levan showed strong scavenging activities on free radicals. Moreover it was suggested that the different structures of the polymer affected the antioxidant activities of levan as the results (Liu et al., 2009).

In a different study, bacterial levan was coated cerium oxide nanoparticles (LCNPs). Cerium oxide has a broad range of applications such as a catalyst, ultraviolet absorber and biomedical material because of its redox activities. According to the results, LCNPs have better radical-scavenging ability compared to other prepared polysaccharides coated with cerium oxide nanoparticles in neutral pH. However, the relationship between the structure of polysaccharide and antioxidant activity is not clear. Therefore, in this study it was suggested that levan has antioxidant activity because it contains many hydroxyl groups that can react with free radicals and terminate radical chain reactions. Additionally, it was reported that the antioxidant activity of LCNPs depend on pH and the concentration. Also the antioxidant and ROS reduction activities LCNPs in hydrogen peroxide-stimulated NIH3T3 (mouse embryonic fibroblast) cells was also examined in this study (Kim et al., 2016). In the light of these results, it can be said that levan coated cerium oxide nanoparticles can be used for the treatment of ROS-related diseases, such as cancer, cardiac diseases and Alzheimer's disease.

In another study about ROS-related diseases, the anti-oxidant effect of non-digestible levan for the possibility of treatment of cardiovascular disease and atherosclerosis was investigated. Cardiovascular diseases is the primary leading cause of death worldwide (Zhao et al. 2014), atherosclerotic disease causes more than 19 million deaths per year and oxidative stress is an important factor for this disease (Dahech et al. 2013). Researches hypothesized that hypercholesterolemia or more specifically high plasma low density lipoprotein cholesterol (LDL-C) is a crucial factor for development and progression of atherosclerosis because of the oxidative modification of low-density lipoprotein (LDL) and studies have shown that hypercholesterolemia generate atherosclerosis and enhance the risk of coronary heart disease (Anderson and Hanna 1999; Festi et al., 2004). Result of the study showed that, non-digestible levan increased the activity of two enzymes involved in oxidative stress; catalase and superoxide dismutase. It was also shown that levan through GI tract has good effect on the glucose metabolism and lipid profile (Dahech et al., 2013).

Several researches in literature have shown that natural or chemically modified polysaccharides containing sulphate groups in their structure exhibit various important biological properties including antioxidant (Xiong et al., 2011; Abdel-Fattah et al., 2012;

Sudharsan et al., 2015), immunomodulating (Xiong et al., 2011; Bayoumi 2012) antithrombotic and anticoagulant activities (Pomin and Mourão 2014; Erginer et al. 2016). Heparin is a highly sulphated glycosaminoglycan that has use as a therapeutic agent. Proteoglycans, such as heparin and heparan sulphate, are known to affect basic biological processes by binding to growth factors, to the sugar recognition domains on selectins or by interacting with extracellular matrix proteins (Silva et al. 2015). Furthermore heparin reduce inflammatory response, facilitate angiogenesis and moderate smooth muscle cell proliferation (Fu et al., 2016; Casu et al., 2015).

The study of sulphated levan from *Halomonas smyrnensis* showed that the levan had heparin mimetic activity and suggested that the sulphated levan can be used in cardiac tissue engineering as a promising biomaterial. In this work, anticoagulation activity of levan was detected by prothrombin time (PT), activated partial thromboplastin time (APTT), and thrombin time (TT) kits and low molecular weight heparin was used as control. By results of the study, native levan did not display any anticoagulation activity whereas sulphated levan was dose-dependent. Additional findings in this study, different sulfation degrees of levan samples have affected the anticoagulant activity. In other terms, the anticoagulant activity had a positive correlation with increasing sulfation degrees of levan. Also heparin equivalency of sulphated *Halomonas* levan compared with the previous studies of other polysaccharide in literature, was found to be close to the sulphated fructan derivatives of *Punctaria plantaginea* (Ustyuzhanina et al., 2016), galactans of *Gracilaria debilis* (Sudharsan et al., 2015) and rhamnan of *Monostroma latissimum* (X. X. Liu et al. 2011). Sulfated *Halomonas* Levan (SHL) exhibited better results than most of other polysaccharides from different origins. Antithrombin III inactivates thrombin and other coagulation proteases like factors Xa, IX, XIa and XIIa in the coagulation cascade as a plasma serine protease and heparin shows its anticoagulation activity *via* interaction with this protease. This study also investigated that the possible action mechanism of SHL. As a result, SHL and heparin have decreased Factor Xa activity with increasing concentration. In addition, the docking study results proposed that high thrombin inhibition of SHL at high concentrations, most likely to binding multiple sites on thrombin and pentasaccharide binding site (PBS) of antithrombin. Accordingly, the cytotoxicity assay of this study, SHL was exhibited high cell viability in mouse fibroblasts L929. (Erginer et al., 2016).
In conclusion, sulphated levan may be considered as an outstanding biomaterial for skin tissue engineering, wound dressing and drug delivery applications having basic properties such as anti-oxidant, anti-inflammatory, immunomodulatory and heparin mimic activity.

Levan and immunomodulatory activity

A study has been published based on purified levan from variety of sources to induce an immune response in human subjects and also the immunochemical behaviour defining structural differences among levans was reported (Allen 1957). *Aerobacter* levan indicated immunostimulatory activities in human body. Lately, the study reported that, while the levan synthesized by *Bacillus licheniformis* 8-37-0-1 significantly stimulated the spleen lymphocytes proliferation, which is a crucial step in the activation cascade of both cellular and humoral immune responses. Another study observed that the levan synthesized by *Paenibacillus bovis* sp. nov BD3526 significantly stimulated the spleen cells proliferation and induced the expression of TNF-α, which is a cell signalling protein. Its main role is in the regulation of immune cells, and it can induce apoptotic cell death, inflammation and can inhibit tumorigenesis by defining its potential as a natural immunomodulator. (Liu et al. 2010; Xu et al. 2016). Another study on this topic showed that no significantly elevation in TNF-α level in *Bacillus subtilis* sulphated Levan compared with the control (Bayoumi 2012).

The anti-microbial effect of levan is an important property for developing levan polysaccharide-based biomaterials in medical applications such as tissue engineering and wound dressing.

Levan and anti-diabetic activity

Nowadays, diabetes mellitus especially type 2, is a global health problem causing significant morbidity and mortality. Mechanisms underlying diabetes are not fully understood despite many investigations hinting towards the importance of inflammation and oxidative stress. The oral drugs have common side effects including hypoglycaemia, weight gain, and gastrointestinal side effects such as abdominal pain, nausea, vomiting and bloating. However, the polysaccharide derived from natural sources like animal, plant and microorganism, are considered less toxic than synthetic drugs. Oxidative process plays a crucial role in the development of some biological process like β-cell dysfunction and impairment of glucose tolerance, which can cause diabetes. Due to the relationship of diabetes and oxidative process, researches suggested that the polysaccharides having antioxidant activity might be used for diabetes treatment as an anti-diabetic agent (Dahech et al. 2011b; J. Wu et al. 2016).

For this purpose, Belghith et al., examined the presence of antidiabetic effect of microbial levan. One study investigated the effects on diabetic rats with treated with *Bacillus licheniformis* levan. Application of levan in rats provoked a significant increase in glycogen level by 52% and a decrease in glucose level in plasma by 52%. In addition, treatment of rats with levan caused increasing the activities of superoxide dismutase (SOD) and catalase (CAT) and the study showed that levan restored abnormal indicator of oxidative stress. This study has verified that levan has anti-diabetic activity for the first time (Dahech et al. 2011a) Another study on this topic investigated the possibility of hypolipidemic and anti-oxidative effects of levan in rats fed with a high-cholesterol diet. According to the results, after the treatment period (60 days), the determined plasma antioxidant enzymes and lipid profile (total cholesterol, HDL-cholesterol, LDL-cholesterol and triglycerides) activities was positively changed by levan treatment. Results of the study also showed that the levan supplemented-diet increased the concentrations of serum HDL-C and decreased the concentrations of serum LDL-C. One possible mechanism of hypocholesterolemic effect of levan is that reduced intestinal absorption of steroids could lead to a lowering of the serum cholesterol concentration. Some studies suggest that levan may bind or entrap the steroids in the intestine and disturb their reabsorption; this mechanism was supported in this study. (Belghith et al. 2012).

In a different work investigated that the effect of bacterial levan on oxidative stress and hyperglycaemia in alloxan-induced diabetic rats. The results showed that the oral approach of levan in diabetic rats caused a decrease in glucose level in plasma and an increase of (SOD), (CAT) and glutathione peroxidase (GPX) activities in both pancreas and liver. Moreover, this study demonstrated that treatment of levan caused a decrease in the indicators of hepatic and pancreatic toxicity including alkaline phosphatases (ALP), aspartate and lactate transaminases (AST and ALT), lactate dehydrogenases (LDH) activities and the thiobarbituric acid-reactive substances (TBARs) in diabetic rats. In this study, three mechanisms were suggested for the ability of levan polysaccharide to reduce the blood glucose level: (1) decrease β-cells damage as more insulin secretion due to its ability to modulate immune system, (2) prevent and decrease the toxicity caused by oxygen-derived free radicals in pancreatic β-cells and (3) improve the intestinal microflora production of short-chain fatty acids (SCFA), that inhibited glucose absorption in intestine (Dahech et al. 2011b).

Levan and anti-virus and anti-microbial activity

Esawy et al. investigated 16 bacterial honey isolates for levansucrase production and reported novel *Bacillus* spp honey isolates as a new source of very important enzymes such as levansucrase. Some of the six *Bacillus* spp had weak effect on adenovirus 40 while each of these samples showed an antiviral effect on H5N1 virus (Esawy et al., 2012).

Antibiotic susceptibility of bacteria, which is another parameter for bacterial safety, were assessed and results showed that the three isolates were sensitive to different kinds of antibiotics.

Three bacterial strains displayed different antimicrobial activity against 13 indicator microorganisms, which are pathogens responsible from human and agricultural diseases. The results indicated a wide range of antimicrobial activity by the bacterial isolates which indicates their potential as sources of novel antimicrobial compounds (Abdel Wahab et al., 2018). Another work focused on anti-bacterial effect of levan produced by *Bacillus polymyxa* reported that the produced nanocomposites containing magnetic nanoparticles of iron and levan biopolymer had antibacterial effects on both *Escherichia coli* and *Staphylococcus aureus* (Taran et al., 2017). *Escherichia coli* and *Staphylococcus aureus* are two main bacteria which cause urinary tract, blood stream infections and wound, bloodstream infections, respectively (World Health Organization, WHO Press, Geneva, n.d.).

Levan and tissue engineering

The main purpose of tissue engineering is to create tissue-like structures by using living cells, biochemicals, biocompatible materials and also a combination of these components. These materials are used to repair injuries or replace malfunctional body parts such as failing organs (Berthiaume et al., 2011). It is important to choose suitable biocompatible materials wisely for tissue engineering applications and levan is one of the important polysaccharides to be used in such applications (Jahangirian et al., 2018).

A group of scientists produced levan films on the nanometric scale. Nanometric scale film method reduces the cost and increases the specific surface area. In their research, they produced films where matrix-assisted pulsed laser evaporation was extended to obtain nanostructured thin films of levan and oxidized levan under vacuum conditions. Matrix-assisted pulsed laser evaporation is the technique where laser-induced material drop out is made backward from solid cryogenic targets, and the ejected material is assembled on specific collectors. Targets are prepared from organic complex dissolved into a laser wavelength absorbing solvent when frozen. This method allows production of microarrays and nanoarrays of biomaterials. Their applications include biochemical analysis, pharmaceutical industry, biosensing, etc. Material that they produced, gave good results with a compact structure, good adhesion to substrate, and a uniform, homogeneous nanostructured surface. High specific surface area was exhibited full compatibility with their potential use in biology or medicine. Furthermore, cell viability and proliferation were increased indicating its suitability for tissue engineering applications (Sima et al., 2011).

Sturzoiu et al. investigated the relation between *Zymomonas mobilis* levan and matrix metalloproteinase (MMP) activity. MMPs play a significant role in tissue modelling and wound healing. MMPs are being thought to be involved in cell proliferation, migration, angiogenesis and host defence. In this research effect of levan polysaccharide on Wistar rat lesions caused by mechanical process on wound and burn were studied. As a result, it was shown that MMP-2 and MMP-9 activity increased after levan treatment with speleothreapy for two weeks. It was a significant study that showed levan to be a good candidate for wound healing in skin tissue engineering due to its ability in preventing infections above epithelial tissue and cellular epithelial proliferation. According to these effects levan treatments increase the wound healing effect and decrease the healing time. (Sturzoiu et al., 2011).

Researchers studied on the adhesive properties of phosphonated levan in order to see if it can be used in electrostatic LbL adsorption to construct nanostructured films. Phosphonated levan and chitosan were assembled onto glass using in-house developed dipping robot. Phosphorylation is a good method for modifying properties and bioactivities of polysaccharides. Phosphorylated polysaccharides have found use in various applications. Such modified levan has shown good properties for the fabrication of adhesive multilayer coatings so they have been investigated within the scope of this study. Relying on electrostatic interactions, hybrid films produced from these two biopolymers were compared with a formulation of chitosan and alginate, two natural ingredients widely regarded as good adhesives. Furthermore, cells were sensitive to this polysaccharide and adhered in high number. Researchers claimed that produced films provide enhanced adhesiveness and protection (Costa et al., 2013).

Another group of researchers studied interaction of chitosan, PEO and levan by producing ternary blend films. Solutions were prepared by casting method and their phase behaviour, miscibility; thermal and mechanical properties as well as their surface energy and morphology were characterized by different techniques. FTIR spectroscopy showed that blend components exhibited significant interactions with each other via hydrogen bonding when PEO component in blends was less than 26.1%. Furthermore, chitosan and levan suppress PEO crystallinity enhancing mechanical and thermal properties of the blend samples. Cell viability and proliferation studies confirmed the biocompatible behaviour of films (Bostan et al., 2014).

Levan polysaccharides combined with methacrylate groups via ester or urethane bonds and photochemically crosslinked with different photoinitiator systems were also investigated. It was concluded that these systems synthesized mechanically stable and slightly degradable cytocompatible levan based hydrogels. Due to their solubility, cytocompatiblity and stability these hydrogels were considered as a new functional biomaterial in advanced (Berg et al., 2017).

There are only two studies regarding fiber formation of levan. One of these studies investigated the parameters for fibers by electrospinning of levan dissolved in water. According to results, the fiber formation was achieved at high concentration of levan and showed that fiber diameter decreased with increased voltage and decreased pump flow rate (Manandhar et al., 2009). In other study, researchers examined levan fibrous matrix by two different electrospinning techniques. Hydrolyzed (hHL) and sulphated hydrolysed (ShHL) *Halomonas* levan were chemically synthesized and used together with polycaprolactone (PCL) and polyethyleneoxide (PEO). PEO and PCL were used for overcoming the poor molecular flexibilities and insufficient entanglement of levan in aqueous solution. According to the results of the study, hHL fibrous matrix showed an increase in cell viability for both L929 (murine fibroblasts) and HUVECs (human umbilical vein endothelial cells). As ShHL was compared with hHL, ShHL exhibited higher cell viability for HUVECs. On the contrary, the presence of ShHL matrix inhibited L929 fibroblast growth. The reason of this inhibition is the heparin like activity of ShHL. Moreover, the result showed that antithrombogenic activity of ShHL fibrous matrices was dependent on ShHL concentration. According to all these results that cell viability assay and fluorescence imaging of L929 fibroblasts and HUVECs as well as heparin mimetic activity of ShHL electrospun fibers indicated their high potential to be used in decreasing neointimal proliferation and thrombogenicity of grafts and prosthesis and also cardiac tissue engineering (Avsar et al. 2018).

Modified levan has also been used to try to produce films. Films based on alginate, chitosan and sulfated levan were produced using layer by layer technique (Gomes et al., 2018). The introduction of sulfated levan gave a chance to produce films with enhanced mechanical properties that required a higher tensile stress in order to rupture and higher shear strength for detachment when compared to the widely used control membranes composed of chitosan and alginate. As a result, it was stated that sulphated levan-containing films have improved properties, such as higher stiffness, surface roughness, higher mechanical resistance and enhanced cell viability and proliferation. Such films are very promising, and the use of a levan-based polymer is exciting for film production in tissue engineering.

Despite these exciting activities and properties, there has been no report about dermal filler hydrogels based on levan until 2018 and this system is thought to be an excellent alternative to hyaluranic acid (HA) based dermal fillers (Choi et al., 2018). Due to high cost of HA, low cost production of levan made it the centre of attention for hydrogel formations. In this study, the researchers developed a levan based dermal filler hydrogel for soft tissue augmentation. They prepared injectable levan hydrogels with pluronic and carboxymethyl cellulose. Hydrogels were formed after increasing temperature up to 37°C from 4°C sol state. It was concluded that elastic modulus of levan was higher than HA-based hydrogel and interconnected porous structures were similar to each other. *In vivo* studies showed that levan enhances the collagen synthesis, is more stable than HA and Pluronic F127 hydrogel and its anti-wrinkle efficiency is higher than standard HA hydrogel. Therefore, levan based dermal fillers are a very promising alternative for overpriced HA-based hydrogels (Choi et al., 2018).

In another study, Combinatorial Matrix-Assisted Pulsed Laser Evaporation (C-MAPLE) technique was applied to levan and chitosan polymers in order to produce compositional libraries of these biopolymers in a single-step process. Scientists used two distinct cryogenic targets, one consisting of sulphated HL (SHL) and the other of quaternized low molecular weight chitosan to produce compositional gradient coatings for surface functionalization. Morphological study demonstrated that surface properties have been enhanced by these thin polymer coatings. Polymer coatings has shown antimicrobial effect on infections caused by *Escherichia coli* and *Staphylococcus aureus*. This nanostructured coating prevented the adherence properties of bacterial biofilms, enhanced the surface properties and proliferation assays demonstrated that the obtained thin films are biocompatible making this polymer blend suitable for broad range of biomedical and tissue engineering applications (Mihailescu et al., 2018).

Recent study was focused on water soluble levan micro-gel formation and levan was obtained from recombinant levansucrase production by *Erwinia amylovora.* Hight molecular weight levan (2.017 x107 g/mol) showed gel forming strength. Gelation of *Erwinia* levan compared with xanthan, carrageenan and guar gums for their textural property. Gel forming ability was inadequate than other polymers but %8 concentrated levan was shown similar properties to 2% guar and 4% xanthan gum. Thereby, water soluble levan hydrogel is thought to be used in food industry for gelling and thickening material. Until 2% concentration, levan showed Newtonian fluid characteristics but on the contrary, it showed non-Newtonian pseudoplastic fluid properties at 8%. Overall, after these significant results levan micro-gels revolved to be used in tissue engineering and cosmetic industry with further studies (Peng et al., 2018).

These studies have shown that high adherence and biocompatible properties of levan make this polysaccharide a very significant candidate for tissue engineering applications.

Levan and Drug Delivery Systems

As smarter technologies emerge, smarter drug delivery methods are being developed. Using biological exopolysaccharides in such systems have numerous advantages such as biodegradability, biocompatibility, ease of production with alterable parameters, bio-waste production, etc. Levan itself has additional beneficial properties to the aforementioned advantages such as high solubility in water, strong adhesiveness and film forming. Nanocarriers obtained from levan with varying particular size, charge and release profiles exhibited sufficient prolongation of BSA protein release hinting towards a successful drug delivery system that can be obtained in the future (Sezer et al. 2011). About six years later, Sezer et al encapsulated vancomycin (VANCO) into biodegradable and biocompatible microparticles of *Halomonas* levan. VANCO is a glycopeptide antibiotic that affects gram positive bacteria (2017). There are some limitations like slow intravenous infusion of this drug due to poor absorption from gastrointestinal tract so; researchers develop several drug delivery systems for this antibiotic. Drug encapsulation capacity differs from 47.7% to 74.7% and drug release was between 10% and 20% for the first 5 hours. After 5 hours, the release was rapid and slow release properties were achieved by increasing levan concentrations. All encapsulated drugs were shown to have antimicrobial effect on *Bacillus subtilis* ATCC 6633. It was concluded that levan microparticles can be used for antibiotic drug release systems (Sezer et al., 2017). In same year, levan was used as crosslinker to synthesize levan/N-isopropyl acrylamide (levan/pNIPA) hydrogels for controlled release of 5-Aminosalycylic acid (5-ASA). 5-ASA can be used for treatment of inflammatory bowel diseases and to overcome gastric irritation during oral use with fat absorption in small intestine, this drug was encapsulated within the hydrogels. It was shown that almost all of the drug was released within 6 hours and *in vitro* results indicated that biocompatibility was increased on L929 fibroblasts with increasing of levan concentration in hydrogels (Osman et al., 2017).

In another study, levan polysaccharide was studied as a new candidate for versatile transient device/system applications. Evaluations on *in vivo* tests illustrated that the possibility of potential applications in implantable devices and drug delivery vehicles. Results of *in vitro* and *in vivo* studies gave proof of the biocompatibility and bioresorption of material. Using NIR irradiation programmable transience was determined, suggesting an approach to trigger life span of transient electronic system for relevant promising usages such as a drug delivery vehicle and/or a self-killing system (Kwon et al., 2018). Studies with levan on drug delivery systems have been increasing since 2000s and potential of this natural polysaccharide has been discovered in every field of biomedicine including its potential as a drug coating material.

Drug tablet coatings are generally made of polymer and polysaccharide, containing pigments and plasticizers. Researchers have made thick films of levan films by extrusion and molding, by adding glycerol as a plasticizer. Levan was blended with glycerol and then was thermally processed and extruded to create a film. It was seen that a 20% weight of glycerol is needed to achieve an efficient plasticization to obtain films on micrometric scale (Barone and Medynets, 2007). A different group of researchers created levan films by using montmorillonite and these films had a significant potential as drug coatings. In this case, the montmorillonite surface oxygen was attached to the levan hydroxyl groups, conferring a thermal stability to the polymer and an increase in the glass transition temperature depending on the montmorillonite composition. Results showed that 1% of montmorillonite increases the glass transition temperature up to 33°C. Young's modulus also exhibited an improvement with the mineral composition, providing up to a 480% increase in comparison with the original levan polymer and study showed that levan film is a promising coating material (Chen et al., 2014). These levan coatings can be used for different applications aside drug coating by forming desired shapes and areas. This can be achieved by different methods such as solvent casting or thermal processing. Solvent casting is a process of making thermoplastic polymer samples by dipping a shape into a solution of the polymer and drawing off the solvent to leave a polymer film adhering to the mould (Prasad et al., 2017).

Levan and cosmetic

Levan can be used in cosmetic industry beside all the mentioned biomedical applications due to its properties like film forming, biocompatibility, moisturizing, cell proliferation and skin protection. It was concluded that moisturizing effect of levan is as same as hyaluronic acid and cell proliferation effect has similar with in human fibroblast and keratinocyte cell lines (Kim et al., 2005). According to patent in 2005, combination of levan and ascorbic acid improves oxidation stability of ascorbic acid and this result is very important for long shelf life of products with vitamin C (Kim et al., 2005b). Darkening of the skin is called hyperpigmentation and it is occurred due to high amount of dark brown melanin. According to another patent in 2006, it was concluded that levan has a 'whitening effect', as it prevents high melanin formation by slowing the tyrosinase activity (Furukawa and Tsuboi, 2006). Levan is also a non-ionic polymer so it is compatible with other chemical ingredients and polymers therefore, surfactants based on levan with cationic substituents can be developed to use in conditioning shampoo (Gunn et al., 2009). In one study, tensile strength of damaged hair was enhanced by using carboxymethylated levan and lifting of cuticle layers caused by bleaching has been decreased by using derivatized levan (Kim et al., 2012). Levan can be used as hair holding ingredient in hair care products due to film forming properties. As levan has low viscosity, it does not affect the physical properties of finished products and it can be used in pump spray and aerosol technology as a hair spray resin (Carson, 2015).

Also PoLevan S and PROTEOLEA® are the commercial products that contain levan polysaccharide on the market nowadays. PROTEOLEA® acts as an anti-aging product and activates the proteasome as it enhances the stress resistance (Choi et al., 2018). It slows down the skin aging and renews the eye area within 4 weeks.

Levan in food applications

Another application area of levan is food industry and there are still many unexplored potential application areas of this polysaccharide. In fields of food levan can be used as a sweetener, emulsifier, stabilizer and thickener and it can enhance the shelf life of food products. Levan can also be used to develop mouth feel taste and spreadability of dairy products as fructans have fat-like properties (Booten et al., 1998). Due to high molecular weight of levan, it is hard to detect levan by tasting, and it can be replaced by fat because it has low volatility to exude detectable odor. Levan is used as a prebiotic in functional food as yoghurt (Xiao et al., 2014). It is suggested that, using levan is favorable to obtain more stable foam, aerosol and emulsion in colloidal food systems, such as fermented cereal, milk and juices (Mays and Dally, 1989). Levan and levan type fructooligosaccharides have a role on beneficial bacteria and complex intestinal microbiota proliferation (Öner et al., 2016). These qualities make levan a very significant prebiotic. Conversion description of levan in cooperative metabolism system with gut bacteria is performed by supplying metabolic scheme with enriching levan with *Bacteroids, Escherichia, Streptococcus and Faecalibacterium* (Adamberg et al., 2015). According to a study performed in 2016, levan in large quantities and with high molecular weight, from acetic acid bacteria, can be useful for gluten free bread improvement. Levan producing acetic acid bacteria on buckwheat sourdough breads, organic acid and pH affected specific volume and crumb hardness while levan was affecting structural features of buckwheat breads (Ua-Arak et al., 2016). Studies concluded that beneficial properties of levan is applicable in many industries especially biomedical and other applications are also progressing day by day including food industry.

Conclusion

As a biopolymer, levan has come a long way; when it was first discovered it was considered as a mere unwanted side product but today numerous applications of levan is found and even more are being explored. Synthesis of levan through microorganisms may be a potential application in a vast majority of industries in the future but today it is not completely developed. Selection of microorganism type is a crucial parameter in obtaining the required amount of levan. Production yield should be efficient enough to obtain the desired value in the shortest amount of time possible, but the organism should not be hard to obtain and/or cultivate. For this reason, scientists are researching microorganisms and optimization methods to obtain levan and modify it in different ways so that it can have additional properties. Anti-coagulant, adhesiveness, prebiotic effect, anti-tumour, anti-irritant, anti-inflammatory, anti-cancer, anti-bacterial, anti-viral properties make it possible to apply levan into smart drug delivery systems, wound-dressings, film packaging and as food and drugs (Öner et al., 2016). Many more can be found as scientists keep exploring different ways to synthesise levan and modify it depending on their objectives for numerous industry applications.

References

Abdel-Fattah, A. M. *et al.* (2012) 'Antitumor and antioxidant activities of levan and its derivative from the isolate Bacillus subtilis NRC1aza', *Carbohydrate Polymers*. Elsevier, 89(2), pp. 314–322. doi: 10.1016/J.CARBPOL.2012.02.041.

Abdel Wahab, W. A., Saleh, S. A. A., Karam, E. A., Mansour, N. M., & Esawy, M. A. (2018). Possible correlation among osmophilic bacteria, levan yield, and the probiotic activity of three bacterial honey isolates. Biocatalysis and Agricultural Biotechnology, 14, 386–394. https://doi.org/10.1016/j.bcab.2018.04.006

Adamberg, K., Tomson, K., Talve, T., Pudova, K., Puurand, M., Visnapuu, T., Alamäe, T., and Adamberg, S. (2015). Levan enhances associated growth of Bacteroides, Escherichia, Streptococcus and Faecalibacterium in fecal microbiota. PloS one *10*, e0144042. https://doi.org/10.1371/journal.pone.0144042

Allen, P. Z. (1957). STUDIES ON THE CAPACITY OF SOME POLYSACCHARIDES TO ELICIT ANTIBODY FORMATION IN MAN. Journal of Experimental Medicine, 105(5), 383–394. https://doi.org/10.1084/jem.105.5.383

Anderson, J. W., & Hanna, T. J. (1999). Impact of Nondigestible Carbohydrates on Serum Lipoproteins and Risk for Cardiovascular Disease. The Journal of Nutrition, 129(7), 1457S–1466S. https://doi.org/10.1093/jn/129.7.1457S

Avsar, G., Agirbasli, D., Agirbasli, M.A., Gunduz, O., and Oner, E.T. (2018). Levan based fibrous scaffolds electrospun via co-axial and single-needle techniques for tissue engineering applications. Carbohydrate Polymers *193*, 316-325. https://doi.org/10.1016/j.carbpol.2018.03.075

Barone, J.R., and Medynets, M. (2007). Thermally processed levan polymers. Carbohydrate polymers *69*, 554-561. https://doi.org/10.1016/j.carbpol.2007.01.017

Bayoumi, F. (2012). Immunomodulatory and Chemo preventive activity of Bacillus subtilis sulphated Levan. Life Science Journal, 9(1097–8135).

Belghith, K.S., Dahech, I., Hamden, K., Feki, A., Mejdoub, H., and Belghith, H. (2012). Hypolipidemic effect of diet supplementation with bacterial levan in cholesterol-fed rats. International journal of biological macromolecules *50*, 1070-1074. https://doi.org/10.1016/j.ijbiomac.2012.02.024

Berg, A., Oner, E.T., Combie, J., Schneider, B., Ellinger, R., Weisser, J., Wyrwa, R., and Schnabelrauch, M. (2017). Formation of new, cytocompatible hydrogels based on photochemically crosslinkable levan methacrylates. International journal of biological macromolecules. https://doi.org/10.1016/j.ijbiomac.2017.10.120

Berthiaume, F., Maguire, T. J. & Yarmush, M. L. (2011). Tissue engineering and regenerative medicine: history, progress, and challenges. *Annu Rev Chem Biomol Eng,* 2**,** 403-30.

Bhardwaj, N., Chouhan, D., and Mandal, B.B. (2018). 3D functional scaffolds for skin tissue engineering. In Functional 3D Tissue Engineering Scaffolds (Elsevier), pp. 345-365. https://doi.org/10.1016/B978-0-08-100979-6.00014-8

Birben, E., Murat, U., Md, S., Sackesen, C., Erzurum, S., & Kalayci, O. (2012). Oxidative Stress and Antioxidant Defense. WAO Journal, 5(January), 9–19. https://doi.org/10.1097/WOX.0b013e3182439613

Bondarenko, O. M., Ivask, A., Kahru, A., Vija, H., Titma, T., Visnapuu, M., … Alamäe, T. (2015). Bacterial polysaccharide levan as stabilizing, non-toxic and functional coating material for microelement-nanoparticles. Carbohydrate Polymers, 136, 710–720. https://doi.org/10.1016/j.carbpol.2015.09.093

Booten, K., De Soete, J., and Frippiat, A. (1998). Fructan-and/or polydextrose-containing dairy powders, process for preparing same and their use. Eur patent appl EP0821885A1.

Bostan, M. S. *et al.* (2014) 'Comprehensive characterization of chitosan/PEO/levan ternary blend films', *Carbohydrate Polymers*. Elsevier, 102, pp. 993–1000. doi: 10.1016/J.CARBPOL.2013.09.096.

Calazans, G. M. T., Lopes, C. E., Lima, R. M. O. C., & De França, F. P. (1997). Antitumour activities of levans produced by Zymomonas mobilis strains. Biotechnology Letters, 19(1), 19–21. https://doi.org/10.1023/A:1018350617120

Calazans, G. M. . *et al.* (2000) 'Molecular weight and antitumour activity of Zymomonas mobilis levans', *International Journal of Biological Macromolecules*. Elsevier, 27(4), pp. 245–247. doi: 10.1016/S0141-8130(00)00125-2.

Carson, J. 2015. Available from URL: http://carsonproductdevelopment.com/

Casu, B., Naggi, A., & Torri, G. (2015). Re-visiting the structure of heparin. Carbohydrate Research, 403, 60–68. https://doi.org/10.1016/j.carres.2014.06.023

Choi, W.I., Hwang, Y., Sahu, A., Min, K., Sung, D., Tae, G., and Chang, J.H. (2018). An injectable and physical levan-based hydrogel as a dermal filler for soft tissue augmentation. Biomaterials science. doi:
10.1039/C8BM00524A

Costa, R.R., Neto, A.I., Calgeris, I., Correia, C.R., Pinho, A.C., Fonseca, J., Öner, E.T., and Mano, J.F. (2013). Adhesive nanostructured multilayer films using a bacterial exopolysaccharide for biomedical applications. Journal of Materials Chemistry B *1*, 2367-2374.doi:
10.1039/C3TB20137F

Dahech, I., Belghith, K. S., Hamden, K., Feki, A., Belghith, H., & Mejdoub, H. (2011a). Antidiabetic activity of levan polysaccharide in alloxan-induced diabetic rats. International Journal of Biological Macromolecules, 49(4), 742–746. https://doi.org/10.1016/j.ijbiomac.2011.07.007

Dahech, I., Belghith, K. S., Hamden, K., Feki, A., Belghith, H., & Mejdoub, H. (2011b). Oral administration of levan polysaccharide reduces the alloxan-induced oxidative stress in rats. International Journal of Biological Macromolecules, 49(5), 942–947. https://doi.org/10.1016/j.ijbiomac.2011.08.011

Dahech, I., Belghith, K. S., Belghith, H., & Mejdoub, H. (2012). Partial purification of a Bacillus licheniformis levansucrase producing levan with antitumour activity. International Journal of Biological Macromolecules, 51(3), 329–335. https://doi.org/10.1016/j.ijbiomac.2012.04.030

Dahech, I., Harrabi, B., Hamden, K., Feki, A., Mejdoub, H., Belghith, H., & Belghith, K. S. (2013). Antioxidant effect of nondigestible levan and its impact on cardiovascular disease and atherosclerosis. International Journal of Biological Macromolecules, 58, 281–286. https://doi.org/10.1016/j.ijbiomac.2013.04.058

El-Sherbiny, I.M., and Yacoub, M.H. (2013). Hydrogel scaffolds for tissue engineering: Progress and challenges. Global Cardiology Science and Practice, 38. https://doi.org/10.5339/gcsp.2013.38

Erkorkmaz, B.A., Kırtel, O., Ateş Duru, Ö., and Toksoy Öner, E. (2018). Development of a cost-effective production process for Halomonas levan. Bioprocess and Biosystems Engineering. https://doi.org/10.1007/s00449-018-1952-x

Erginer, M. *et al.* (2016) 'Sulfated levan from Halomonas smyrnensis as a bioactive, heparin-mimetic glycan for cardiac tissue engineering applications', *Carbohydrate Polymers*. Elsevier, 149, pp. 289–296. doi: 10.1016/J.CARBPOL.2016.04.092.

Esawy, M. A. *et al.* (2011) 'Production of levansucrase from novel honey Bacillus subtilis isolates capable of producing antiviral levans', *Carbohydrate Polymers*. Elsevier, 86(2), pp. 823–830. doi: 10.1016/J.CARBPOL.2011.05.035.

Esawy, M. A., Ahmed, E. F., Helmy, W. A., Mansour., N. M., El-Senousy, W. M., & El-Safty, M. M. (2012). Antiviral Levans from Bacillus spp. Isolated from Honey. The Complex World of Polysaccharides, 195–214. https://doi.org/10.5772/52483

Esawy, M. A. *et al.* (2013) 'Scaling up, characterization of levan and its inhibitory role in carcinogenesis initiation stage'', *Carbohydrate Polymers*. Elsevier, 95(1), pp. 578–587. doi: 10.1016/J.CARBPOL.2013.02.079.

Festi, D., Colecchia, A., Sacco, T., Bondi, M., Roda, E., & Marchesini, G. (2004). Hepatic steatosis in obese patients: clinical aspects and prognostic significance. Obesity Reviews, 5(1), 27–42. https://doi.org/10.1111/j.1467-789X.2004.00126.x

Franken, J. *et al.* (2013) 'Biosynthesis of Levan, a Bacterial Extracellular Polysaccharide, in the Yeast Saccharomyces cerevisiae', *PLoS ONE*. Edited by P. A. Cobine. Public Library of Science, 8(10), p. e77499. doi: 10.1371/journal.pone.0077499.

Frenot, A., & Chronakis, I. S. (2003). Polymer nanofibers assembled by electrospinning. Current Opinion in Colloid & Interface Science, 8(1), 64–75. https://doi.org/10.1016/S1359-0294(03)00004-9

Fu, L., Suflita, M., & Linhardt, R. J. (2016). Bioengineered Heparins and Heparan Sulfates HHS Public Access. Adv Drug Deliv Rev, 97, 237–249. https://doi.org/10.1016/j.addr.2015.11.002

Furukawa, M., and Tsuboi, T. (2006). Beautifully whitening agent. Japanese patent *2006052146*.

Gomes, T. D. *et al.* (2018) 'Adhesive free-standing multilayer films containing sulfated levan for biomedical applications', *Acta Biomaterialia*, 69. doi: 10.1016/j.actbio.2018.01.027.

González-Garcinuño, Á., Tabernero, A., Domínguez, Á., Galán, M.A., and Martin del Valle, E.M. (2018). Levan and levansucrases: polymer, enzyme, micro-organisms and biomedical applications. Biocatalysis and Biotransformation *36*, 233-244

Hamid, A. *et al.* (2018) *Bioprocess Optimization for Levan Production by Bacillus subtilis B58, Journal of Scientific & Industrial Research.* Available at: http://nopr.niscair.res.in/bitstream/123456789/44675/1/JSIR 77%287%29 386-393.pdf (Accessed: 18 September 2018).

Han, Y. W. (1989) 'Levan production byBacillus polymyxa', *Journal of Industrial Microbiology.* Springer-Verlag, 4(6), pp. 447–451. doi: 10.1007/BF01569641.

Han, Y., and Watson, M. (1992). Production of microbial levan from sucrose, sugarcane juice and beet molasses. Journal of industrial microbiology *9*, 257-260. doi: https://doi.org/10.1007/BF01569633

Hill, A., Karboune, S., and Mateo, C. (2017). Investigating and optimizing the immobilization of levansucrase for increased transfructosylation activity and thermal stability. Process Biochemistry *61*, 63-72. https://doi.org/10.1016/j.procbio.2017.06.011

Ishida, R., Sakaguchi, K., Matsuzaki, C., Katoh, T., Ishida, N., Yamamoto, K., and Hisa, K. (2016). Levansucrase from Leuconostoc mesenteroides NTM048 produces a levan exopolysaccharide with immunomodulating activity. Biotechnology letters *38*, 681-687. https://doi.org/10.1007/s10529-015-2024-9

Jahangirian, H., Lemraski, E.G., Rafiee-Moghaddam, R., and Webster, T.J. (2018). A review of using green chemistry methods for biomaterials in tissue engineering. International Journal of Nanomedicine *13*, 5953-5969. doi: [10.2147/IJN.S163399]

Jathore, N. R. *et al.* (2012) 'Microbial levan from Pseudomonas fluorescens: Characterization and medium optimization for enhanced production', *Food Science and Biotechnology.* The Korean Society of Food Science and Technology, 21(4), pp. 1045–1053. doi: 10.1007/s10068-012-0136-8.

Jeong, K.-H., Park, D., and Lee, Y.-C. (2017). Polymer-based hydrogel scaffolds for skin tissue engineering applications: a mini-review. Journal of Polymer Research *24*, 112. https://doi.org/10.1007/s10965-017-1278-4

Kazak Sarilmiser, H. and Toksoy Oner, E. (2014) 'Investigation of anti-cancer activity of linear and aldehyde-activated levan from Halomonas smyrnensis AAD6T', *Biochemical Engineering Journal.* Elsevier, 92, pp. 28–34. doi: 10.1016/J.BEJ.2014.06.020.

Kenry, & Lim, C. T. (2017). Nanofiber technology: current status and emerging developments. Progress in Polymer Science, 70, 1–17. https://doi.org/10.1016/j.progpolymsci.2017.03.002

Kim, K.H., Chung, C.B., Kim, Y.H., Kim, K.S., Han, C.S., and Kim, C.H. (2005). Cosmeceutical properties of levan produced by Zymomonas mobilis. Journal of cosmetic science *56*, 395-406.

Kim, C.H., Rhee, S.K., Kim, K.H., Kim, K.S., Han, C.S., Choi, B.R. Beta-fructosyl-L ascorbic acid and the preparation method thereof. World patent appl. WO2005026183A1, 2005b.

Kim, Y.H., Kim, J.G., Jang, J., Lee, S.Y., Heo, W.B., and Kim, K.H. (2012). Study on the synthesis of modified levan and its properties on the hair. In International Federation of Societies of Cosmetic Chemists, Johannesburg, South Africa, Poster Session 5, Number 302.

Kim, S. J., Bae, P. K., & Chung, B. H. (2015). Self-assembled levan nanoparticles for targeted breast cancer imaging. Chemical Communications, 51(1), 107–110. https://doi.org/10.1039/c4cc07679f

Kim, S. J., & Chung, B. H. (2016). Antioxidant activity of levan coated cerium oxide nanoparticles. Carbohydrate Polymers, 150, 400–407. https://doi.org/10.1016/j.carbpol.2016.05.021

Kwon, K.Y., Lee, J.S., Ko, G.J., Sunwoo, S.H., Lee, S., Jo, Y.J., Choi, C.H., Hwang, S.W., and Kim, T.i. (2018). Biosafe, Eco-Friendly Levan Polysaccharide toward Transient Electronics. Small *14*, 1801332. https://doi.org/10.1002/smll.201801332

Lee, J.K.Y., Chen, N., Peng, S., Li, L., Tian, L., Thakor, N., and Ramakrishna, S. (2018). Polymer-based composites by electrospinning: Preparation & functionalization with nanocarbons. Progress in Polymer Science. https://doi.org/10.1016/j.progpolymsci.2018.07.002

Li, Y. (2013). Techniques for fabrication and construction of three-dimensional scaff | IJN, 337–350. Retrieved from http://www.dovepress.com/techniques-for-fabrication-and-construction-of-three-dimensional-scaff-peer-reviewed-article-IJN

Li, R. *et al.* (2015) 'Purification and characterization of an intracellular levansucrase derived from Bacillus methylotrophicus SK 21.002', *Biotechnology and Applied Biochemistry*, 62(6), pp. 815–822. doi: 10.1002/bab.1334.

Li, W., Yu, S., Zhang, T., Jiang, B., and Mu, W. (2017). Synthesis of raffinose by transfructosylation using recombinant levansucrase from Clostridium arbusti SL206. Journal of the Science of Food and Agriculture *97*, 43-49. https://doi.org/10.1002/jsfa.7903

Liberti, M. V, & Locasale, J. W. (2016). The Warburg Effect: How Does it Benefit Cancer Cells? Trends in Biochemical Sciences, 41(3), 211–218. https://doi.org/10.1016/j.tibs.2015.12.001

Lim, F., and Sun, A.M. (1980). Microencapsulated islets as bioartificial endocrine pancreas. Science *210*, 908-910. doi: 10.1126/science.6776628

Liu, J., Luo, J., Ye, H., Sun, Y., Lu, Z., & Zeng, X. (2009). Production, characterization and antioxidant activities in vitro of exopolysaccharides from endophytic bacterium Paenibacillus polymyxa EJS-3. Carbohydrate Polymers, 78(2), 275–281. https://doi.org/10.1016/j.carbpol.2009.03.046

Liu, C. *et al.* (2010) 'Isolation, structural characterization and immunological activity of an exopolysaccharide produced by Bacillus licheniformis 8-37-0-1', *Bioresource Technology*. Elsevier, 101(14), pp. 5528–5533. doi: 10.1016/J.BIORTECH.2010.01.151.

Liu, J. *et al.* (2010) 'In vitro and in vivo antioxidant activity of exopolysaccharides from endophytic bacterium Paenibacillus polymyxa EJS-3', *Carbohydrate Polymers*. Elsevier, 82(4), pp. 1278–1283. doi: 10.1016/J.CARBPOL.2010.07.008.

Liu, X. X., Wan, Z. J., Shi, L., & Lu, X. X. (2011). Preparation and antiherpetic activities of chemically modified polysaccharides from Polygonatum cyrtonema Hua. Carbohydrate Polymers, 83(2), 737–742. https://doi.org/10.1016/j.carbpol.2010.08.044

Liu, J. *et al.* (2012) 'Preparation, antioxidant and antitumor activities in vitro of different derivatives of levan from endophytic bacterium Paenibacillus polymyxa EJS-3', *Food and Chemical Toxicology*. Pergamon, 50(3–4), pp. 767–772. doi: 10.1016/J.FCT.2011.11.016.

Ma, Z., Kotaki, M., Inai, R., & Ramakrishna, S. (2005). Potential of Nanofiber Matrix as Tissue-Engineering Scaffolds. Tissue Engineering, 11(1–2), 101–109. https://doi.org/10.1089/ten.2005.11.101

Manandhar, S., Vidhate, S., & D'Souza, N. (2009). Water soluble levan polysaccharide biopolymer electrospun fibers. Carbohydrate Polymers, 78(4), 794–798. https://doi.org/10.1016/j.carbpol.2009.06.023

Mays, T.D., Dally, E.L. (1989). Microbial production of polyfructose. US patent 4,879,228.

Mihailescu, N., Haskoylu, M.E., Ristoscu, C., Bostan, M.S., Sopronyi, M., Eroğlu, M.S., Chifiriuc, M.C., Mustaciosu, C.C., Axente, E., and Oner, E.T. (2018). Gradient multifunctional biopolymer thin film assemblies synthesized by Combinatorial MAPLE. Applied Surface Science. https://doi.org/10.1016/j.apsusc.2018.10.077

Molinari, M. L. and Boiardi, J. L. (2013) 'Levans production by Gluconacetobacter diazotrophicus', *Electronic Journal of Biotechnology*, 16(3). doi: 10.2225/vol16-issue3-fulltext-9.

Osman, A., Oner, E. T. and Eroglu, M. S. (2017) 'Novel levan and pNIPA temperature sensitive hydrogels for 5-ASA controlled release', *Carbohydrate Polymers*. Elsevier, 165, pp. 61–70. doi: 10.1016/J.CARBPOL.2017.01.097.

Öner, E.T., Hernández, L., and Combie, J. (2016). Review of levan polysaccharide: from a century of past experiences to future prospects. Biotechnology advances *34*, 827-844. https://doi.org/10.1016/j.biotechadv.2016.05.002

Park, S.-B., Lih, E., Park, K.-S., Joung, Y. K., & Han, D. K. (2017). Biopolymer-based functional composites for medical applications. Progress in Polymer Science, 68, 77–105. https://doi.org/10.1016/j.progpolymsci.2016.12.003

Peng, J., Xu, W., Ni, D., Zhang, W., Zhang, T., Guang, C., and Mu, W. (2018). Preparation of a novel water-soluble gel from Erwinia amylovora levan. International journal of biological macromolecules. https://doi.org/10.1016/j.ijbiomac.2018.10.093

Pomin, V. H., & Mourão, P. A. S. (2014). Specific sulfation and glycosylation-a structural combination for the anticoagulation of marine carbohydrates. Frontiers in Cellular and Infection Microbiology, 4, 33. https://doi.org/10.3389/fcimb.2014.00033

Prasad, A., Sankar, M.R., and Katiyar, V. (2017). State of art on solvent casting particulate leaching method for orthopedic scaffoldsfabrication. Materials Today: Proceedings *4*, 898-907. https://doi.org/10.1016/j.matpr.2017.01.101

Queiroz, E. A. I. F. *et al.* (2017) 'Levan promotes antiproliferative and pro-apoptotic effects in MCF-7 breast cancer cells mediated by oxidative stress', *International Journal of Biological Macromolecules*. Elsevier, 102, pp. 565–570. doi: 10.1016/J.IJBIOMAC.2017.04.035.

Santos-Moriano, P., Fernandez-Arrojo, L., Poveda, A., Jimenez-Barbero, J., Ballesteros, A.O., and Plou, F.J. (2015). Levan versus fructooligosaccharide synthesis using the levansucrase from Zymomonas mobilis: effect of reaction conditions. Journal of molecular catalysis B: Enzymatic *119*, 18-25. https://doi.org/10.1016/j.molcatb.2015.05.011

Senthilkumar, V. and Gunasekaran, P. (2005) *Influence of fermentation conditions on levan production by Zymomonas mobilis CT2*, *Indian Journal of Biotechnology*. Available at: http://nopr.niscair.res.in/bitstream/123456789/5777/1/IJBT 4%284%29 491-496.pdf (Accessed: 26 September 2018).

Sezer, A. D. *et al.* (2011) 'Levan-based nanocarrier system for peptide and protein drug delivery: Optimization and influence of experimental parameters on the nanoparticle characteristics', *Carbohydrate Polymers*. Elsevier, 84(1), pp. 358–363. doi: 10.1016/J.CARBPOL.2010.11.046.

Sezer, A.D., Kazak Sarılmışer, H., Rayaman, E., Çevikbaş, A., Öner, E.T., and Akbuğa, J. (2017). Development and characterization of vancomycin-loaded levan-based microparticular system for drug delivery. Pharmaceutical development and technology *22*, 627-634. https://doi.org/10.3109/10837450.2015.1116564

Silva, A. K. A., Juenet, M., Meddahi-Pellé, A., & Letourneur, D. (2015). Polysaccharide-based strategies for heart tissue engineering. Carbohydrate Polymers, 116, 267–277. https://doi.org/10.1016/j.carbpol.2014.06.010

Sima, F., Mutlu, E.C., Eroglu, M.S., Sima, L.E., Serban, N., Ristoscu, C., Petrescu, S.M., Oner, E.T., and Mihailescu, I.N. (2011). Levan nanostructured thin films by MAPLE assembling. Biomacromolecules *12*, 2251-2256. doi: 10.1021/bm200340b

Simms, P.J., Boyko, W.J., and Edwards, J.R. (1990). The structural analysis of a levan produced by Streptococcus salivarius SS2. Carbohydrate research *208*, 193-198. doi: 10.1021/bm200340b

Smith, L. A., & Ma, P. X. (2004). Nano-fibrous scaffolds for tissue engineering. Colloids and Surfaces B: Biointerfaces, 39(3), 125–131. https://doi.org/10.1016/j.colsurfb.2003.12.004

Srikanth, R., Siddartha, G., *et al.* (2015a) 'Antioxidant and anti-inflammatory levan produced from Acetobacter xylinum NCIM2526 and its statistical optimization', *Carbohydrate Polymers*. Elsevier, 123, pp. 8–16. doi: 10.1016/J.CARBPOL.2014.12.079.

Srikanth, R., Reddy, C. H. S. S. S., *et al.* (2015b) 'Review on production, characterization and applications of microbial levan', *Carbohydrate Polymers*. Elsevier, 120, pp. 102–114. doi: 10.1016/J.CARBPOL.2014.12.003.

Sturzoiu, C., Petrescu, M., Galateanu, B., Anton, M., Nica, C., Simionca, G.I., Dinischiotu, A., and Stoian, G. (2011). Zymomonas mobilis Levan is Involved in Metalloproteinases Activation in Healing of Wounded and Burned Tissues. Scientific Papers Animal Science and Biotechnologies *44*, 453-458.

Sudharsan, S., Subhapradha, N., Seedevi, P., Shanmugam, V., Madeswaran, P., Shanmugam, A., & Srinivasan, A. (2015). Antioxidant and anticoagulant activity of sulfated polysaccharide from Gracilaria debilis (Forsskal). International Journal of Biological Macromolecules, 81, 1031–1038. https://doi.org/10.1016/J.IJBIOMAC.2015.09.046

Tabernero, A. *et al.* (2017) 'Development of a nanoparticle system based on a fructose polymer: Stability and drug release studies', *Carbohydrate Polymers*. Elsevier, 160, pp. 26–33. doi: 10.1016/J.CARBPOL.2016.12.025.

Taran, M., Etemadi, S., & Safaei, M. (2017). Microbial levan biopolymer production and its use for the synthesis of an antibacterial iron(II,III) oxide–levan nanocomposite. Journal of Applied Polymer Science, 134(12), 2–6. https://doi.org/10.1002/app.44613

Tian, F., Inthanavong, L., and Karboune, S. (2011). Purification and characterization of levansucrases from Bacillus amyloliquefaciens in intra-and extracellular forms useful for the synthesis of levan and fructooligosaccharides. Bioscience, biotechnology, and biochemistry *75*, 1929-1938. https://doi.org/10.1271/bbb.110315

Ua-Arak, T., Jakob, F., and Vogel, R.F. (2016). Characterization of growth and exopolysaccharide production of selected acetic acid bacteria in buckwheat sourdoughs. International journal of food microbiology *239*, 103-112

Ustyuzhanina, N. E., Bilan, M. I., Gerbst, A. G., Ushakova, N. A., Tsvetkova, E. A., Dmitrenok, A. S., Nifantiev, N. E. (2016). Anticoagulant and antithrombotic activities of modified xylofucan sulfate from the brown alga Punctaria plantaginea. Carbohydrate Polymers, 136, 826–833. https://doi.org/10.1016/j.carbpol.2015.09.102

Vasita, R., & Katti, D. S. (2006). Nanofibers and their applications in tissue engineering. International Journal of Nanomedicine, 1(1), 15–30. https://doi.org/10.2147/nano.2006.1.1.15

Vina, I., Karsakevich, A., & Bekers, M. (2001). Stabilization of anti-leukemic enzyme L-asparaginase by immobilization on polysaccharide levan. Journal of Molecular Catalysis - B Enzymatic, 11(4–6), 551–558. https://doi.org/10.1016/S1381-1177(00)00043-6

Wang, Y., Xin, D., Liu, K., Zhu, M., & Xiang, J. (2009). Heparin−Paclitaxel Conjugates as Drug Delivery System: Synthesis, Self-Assembly Property, Drug Release, and Antitumour Activity. Bioconjugate Chemistry, 20(12), 2214–2221. https://doi.org/10.1021/bc8003809

Wichterle, O., and Lim, D. (1960). Hydrophilic gels for biological use. Nature *185*, 117. https://doi.org/10.1038/185117a0

World Health Organization, WHO Press, Geneva, S. (n.d.). full-text (Vol. 2).

Wu, J., Shi, S., Wang, H., & Wang, S. (2016). Mechanisms underlying the effect of polysaccharides in the treatment of type 2 diabetes: A review. Carbohydrate Polymers, 144, 474–494. https://doi.org/10.1016/j.carbpol.2016.02.040

Xiao, M., Feng, F., and Lu, L. (2014). Preparation method of levan-contained yoghourt (Google Patents).

Xiong, S. li, Li, A., Huang, N., Lu, F., & Hou, D. (2011). Antioxidant and immunoregulatory activity of different polysaccharide fractions from tuber of Ophiopogon japonicus. Carbohydrate Polymers, 86(3), 1273–1280. https://doi.org/10.1016/J.CARBPOL.2011.06.025

Xu, X., Gao, C., Liu, Z., Wu, J., Han, J., Yan, M., & Wu, Z. (2016). Characterization of the levan produced by Paenibacillus bovis sp. nov BD3526 and its immunological activity. Carbohydrate Polymers, 144, 178–186. https://doi.org/10.1016/j.carbpol.2016.02.049

Yoo, S.-H. *et al.* (2004) 'Antitumor activity of levan polysaccharides from selected microorganisms', *International Journal of Biological Macromolecules*. Elsevier, 34(1–2), pp. 37–41. doi: 10.1016/J.IJBIOMAC.2004.01.002.

Yoon, E. J., Yoo, S. H., Cha, J., & Lee, H. G. (2004). Effect of levan's branching structure on antitumour activity. International Journal of Biological Macromolecules, 34(3), 191–194. https://doi.org/10.1016/j.ijbiomac.2004.04.001

Zhao, Y., Feric, N. T., Thavandiran, N., Nunes, S. S., & Radisic, M. (2014). The role of tissue engineering and biomaterials in cardiac regenerative medicine. Can J Cardiol, 30(11), 1307–1322. https://doi.org/10.1016/j.cjca.2014.08.027

Zong, A., Cao, H., & Wang, F. (2012). Anticancer polysaccharides from natural resources: A review of recent research. Carbohydrate Polymers, 90(4), 1395–1410. https://doi.org/10.1016/J.CARBPOL.2012.07.026

Molecular Modification of Levan and Biotechnological Applications of its Derivatives

10

Edmilson Clarindo de Siqueira[1], Bogdan Doboszewski[2], Juliana de Souza Rebouças[3], Irapuan Oliveira Pinheiro[1], Ebru Toksoy Öner[4] and Fabio Rocha Formiga[1,5,*]

[1] Laboratório de Biotecnologia, Programa de Pós-Graduação em Biologia Celular e Molecular Aplicada, Instituto de Ciências Biológicas, Universidade de Pernambuco (UPE), 50100-130, Recife/PE, Brazil.

[2] Departamento de Química, Universidade Federal Rural de Pernambuco (UFRPE), 52171-900, Recife/PE, Brazil.

[3] Instituto de Ciências Biológicas, Universidade de Pernambuco (UPE), 50100-130, Recife/PE, Brazil.

[4] IBSB, Department of Bioengineering, Marmara University, Goztepe Campus, 34722 Istanbul, Turkey.

[5] Departamento de Imunologia, Instituto Aggeu Magalhães (IAM), Fundação Oswaldo Cruz (FIOCRUZ), Recife/PE, Brazil.

*Correspondinf author: fabio.formiga@upe.br

DOI: https://doi.org/10.21775/9781912530267.10

Abstract

Levan is a fructose-based homopolymer, a fructan, with emerging potential in different fields, including biotechnology, food and health. It is obtained from sucrose by a wide range of microorganisms. Levan has become a versatile biopolymer due to its physicochemical and biological properties stemming from its molecular weight, functional groups and degree of substitution. Levan can undergo changes in its molecular structure to broaden its functional properties and consequently its applications. Molecular modification methods mainly involve chemical, physical and biological changes. Chemical modification is the most widely used method to enhance the bioactivity of levan, which occurs through the addition or substitution of functional groups. Physical and biological modifications alter only the molecular weight into varied sizes. This review offers a description of the main molecular modification methods of levan and the contributions to its physicochemical and biological properties.

Introduction

Levan is a fructose-based homoglycan (fructan) and biologically active polysaccharide found broadly and diversely distributed among plants, fungi and bacteria (Oner, Hernández and Combie, 2016). Microbial levan is synthesized in the extracellular environment through the action of levansucrase (Han, 1990; Li et al., 2015), which is an enzyme that acts on sucrose, catalyzing transglycosylation reactions to form a polymer. Levan is one of the few biopolymers in which carbohydrate is found in its furanosidic form (Gupta et al., 2011) besides inulin (Porras-Domínguez 2014). Structural characteristics of levan such as molecular weight and degree of branching are affected by the nature of its origin (if vegetal or microbial) as well as growth conditions. Importantly, the biocompatibility and biodegradability of levan make it a functional biopolymer with considerable potential in food, cosmetic and pharmaceutical industries (Kang et al., 2009; Oner, Hernández and Combie, 2016; Eroglu, et al., 2017).

The majority of microbial polysaccharides have been used in a modified form to enhance their functional properties. The term "molecular modification" regards a change in the structure of a polymer through chemical, physical and biological means to obtain a greater number of structural derivatives (Li et al., 2016). Chemical modification is the most widely used method and is considered an important tool for enhancing the bioactive properties of polysaccharides and significantly broadening their applications (Kagimura et al., 2014). This method consists of altering the structures of polysaccharides through the introduction of substituent groups to enhance the original bioactivity as well as generate new functional bioactivities (Li et al., 2016). In turn, physical and biological modification methods only alter the molecular weight of a polysaccharide, affecting its conformation through a change in its spatial structure rather than its primary structure (Wang et al., 2017).

The presence of hydroxyl units in the levan molecule has been chemically explored for the preparation of polymeric derivatives with exclusive properties (Xue, 2009; Oner, Hernández, and Combie, 2016). Levan derivatives, such as sulfates, phosphates and acetates, have been investigated as antiviral drugs (Roberts and Garegg, 1998) and as food additives (Kang et al., 2009). Moreover, the modification of levan by acetylation, phosphorylation and benzoylation significantly improves its antioxidant and anti-tumor activities (Liu et al., 2012). A recent study reports that sulfated levan exhibits anti-coagulant activity similar to that found for heparin (Erginer et al., 2016). Physical and biological methods have been used mainly to obtain fructooligosaccharides, which are short-chain fructans (Bakar and Kaplan-Türköz, 2017). The most commonly used physical methods for the modification of levan include microwaves (De Paula et al., 2008; Coimbra et al., 2010) and radiation-induced reactions (Jalan et al., 2013). Regarding biological methods, enzymatic hydrolysis is the major procedure leading to degradation of the polymer by enzymes (Porras-Domingues et al., 2014).

Recent applications of levan in industry, biotechnology and medicine require molecular structures with unique properties. Therefore, the present review offers a summary of the most important levan modification methods as well as the impacts on the physicochemical properties and bioactivity of this fructan.

Chemical structure and properties of levan

Levan is an exopolysaccharide (EPS) chemically composed of D-fructofuranosyl residues united by β-(2,6) bonds (Figure 1) (Oner, Hernández and Combie, 2016). However, some microbial cultures, such as *Bacillus* and *Zymomonas*, produce branched levan with β-(2,1) bonds that occur in C-1 (Eroglu et al., 2017).

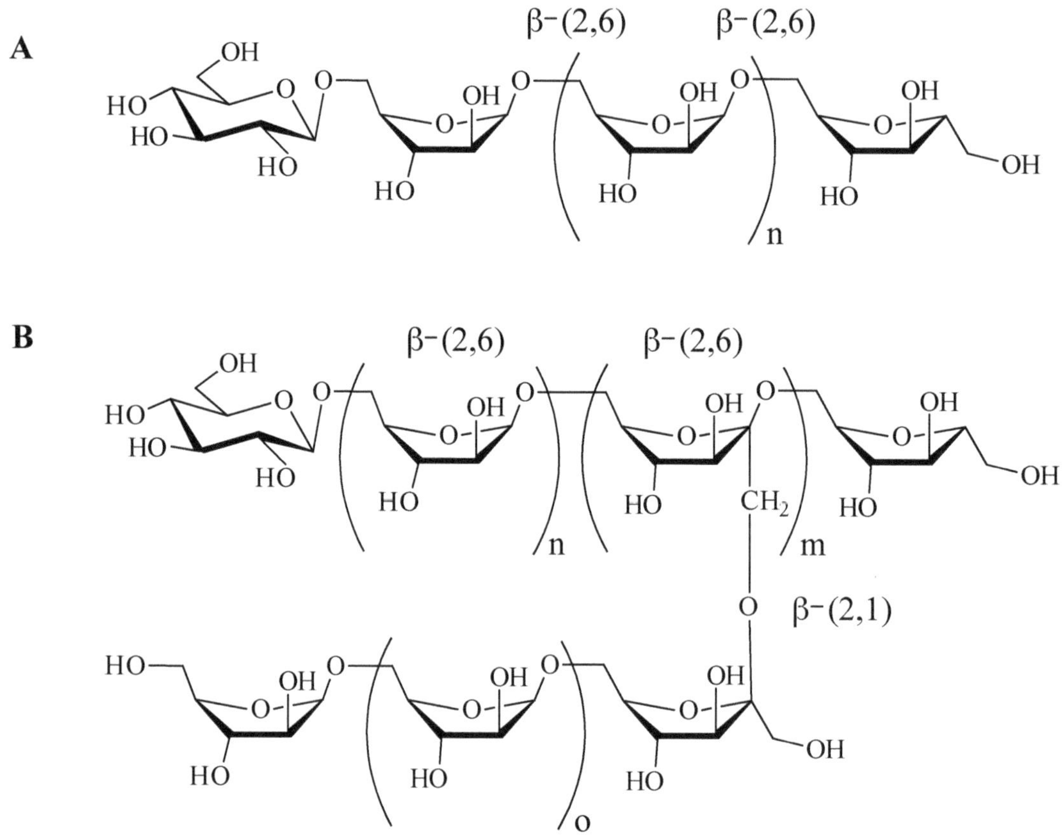

Figure 1. Molecular structure of levan $(C_6O_{10}H_5)_n$, linear (A) and branched (B).

Levan has the characteristic of an amorphous white powder or microcrystalline material with typical properties that distinguish it from other EPSs (Gupta et al., 2011). It is an amphiphilic polymer due to the CH_2 groups in the furanose residues (Sezer et al., 2011), making it soluble in water and oil, but it is insoluble in the majority of organic solvents, except dimethyl sulfoxide (DMSO) (Manandhar et al., 2009; Freitas et al., 2011; Sarilmiser and Oner, 2014; Sarilmiser et al., 2015). Levan presents low intrinsic viscosity (~0.14 dl/g) compared to other polysaccharides with a similar molecular weight (Kasapis et al., 1994). For concentrations of up to 30% in mass, levan exhibits Newtonian behavior, meaning that viscosity is constant for different shearing rates and does not vary with time. Between 30 and 55%, it exhibits pseudoplastic behavior, which is a reduction in viscosity with the increase in shearing rate, likely due to the participation of the lateral branches (Arvidson et al., 2006; Kang et al., 2009).

Other properties include thermal stability (Barone and Medynets, 2008; Manandhar et al., 2009), chemical stability (Sarilmiser and Oner, 2014), traction resistance (Freitas et al., 2011) and compatibility with mineral salts and surfactants (Silbir et al., 2014). Levan is also highly adhesive (Sarilmiser et al., 2015; Gonzalez-Garcinuno 2017) and forms nanoparticles in aqueous media (224.3 nm) and ethanol (251.8 nm) (Report - CIR, 2012; Tomulescu et al., 2016). Moreover, levan has low permeability to oxygen and can be used as an edible lining material for foods and pharmaceutical products (Vijayendra and Shamala, 2014).

Levan has been found to have many excellent functional properties and consequently various applications. In industry, it is used as a surfactant for home use due to its surface properties (Gupta 2011). In the food industry, it is used as a stabilizer, emulsifier, additive, encapsulating agent and transporter of aromas and fragrances (Han, 1990). Moreover, a biphasic aqueous system formed by polyethylene glycol (PEG 60%, w/w) and levan (6.77%, w/w) is used in the partition of proteins, such as pectin (Kang, 2009).

Due to its moisturizing and cell proliferation-promoting effects, levan produced by *Z. mobilis* has been studied for its cosmetic properties and Kim et al. (2005) found it to be a safe, useful cosmetic agent. These properties may be due to the size of its particles, which are much smaller compared to those of other polysaccharides, improving the skin penetration capacity.

Levan also has immune-stimulating, anti-diabetic, anti-tumor and anti-inflammatory activities. Its anti-tumor activity is dependent on the molecular weight (Calazans et al., 2000; Yoo et al., 2004) and is directly proportional to its degree of branching, with less branching leading to lower anti-tumor activity (Yoon et al., 2004; Sarilmiser and Oner, 2014). Moreover, the consumption of levan with a high molecular weight (1 to 5%, w/v) reduces total cholesterol in the serum in a dose-dependent manner (Yamamoto et al., 1999). Levan with a low molecular weight has a more accentuated antibacterial effect *in vitro* (Byun, Lee, and Mah, 2014). In addition, the properties of levan have gained greater attention with regard to use as a promising biomaterial in nanosystems for the release of bioactive molecules (Sezer et al., 2011; Sezer, 2015; Kim, Bae and Chung, 2015).

The physicochemical properties of a polymer exert an influence on its biological properties and both are favored by modifications in the molecular structure (Ma, 2012; Ahmad, Mustafa and Man, 2015; Li et al., 2016). Accordingly, these characteristics should be examined in levan when modified by appropriate methods.

Molecular modification of levan

Polysaccharide molecular modification methods mainly involve chemical, physical and biological alterations (Figure 2).

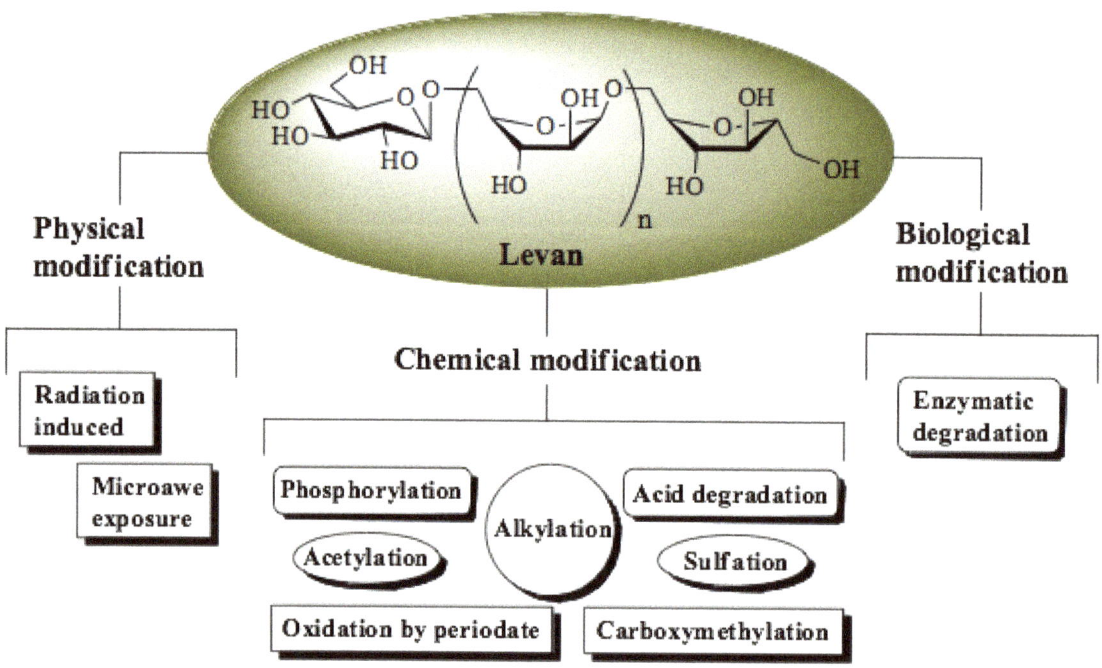

Figure 2. Molecular modification approaches for microbial levan. Adapted from Li et al. (2016).

Chemical modification is the most widely method used to enhance bioactivity and occurs by the introduction or substitution of functional groups. Physical modifications are evidenced by the change in molecular weight and destruction of the glycosidic chain through extra energy. Biological modifications only alter the molecular weight in varied sizes through enzymatic hydrolysis of the glycosidic chain (Li et al., 2016).

Chemical modifications of levan and applications

The majority of chemical modification processes of levan depend on the multiplicity of nucleophilic groups found throughout its structure in the form of hydroxyl groups Chatterjee, Pong and Sen, 2015). Such processes include oxidation with periodates ($NaIO_4$ or KIO_4), sulfation, carboxymethylation, phosphorylation and acetylation (Kagimura et al., 2014; Wang et al., 2016).

Oxidation by periodate

The oxidation of levan with periodate is seen as a way to introduce aldehyde groups in its structure, which are more reactive to *N*-nucleophiles, such as carbamates, hydrazines and amines in general (Vına, Karsakevich and Bekers, 2001). The reaction process occurs in a 3,4-diol system of each D-fructosyl unit by the selective cleaving of chemical bonds between carbon atoms C3 and C4, resulting in the formation of reactive dialdehydes (Gonta, Karsakevida, and Vina, 1999; Sarilmiser and Toksoy Oner, 2014), which can be easily reduced with sodium borohydride ($NaBH_4$) (Figure 3).

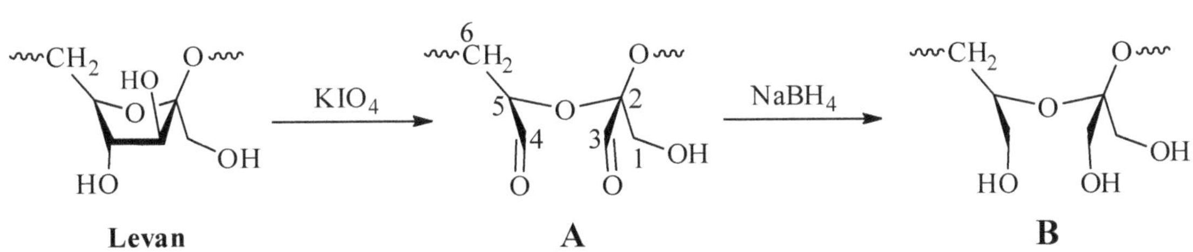

Figure 3. Modification of levan with periodate: A) oxidized form; B) reduced form.

Oxidation with periodate is a well-established method in the structural determination of a polysaccharide. The effect of this chemical modification of levan was established by Gonta et al. (2004) through simulations and computational analysis. The authors found that the spatial structure of oxidized levan is expanded by the increase in the distance between the atoms C3 and C4 and the C5-O-C2 angle in the furanosyl residue, implying a region of greater affinity for nucleophilic reagents and, more importantly, a strategic site of the biological activity of levan (Gonta et al., 2004; Sarilmiser and Toksoy Oner, 2014). For instance, Gonta, Karsakevida and Vina (1999) studied the immunomodulating activity of levan produced by *Zymomonas mobilis* in the oxidized form and found that only the reducing extremities of the polymer chain had predominance for this activity.

Levan in the oxidized form has been also used to immobilize enzymes through the condensation reaction (Ulu and Ates, 2017), as shown in Figure 4.

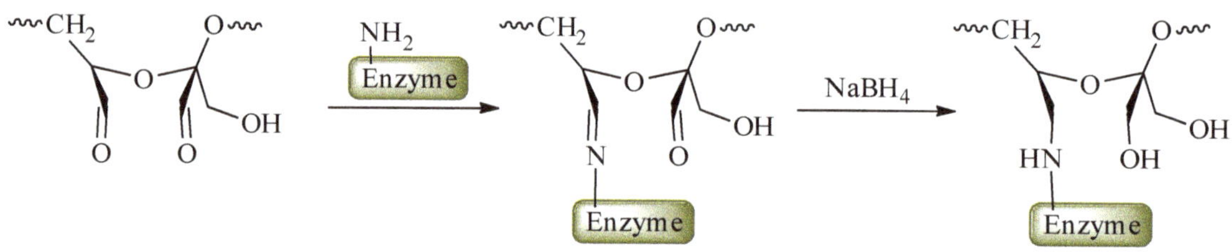

Figure 4. Condensation reaction of dialdehyde levan with protein/enzymes.

By using the condensation reaction, a Schiff base is produced by a reaction between oxidized levan and the amino groups of molecules, such as enzymes, lectins, immunoglobins and other proteins. The Schiff base is relatively stable and can be reduced with the addition of $NaBH_4$, resulting in the formation of a stable bond between the protein and the polysaccharide (Gonta et al., 2004; Sarilmiser and Toksoy Oner, 2014).

Vına, Karsakevich and Bekers (2001) investigated the activity of the enzyme L-asparaginase (L-ASNase) immobilized covalently on levan. The immobilized L-

ASNase exhibited optimized stability and biocatalytic activity in comparison to the free enzyme under the same assay conditions (Vına, Karsakevich and Bekers, 2001; Ulu and Ates, 2017). Tabandeh and Aminlari (2009) obtained similar results with the same immobilized enzyme and oxidized inulin. Like levan, inulin is a fructan, differing only in the glycosidic bonds of the principal chain (β-2,1 type in inulin and β-2,6 type in levan) (Ortiz-Soto and Seibel 2014; Raga-Carbajal et al., 2015).

In another strategy, Maciel et al. (2012) used oxidized levan in the form of magnetic particles as a support for immobilized trypsin. The magnetic particles were prepared by co-precipitation with iron ions under alkaline conditions. The polymer was then functionalized with $NaIO_4$ and reduced with $NaBH_4$ (Angeli et al., 2009). The partially oxidized magnetic were covalently bound to the trypsin and retrieved with a magnetic field (Maciel et al., 2012).

Bezerra, Silva and Belian (2009) used dialdehyde levan as a drug tracking matrix, reporting the synthesis and characterization of europium cryptate tagged with the levan matrix. This tagging exhibited strong luminescence intensity even in only 1% of signaled fructosidic groups.

New polymeric platforms have been created by employing levan in its oxidized form, such as thin nano-structured films. Thin nano-structured levan (L) films and oxidized levan (OL) are obtained by Matrix-Assisted Pulsed Laser Evaporation (MAPLE) (Sima et al., 2011; 2012). However, OL has a higher potential for cellular proliferation compared to L, exhibiting a greater positive interaction with human osteoblasts (SaOs2) (Axente et al., 2014).

The degree of levan oxidation depends greatly on its molecular mass and plays an essential role in the bioactivity of the resulting structures (Gonta, Karsakevida and Vina, 1999). Sarilmiser and Toksoy Oner (2014) investigated the anti-carcinogen activity of L and OL (with an increased rate of aldehydes) in human lung adenocarcinoma (A549), human liver hepatocellular carcinoma (HepG2/C3A), human gastric adenocarcinoma (AGS) and human breast adenocarcinoma (MCF-7). The authors found that OL exhibited greater anti-cancer activity than L against all cancer cell lines and that this increase was directly proportional to the degree of oxidation (Sarilmiser and Toksoy Oner, 2014).

Carboxymethylation

Carboxymethylation is generally achieved by suspending the biopolymer in monochloroacetic acid under alkaline conditions (Kagimura et al., 2014) (Figure 5-A). The reaction follows the Williamson etherification protocol, in which the sample is treated with a strong base, such as sodium hydroxide, which deprotonates free hydroxyl groups to form alkoxide. The generated adduct undergoes nucleophilic substitution with monochloroacetic acid to yield the carboxymethyl derivative (Figure 5-B) (Rahul et al., 2014).

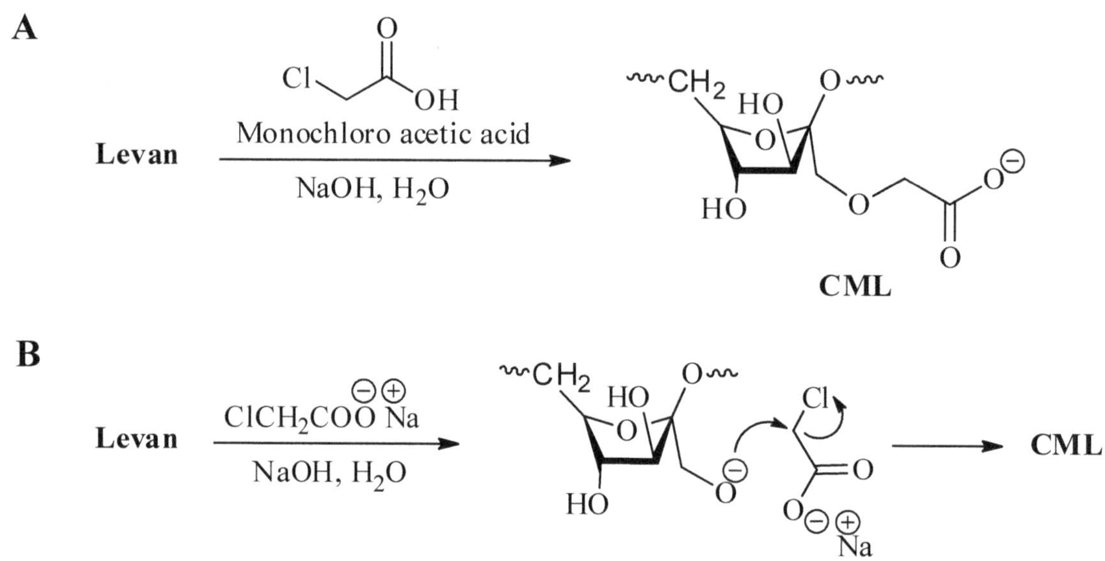

Figure 5. Synthetic method (A). Carboxymethylation mechanism (B).

Carboxymethyl levan (CML) has recently been used in a drug-delivery system (DDS). Tabernero et al. (2017) used levan and CML as vehicles to deliver the anti-tumor drug 5-fluorouracil (5-FU). The release of the drug was 45-75% in 23 h with the levan/5-FU system, whereas only a small percentage of the drug was released from the CML/5-FU system. The low percentage released by the CML/5-FU system was due to the formation of strong amide bonds, which slows the release of the drug (Tabernero et al., 2017).

Osman et al. (2017) also reported the synthesis of CML for use in a DDS. In this case, CML was obtained first and then set to react with methacrylic anhydride (MA) to obtain methacrylate levan (MAL), as shown in Figure 6.

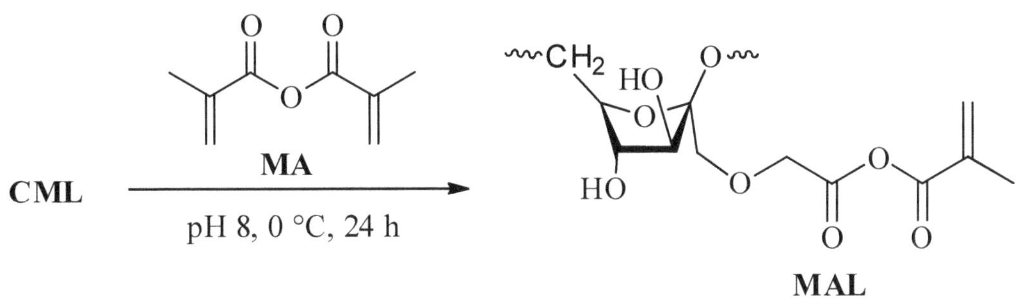

Figure 6. Synthesis of carboxymethyl levan (CML) and methacrylated levan (MAL).

MAL was used as a cross-linking agent with *N*-isopropylacrylamide to obtain hydrogels, which were used as a DDS for 5-aminosalycilic acid (5-ASA). The methacrylate groups were found to react more strongly with carboxymethyl groups than hydroxyl groups of the fructose ring due to the greater nucleophilic power of the former (Osman et al., 2017).

Curiously, MAL has been synthesized and characterized for use in the preparation of hydrogels (Leite et al., 2014; Wyrwa et al.; 2014; Osman, Oner and Eroglu, 2017; Berg et al., 2018). Leite et al. (2014) produced MAL-based hydrogels and ternary glass nanoparticles (SiO_2:CaO:P_2O_5) and found greater bond strength (1.09 ± 0.21 kPa) compared to hydrogels prepared with alginate (null). Moreover, live/dead assays using SaOs2 osteoblasts revealed the successful encapsulation of the cells, which remained alive after 24 h. These hydrogels are an option in tissue engineering to replace adhesive membranes (Leite et al., 2014).

Berg et al. (2018) recently prepared novel MAL derivatives united by ester or urethane bonds to levan fructosyl residues, with the aim of using MALs as reticulating agents to obtain hydrogels. For such, levan and different methacrylating agents, such as methacrylic anhydride (MAA), glycidyl methacrylate (GMA) and 2-isocyanatoethyl methacrylate (IEM) (Figure 7), were submitted to different reaction conditions and photo initiating systems (Berg et l., 2018).

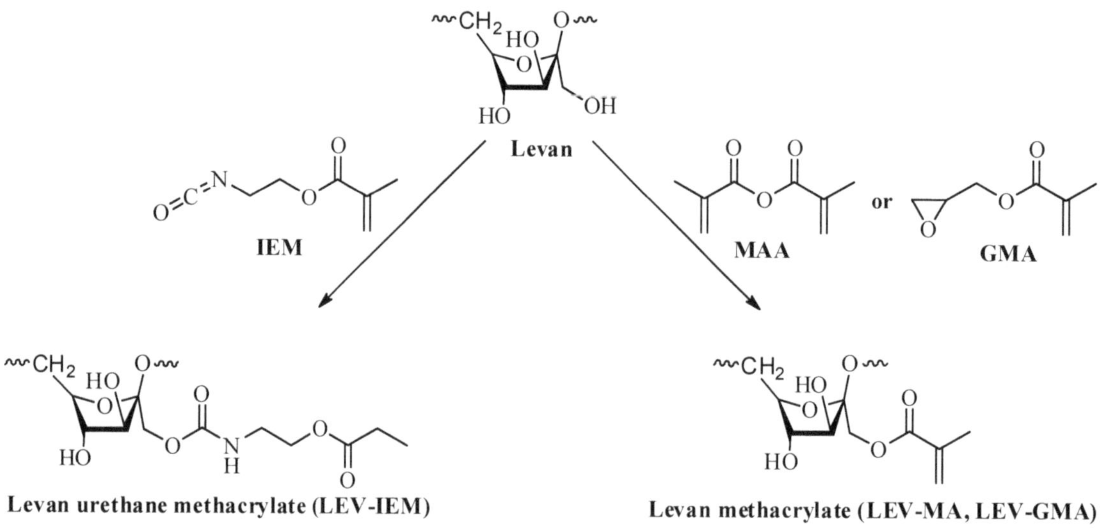

Figure 7. Synthesis of novel methacrylate levan derivatives.

It should be pointed out that two different donors of the methacrylate moiety, namely the methacrylate anhydride (MMA) and the glycidyl methacrylate (GMA) both furnished the same product. It is clear that the glycidyl moiety acted as a leaving group only rather than to act as an alkylator and to react via ring opening.

The hydrogels exhibited relatively stable behavior against hydrolytic degradation in an aqueous medium due to reticulation consistency. The swelling rate increased in the following order: LEV-IEM << LEV-MA << LEV-GMA. In contrast, the amount of available reticulation units exhibited the opposite behavior (Berg et al., 2018).

The authors also studied the cytotoxicity of the hydrogels based on the relative activity of dehydrogenase (WST1 test) using fibroblast cells for a period of eight days. The LEV-MA and LEV-GMA hydrogels did not have cytotoxic effects during the elution period, whereas the LEV-IEM hydrogel exhibited strong cytotoxicity in the initial days, which was attributed to the slow release of DMSO into the compact samples. However, after five to six days of elution, this cytotoxic effect was attenuated. These novel methacrylates represent a new class of hydrogel-forming materials that can be used in the fabrication of functional biomaterials (Berg et al., 2018).

CML also has significantly greater prebiotic activity compared to free levan. Husseiny et al. (2015) found that the amount of *Lactobacillus lactis* MCAIN B01357 grown in a medium with CML was greater than that grown in a medium with free levan (Husseiny et al., 2015). Therefore, chemical modification by carboxymethylation is considered an important way to enhance the properties and biological activities of levan (Wang et al., 2017).

Sulfation

There are several methods for the modification of polysaccharides by sulfation. The most widely employed methods are chlorosulfonic acid in pyridine (CSA-Py), SO_3 complex with pyridine (Py•SO_3) and concentrated sulfuric acid (Li et al., 2016).

Sulfated levan (SL) has been commonly achieved using the CSA-Py method (Figure 8) with levans produced by *Halomonas smyrnensis* AAD6^T (Akçay, 2014; Erginer et al., 2016; Avsar et al., 2018; Gomes et al., 2018) and *Bacillus subtilis* (Abdel-Fattah et al., 2012; Bayoumi et al., 2012; 2013), with the proportion of CSA to Py, temperature and reaction time as determinant factors in this process (Wang et al., 2009; Li et al., 2016).

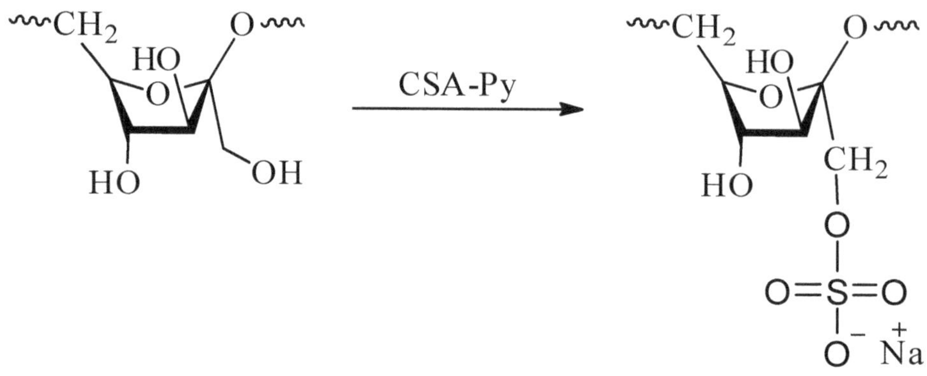

Figure 8. Reaction for production of sulfated levan (SL)

However, Husseiny et al. (2015) report the sulfation of levan produced by *Chromohalobacter salexigens* KT7 using the Py•SO_3 complex in dimethylformamide (DMF). According to the authors, the sulfation method has a transition state with only weak N-S and O-S bonds, in which the transference of the sulfate group between the nitrogen in the Py•SO_3 complex and oxygen in the polysaccharide occurs in "concert line" (Husseiny et al., 2015).

SL is considered to be an adequate biomaterial for designing functionalized, biologically activated films (Ergenir et al., 2016). Avsar, Gunduz and Oner (2018) produced biomaterials by electrospinning using SL, poly-epsilon-caprolactone (PCL) and polyethylene oxide (PEO). The blood clotting index (BCI) of the fibrous membranes with 5% SL was 4-fold higher compared to matrices composed with PCL alone. Moreover, an increase in the concentration of SL resulted in higher BCI values and SL demonstrated activity similar to that of heparin regarding the inhibition of the proliferation of fibroblasts L929 (Avsar, Gunduz and Oner, 2018). Therefore, SL enhances the biocompatibility and bioadhesiveness of PCL, resulting in intelligent biomaterials with applications in tissue engineering (Ergenir et al., 2016; Avsar, Gunduz and Oner, 2018).

Gomes et al. (2018) found that SL can be successfully introduced in multilayer films of chitosan and alginate (CHI/ALG) for the creation of thin and viscoelastic films. The authors first obtained SL using CSA-Py and then used it to make membranes employing the layer-by-layer method based on electrostatic interactions. Free support (FS) films were produced with $(CHI/ALG/CHI/ALG)_{100}$, non-reticulated $(CHI/ALG/CHI/SL)_{100}$ and $(CHI/ALG/CHI/SL)_{100}$ reticulated with genipin and the properties were studied. The inclusion of SL in the CHI/ALG system made the FS smooth, with roughness found after reticulation. Moreover, the presence of SL led to greater resistance to traction in both the non-reticulated (3.80 ± 0.75 MPa) and reticulated (3.73 ± 0.3 MPa) forms compared to $(CHI/ALG/CHI/ALG)_{100}$ (1.25 ± 0.47 MPa). The inclusion of SL also led to a fourfold increase in resistance to shearing in the non-reticulated and reticulated forms in comparison to $(CHI/ALG/CHI/ALG)_{100}$ (Gomes et al., 2018).

Regarding the compatibility and cellular viability assays, the authors found greater dispersion of myoblasts (C2C12) throughout the entire surface of the reticulated $(CHI/ALG/CHI/SL)_{100}$ film. Based on the literature, the authors point out that C2C12 cells depend on rigid substrates for propagation and that SL can also exert an influence on the differentiation of these cells (Gomes et al., 2018).

SL has been studied as an anti-tumor agent, antioxidant and anticoagulant. Abdel-Fattah et al. (2012) studied the anti-carcinogen activity of SL and its mechanism in the apoptotic pathway. SL was found to induce apoptosis through the intrinsic pathway in the HepG2 line. Husseiny et al. (2015) report similar results. Moreover, SL has exhibited strong antioxidant activity, as demonstrated by the elimination of free radicals (Abdel-Fattah et al., 2012), and is a promising fibrinolytic agent (Husseiny et al., 2015).

The immunomodulating and chemopreventive effects of SL have also been studied (Bayoumi et al., 2012; 2013). The administration of SL in rats induced by diethylnitrosoamine (DEN) resulted in a reduction in the level of vascular endothelial growth factor, which may be linked to the inhibiting role of SL with regard to nitric oxide and tumor necrosis factor-alpha (Bayoumi et al., 2012). SL also inhibited DNA fragmentation and demonstrated anti-progressive activity against HepG2 cells (Bayoumi et al., 2013).

SL has been investigated as an anticoagulant as an alternative to heparin (Akçay, 2014; Ergenir et al., 2016). SL demonstrated an inhibitory effect on the intrinsic pathway of blood coagulation as a consequence of the prolongation of the activated partial thromboplastin time (APTT). In the anti-thrombin tests, an increase in the concentration of SL reduced the activity of Factor Xa. Moreover, above a certain concentration, SL presented better inhibitory activity than heparin (Ergenir 2016).

Phosphorylation

There are several methods for the phosphorylation of polysaccharides, such as the use of phosphoric acid, phosphoric anhydride or low-cost phosphates, such sodium hydrogen phophate, sodium di-hidrogen phophate and sodium tripolyphosphate (Li et al., 2016).

Liu et al. (2012) obtained phosphorylated levan (PL) by treating a levan from *Paenibacillus polymyxa* EJS-3 with a mixture of formaldehyde (FA) and tributylamine, followed by polyphosphoric acid (Zhang et al., 2009). The anti-tumor activity of PL was evaluated in human gastric carcinoma cells (BGC-823) at a concentration of 400 μg/ml. PL achieved a much higher inhibition rate compared to free levan (95.50% and 55.37%, respectively). This greater anti-tumor activity was attributed to the high affinity of the phosphate group with immune cell receptors, activating the immune response (Liu et al., 2012).

Costa et al. (2013) synthesized phosphonated levan (PhL) using phosphorus trichloride (PCl_3) and pyridine (Py) (Figure 9-A). PhL was used with chitosan (CHI) to form films on glass slides by electrostatic self-assembly using the layer-by-layer method (Figure 9-B).

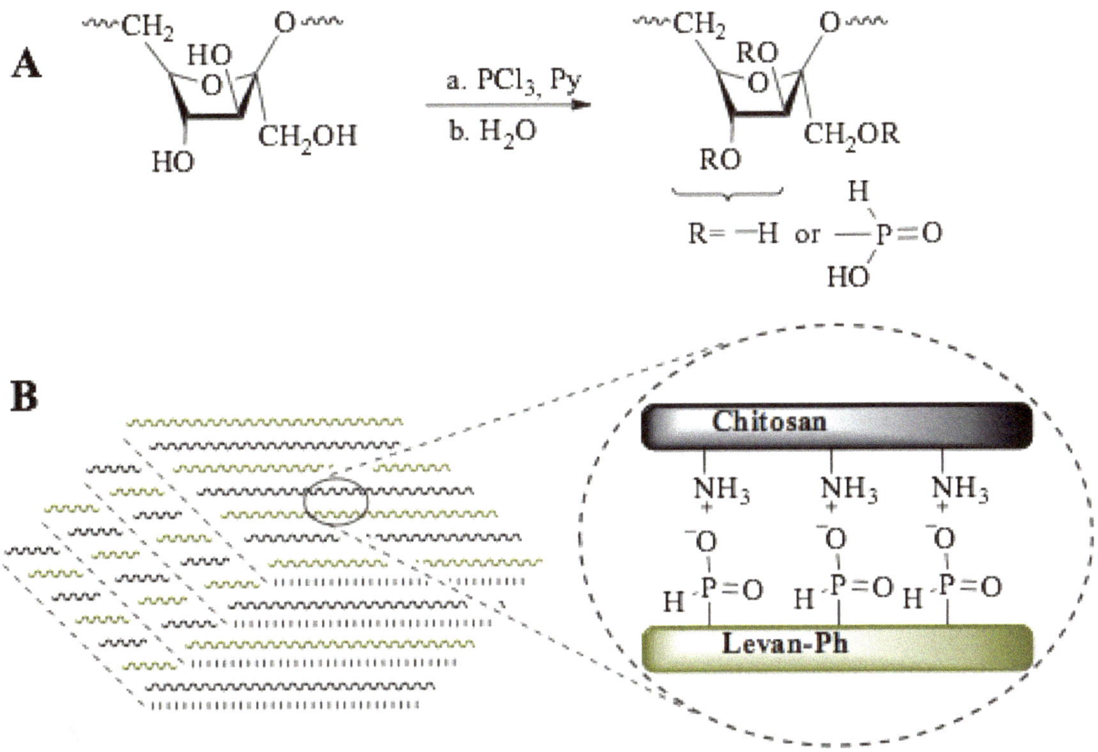

Figure 9. Modification of levan with phosphonate group (A). Preparation of film made of phosphonated levan and chitosan by electrostatic interactions (B).

PhL was obtained with a yield of 50%. The PhL-CHI films exhibited notable bond strength (2.5 MPa) compared to the control formed by alginate and chitosan (ALG-CHI), which was 0.9 MPa. The authors also investigated the adhesiveness of films with a rat fibroblast line (L929) and found that the PhL-CHI film had better adhesion (110 cells/mm^2) compared to the ALG-CHI film (20 cells/mm^2), confirming the use of levan as a base polymer for tissue engineering (Costa et al., 2013; Ergenir et al., 2016).

Acetylation

Chemical modification by acetylation is mainly used for the treatment of the branching of polysaccharide molecules with the aim of enhancing the solubility of the polymer (Li et al., 2016). The essence of this reaction is electrolytic substitution, which consists of preliminary dissolution of the polysaccharide in a solvent, such as

H_2O, DMSO or FA, followed by the drop-by-drop addition of *N*-bromosuccinimide in ethyl acetate (NBS/Ac_2O) or a Py/Ac_2O solution (Li et al., 2016).

Using the NBS/Ac_2O method, Liu et al., (2012) successfully obtained derivatives of acetylated levan (AL) and determined the antioxidant activity *in vitro*. At a concentration of 1 mg/l, the reducing power of AL (0.493) was greater than that of the free levan (0.238) and other derivatives, such as PL (0.359) and benzoyled levan (0.312). Wang et al. (2009) report similar results with acetylated fucoidan in comparison to benzoyled fucoidan. The authors found that the reducing potential diminished in the following order: acetylated fucoidan > oversulfated fucoidan > benzoyled fucoidan (Wang et al., 2009).

Thus, the increased antioxidant activity in AL can be explained by the presence of functional groups, such as COOH, which are generally recognized as good electron donors and therefore exert an influence on the antioxidant activity of the polysaccharide (Liu et al., 2012; Lu et al., 2014).

Acid degradation

The hydrolysis of polysaccharides by acid degradation is one of the traditional methods. In this case, the glycosidic bonds of the polymer are preferentially cleaved by an acidic solution, breaking down the polysaccharides into fragments with a low molecular weight with enhanced solubility and biological activity (Li et al., 2016).

Levan is a non-reducing polymer that is not hydrolysable by amylase or invertase but is very sensitive to hydrolysis by acid and is partially hydrolyzed into fructose and low molecular weight oligosaccharides under acidic conditions and high temperatures (Gupta 2011; Kang et al. 2009). Hydrolysis with organic acids (lactic, acetic and gluconic) ensures the greater formation of fructose monomers, whereas the hydrolysis of levan by HCl ensures large amounts of fructooligosaccharides (FOS) and levan with a low molar mass (Shih et al., 2005).

Among many experimental methods, chemical hydrolysis with nitrous acid is becoming the preferred technique for the production of low molecular weight (LMW) levan due to the fact that it is a fast, economical method that enables control to produce pre-directed sizes (Eroglu et al., 2017). For such, the polymer is dissolved in a solution of acetic acid/water solution (2%) under agitation, treated with sodium nitrite ($NaNO_2$), neutralized with Na_2CO_3 (1 M), dialyzed and lyophilized (Mutlu, 2014). Using these conditions, Osman, Oner and Eroglu (2017) obtained LMW levan (Mn: 16 000 g/mol) with yields of 70%.

Other chemical modification methods for levan

Fructose-based polysaccharides can be explored as chemical building blocks in organic synthesis (Barclay et al., 2012). For instance, Oscarson and Sehgelmeble (2002) prepared a donor of fructofuranosil, 6-*O*-acetyl-3-*O*-benzyl-1,4-*O*-(1,1,3,3-tetraisopropyldisiloxane-1,3-diyl)-2-thio-β-D-fructofuranoside (**1**), designed to produce β-linkages and cause elongation in positions 6 and/or 1 in the synthesis of levan and inulin structures. The glycosylation promoted between the acceptor of methyl-D-fructofuranoside 6-OH (**2**) and **1** resulted in methyl levanobioside (**LVB**), with a yield of 80%, which was once again coupled with donor **1** to obtain methyl levanotrioside (**LVT**), with a yield of 65% (Figure 10).

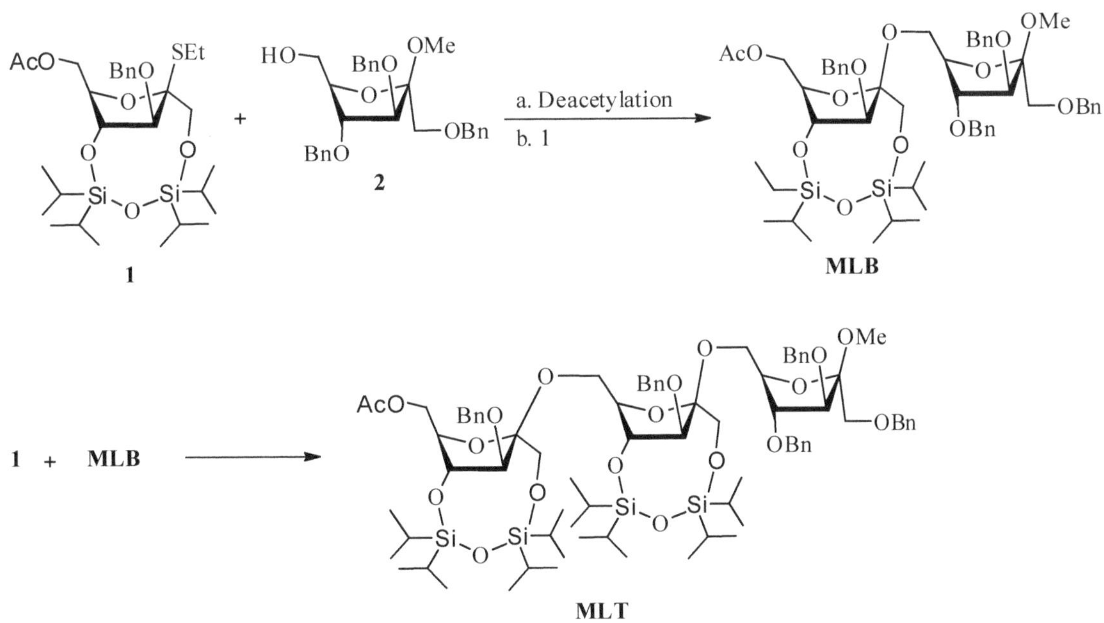

Figure 10. Synthesis of levanobioside (**LVB**) and methyl-D-levanotrioside (**LVT**)

In the future, levan could be an option in the synthesis of renewable products of high value, such as 5-hydroxymethylfurfural (HMF) and its derivatives (Barclay et al., 2012) due to the fact that fructose is an efficient starter material and selective for the production of HMF (Barclay et al., 2012).

Novel biomaterials have been synthesized by reactions of the hydroxyl groups in levan. For instance, Mutlu (2014) used LMW levan for the preparation of quaternized levan (QL), which was used as a sensor and drug nano-support in cancer therapy. QL was obtained by treatment with glycidyltrimethylammonium chloride and used in nanosystems for paclitaxel in the form of nanoparticles (QL-PXSLNPs). The study also included the encapsulation of lectin with QL based on core micelles; the size of the micellar structures was between 20 and 200 nm (Mutlu, 2014).

Esawy et al. (2013) obtained QL from 2,3-epoxypropyltriethylammonium chloride and found anti-tumor activity against some cancer cell lines (MCF-7, HepG2 and A549). A preliminary test of the cytotoxic activity of levan and QL was performed *in vitro* with Ehrlich ascites carcinoma. The test revealed that the QL activity was higher than that of the free levan, killing 80% of the cells at a concentration of 800 g/ml. This superior performance was also found for the MCF-7 and HepG2 cell lines. Moreover, QL exhibited fibrinolytic activity (Esawy et al., 2013).

Kekez et al. (2016) reported the synthesis of novel macromolecular graft co-polymers based on levan produced by *Bacillus licheniformis* and polystyrene (Lev-g-PS). The synthesis was performed by radical reaction using potassium persulfate (PPS) as the primer. The maximum graft percentage (58.1%) was achieved in a reaction temperature of 70 °C (in N_2 atmosphere), using diethanolamine (DEA) as the amine activator. The authors also proposed the reaction mechanisms of the graft.

Physical modifications of levan and its applications

Physical modification methods generally involve the rupture of the polysaccharide by extra energy of mechanical force, thermal treatments or a combination of the two to obtain fragments with a lower molecular weight (Figure 11) (Kaur et al., 2012; Hebelstrup, 2015).

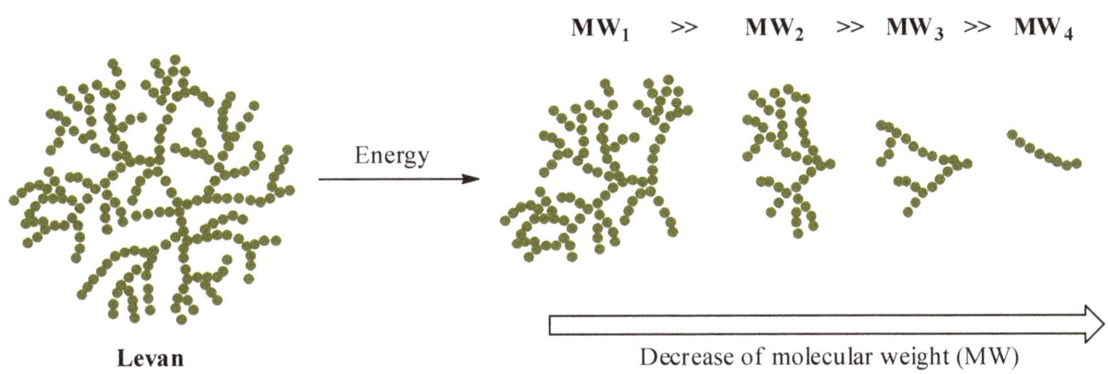

Figure 11. Degraded molecular chains of levan.

The most commonly employed methods are exposure to microwaves, radiation-induced reactions and ultrasound interruption (Li et al., 2016).

Exposures to microwaves

Levan can be hydrolyzed using microwaves to reduce the reaction time. Paula et al. (2008) hydrolyzed levan produced by *Z. mobilis* in a microwave oven at 60% power. The total amount of fructose released after hydrolysis was 73% (0.44 g) in a period of 4.5 minutes. The results show that microwaves were able to reduce the reaction time from a period of hours to minutes. Moreover, this method furnishes an end product without glucose or sucrose, resulting in the industrial production of highly pure FOS (De Paula et al., 2008). Using the same method, Coimbra et al. (2010) produced FOS from the hydrolysis of levan in 3 min and 40 s in a study aimed at the separation of FOS from levan in an aqueous solution using three-phase partitioning.

Avsar, Gunduz and Oner (2018) hydrolyzed a type of linear levan with a high molecular weight using microwave-assisted acidic hydrolysis (De Paula et al., 2008). The aim was to lower the molecular weight of the levan prior to submitting it to sulfation for subsequent use in the fabrication of biomaterials. The fact that levan has a compact globular conformation in an aqueous medium is due to the large number interactions with intermolecular hydrogen (Sima et al., 2012), which impedes a greater range of the sulfate groups. Therefore, the reduction in the molecular weight lessens this entanglement problem and increases the solubility of the polymer (Osman, Oner and Eroglu, 2017).

Radiation-induced treatment

Induction by radiation can cleave polysaccharides and alter their primary structures. Co-60 is often used as a source of gamma rays (Li et al., 2016).

Levan produced by *Bacillus megaterium* type 1 was exposed (in solid form and in a 10% solution) to gamma radiation as an alternative method for producing FOS (Jalan et al., 2013). Levan in powder form demonstrated good stability to degradation by radiation at a dose of up to 150 kGy. In the 10% aqueous solution, gamma radiation of levan at a dose of 250 kGy produced 63.0% FOS with a mean molecular weight of 1250 D in comparison to acidic hydrolysis with HCl 0.5 N for 60 min, which led to a lower yield (23.1%) (Jalan et al., 2013). The authors also found that the viscosity of the levan solution diminished from 2.74 to 1.67 cps when irradiated at 150 kGy and to 1.46 cps when the dose was 250 kGy.

In summary, physical modification methods are normally used on high molecular weight (HMW) polysaccharides with recognized bioactivities and the aim of reducing the molecular weight is to enhance such activities (Li et al., 2016).

Biological modifications of levan and its applications

Biological modifications of polysaccharides occur mainly by enzymatic catalysis (Li et al., 2016). There has been a recent intensification of studies on enzymes that affect the biosynthesis and modification of polysaccharides as well as the production of oligosaccharide derivatives and glycosylated organic compounds, especially with regard to fructosyltransferases (FTFs) due to their wide availability and easy heterologous expression (Ortiz-Soto and Seibel, 2014).

FTFs are multifunctional enzymes grouped by the Carbohydrate-Active enzymes (CAZY) database into the glycosyl hydrolase families GH32 and GH68 (Ortiz-Soto and Seibel, 2014). GH32 encompasses vegetal fructosyltransferases, levanases, invertases, inulinases and fructofuranosidases (Ortiz-Soto and Seibel, 2014), whereas GH68 includes inulosucrase, which synthesizes inulin, and levansucrase, which produces levan (Ortiz-Soto and Seibel, 2014; Raga-Carbajal et al., 2015). In the GH68 family, levansucrase is the main enzyme for the production of levan and derivatives through the transfructosylation of sucrose (Bakar and Kaplan-Türköz, 2017) and is expressed by the supplementation of this sugar during fermentation (Li et al., 2015).

Levansucrase and its mechanism of action

The enzyme levansucrase (sucrose: 2,6-β-D-fructan 6-β-D-fructosyltransferase; EC 2.4.1.10) is widely distributed among a number of bacterial species from the genera *Acetobacter*, *Aerobacter*, *Azotobacter*, *Bacillus*, *Corynebacterium*, *Erwinia*, *Gluconobacter*, *Mycobacterium*, *Pseudomonas*, *Streptococcus*, *Zymomonas* and *Halomonas*, where it cleaves sucrose to generate and transfer fructosyl groups to a variety of acceptor molecules. The acceptors can be monosaccharides, disaccharides or oligosaccharides (transfructosylation), a growing fructan chain (polymerization) or a water molecule, resulting in the hydrolysis of sucrose (Li et al., 2015; Oner, Hernández and Combie, 2016; Eroglu et al., 2017).

In Gram-positive bacteria, such as *B. subtilis, B. megaterium* and *Streptococcus salivarius*, levansucrase tends to produce levan without the accumulation of intermediate FOS (Chambert, Gonzy-Treboul and Dedonder, 1974). The high polymerization of levan occurs by a processing mechanism, in which the growing chain remains linked to the enzyme and is progressively elongated (Ozimek et al., 2006). In Gram-negative species, such as *G. diazotrophicus, E. amylovora* and *Pseudomonas syringae*, levansucrase preferentially produces FOS (Bekers et al., 2002; Visnapuu et al., 2015; Oner, Hernández and Combie, 2016). In this case, the process occurs by a non-processing mechanism, in which the FOS accumulated after each transference of fructosyl are subsequently released from the enzyme (Oner, Hernández and Combie, 2016).

Ozimek et al. (2006) addresses this duality of processing and non-processing mechanisms in the synthesis of fructans with regard to differences in specificities between FTFs that generate HMW fructans and those that only generate FOS. In Gram-negative bacteria, sucrose and its products are immersed in the periplasmic space and there is resistance to mass transfer to the culture medium. In contrast, in Gram-positive bacteria, the enzyme is in direct contact with the culture medium

(Gonzalez-Garcinuño et al., 2017.b). The levansucrase activities in bacteria are depicted in Figure12.

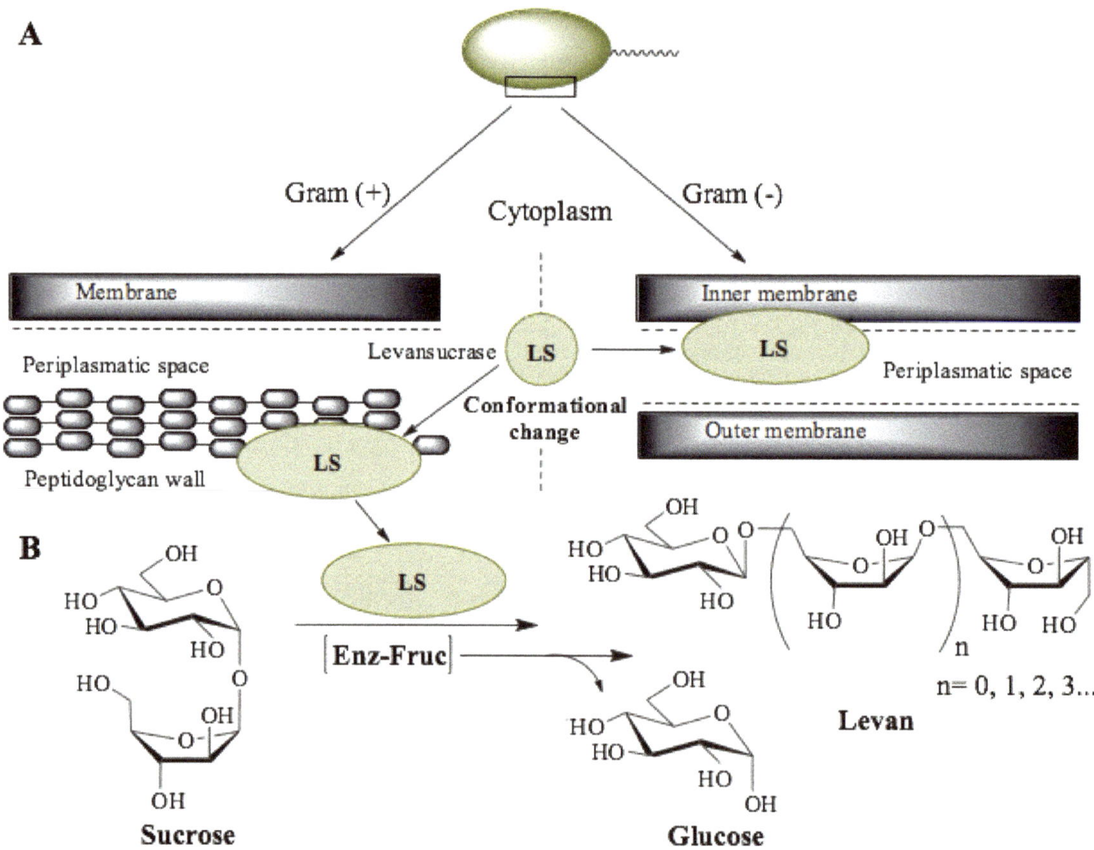

Figure 12. Representation of levansucrase activity in Gram-negative (A, left) and Gram-positive (B, right) bacteria. The biosynthesis of levansucrase occurs in the cell and is accumulated in the periplasmatic spaces, where it adopts its final conformation. In B, levansucrase is excreted outside the cell, where is acts on sucrose molecules to produce levan by transfructosylation reactions.

Levansucrases function through a dual-exchange mechanism (also known as the ping-pong mechanism), in which a fructosyl-enzyme intermediate is formed by covalent bonds (Schematic 1.B) (Chambert et al., 1974; Srikanth et al., 2015; Oner, Hernández and Combie, 2016). Chambert, Gonzy-Treboul and Dedonder (1974) proposed this mechanism, in which the enzyme forms an intermediate complex with fructose (enzyme-fructose) from a sucrose molecule (ping), releasing glucose (pong). For instance, the enzyme-fructose complex may interact with a water molecule (ping) and release a second fructose (pong) (Flores-Maltos et al., 2014; Oner, Hernández and Combie, 2016).

Transferase activity versus levansucrase hydrolysis

The specificity of the levansucrase reaction (*i.e.*, transfructosylation/hydrolysis) is modulated by different factors, such as the concentration of sucrose and enzyme, temperature, pH, the presence of organic solvents in the reaction medium, etc. (Li et al., 2015; Oner, Hernández and Combie, 2016).

A high concentration of sucrose normally promotes a transfructosylation reaction and reduces the hydrolysis of the sucrose (Porras-Domínguez et al., 2015). Ozimek et al. (2006) found that the hydrolase activity of levansucrase isolated from *Lactobacillus reuteri* 121 was much greater with the depletion of sucrose. In turn, high levels of levansucrase isolated from *B. subtilis* resulted in a significant reduction in the levan size (Raga-Carbajal et al., 2016). Adamberg et al. (2015) report similar findings.

The pH is another factor that exerts an influence on the transfructosylation/hydrolysis relationship. According to Goldman et al. (2008), at pH > 7, levansucrase produced by *Z. mobilis* exhibits hydorlase activity, whereas at pH < 6, transferase is much greater for the formation of levan.

Other factors, such as temperature and ionic strength, can determine one type of reaction over the other. Nakapong et al. (2013) found that the molecular weight of the levan produced by levansucrase from *Bacillus licheniformis* RN-01 was controlled by the reaction temperature; levansucrase produced HMW levan (612 kDa) at a high temperature (50 °C) and LMW levan (11 kDa) at 30 °C. The authors also found that ionic strength exerts a considerable impact on levansucrase activity, which is reflected in the molecular weight of the levan. When NaCl (0.5 M) was added, the molecular weight of levan diminished from 612 to 11 kDa (Nakapong et al., 2013). Therefore, an increase in ionic strength can result in changes on LMW of levan (Yamamoto et al., 1985; Chambert and Petit-Glatron, 1989).

The use of organic solvent promotes a transfructosylation reaction over a hydrolysis reaction (Castillo and López-Munguía, 2004; Oner, Hernández and Combie, 2016). Castillo and López-Munguıa (2004) found that the activity of levansucrase from *B. subtilis* was increased by more than 40% when incubated in 40-50% 2-methyl-2-propanol, also increasing transfructosylation by 40% in comparison to the reaction in the buffer alone (85 and 45%, respectively). Vigants et al. (2013) found that ethanol at a concentration of 25% (v/v) at temperatures lower than 40 °C had a positive effect on the transferase activity of levansucrase in the production of fructooligosaccharides.

In summary, the specificity of levansucrase acting as fructosyltransferase or hydrolase depends on the source of the enzyme and the conditions of the medium and is characterized by the ratio of the frutosyltransferase activity to hydrolyzing activity.

Applications of fructosyltransferases/levansucrases

Several enzymes originating from different microorganisms have been identified and used in the production of FOS and fructose polymers using sucrose-rich substrates. The use of enzymes has been proposed as an effective, economical method for the large-scale production of LMW fructooligosaccharides or fructose polymers from sucrose (Bali et al., 2013; Singh, 2017).

Beginning with sucrose as the donor of fructosyl groups and a set of mono-, di- and oligosaccharides as well as sugar and aliphatic alcohols as fructosyl acceptors, fructosyltransferases/levansucrases can synthesize a variety of products with diverse applications (Lu et al., 2014).

Production of levan fructooligosaccharides

Fructooligosaccharides are fructose oligomers containing a single residue of glucose, such as 1-kestose (GF2), nystose (GF3) and ^{1}F-fructofuranosyl nystose (GF4). Depending on the origin, FOS can be of the inulin type (^{1}F-FO), levan type (^{6}F-FO) or a neo-fructo-oligosaccharide (^{6}G-FO). FOS are used as food ingredients due to their prebiotic effects, which enhance the growth of bifidobacteria in the intestinal flora (Kurakake et al., 2010). FOS are calorie free and can be used as alternative sweeteners, as they have one third the sweetness of sucrose (Zambelli et al., 2016).

The production of FOS for human consumption mainly involves the inulin type (Tian et al., 2014). However, levan fructooligosaccharides (LFOS) have been recently described with applications in the pharmaceutical and food industries due to their beneficial properties, such as low intrinsic viscosity and high solubility in water (Kanjanatanin 2016; Bakar and Kaplan-Türköz, 2017).

Experimental results demonstrate that LFOS exhibit superior prebiotic properties in comparison to the inulin type (Zambelli, 2014; Visnapuu, Mardo and Alamae, 2015). A prebiotic is a food ingredient that is not hydrolysable by mammal enzymes (Mutanda et al., 2014; Singh et al., 2017). Marx et al. (2000) tested LFOS prepared by acid hydrolysis of levan from *Z. mobilis* on the growth of bifidobacteria (*Bifidobacterium adolescentis, B. longum, B. breve* and *B. pseudocatenulatum*). The bifidobacteria were able to ferment the LFOS and produce short-chain fatty acids, with the greatest production achieved by *B. adolescentis*.

Porras-Dominguez et al. (2014) recently investigated the prebiotic activity of LFOS for the growth of six strains of probiotic bacteria (*Bifidobacterium longum* ATCC 15707, *B. longum var. infantis* ATCC 17930, *B. longum var. infantis* NRRL 41661, *B. breve* NRRL41408, *B. bifidum* ATCC 29521 and *Lactobacillus paracasei*) and found that *B. bifidum*, *B. longum var. infantis* NRRL 4661 and *B. breve* exhibited the best growth in LFOS (Porras-Dominguez et al., 2014).

Levanases exhibit glycosyl hydrolase activity, leading to the production of LFOS. For instance, endolevanse from *B. subtilis* 168 has high specificity for β(2–6) linkages of the bacterial levan substrate (Jensen et al., 2016). Mardo et al. (2017) studied the biochemical properties of endolevanase BT1760 in six bacterial levans synthesized by levansucrase Lsc3 from *P. syringae* pv. tomato, mutant Asp300Asn, levansucrases from *Z. mobilis*, *Erwinia herbicola*, *H. smyrnensis* and levans isolated from cattail. BT1760 successfully degraded levans into LFOS with a 2 to 13 degree of polydispersity at body temperature (37 °C) and pH between 5 and 6. The authors concluded that levans derived from plants or bacteria serve as a prebiotic for

Bacteroides thetaiotaomicron by promoting the synthesis of short-chain fatty acids by intestinal microbiota (Mardo et al., 2017).

Fructosidases, particularly exo-β-fructosidases, can act on levan to produce fructose (Munoz-Gutierrez 2009). Saito et al. (2000) produced an exoactive enzyme from *Streptomyces exfoliatus* F3-2, which extracellulary converted levan into 1-kestose, nystose, 1-fructosylnystose and levanbiose. Additionally, the enzyme catalyzed the hydrolysis of these LFOS, with the release of fructose. The various physiological functions of these LFOS are related to their respective structures (Saito et al., 2000).

The low yields of short-chain FOS have led researchers to search for alternative enzyme sources, such as fungi. Tian et al. (2014) report a dual-enzymatic system composed of levansucrase from *Bacillus amyloliquefaciens* and the endo-inulinase from *Aspergillus niger* used to produce LFOS and oligolevans from sucrose. Levansucrase catalyzed the synthesis of levan, whereas endo-inulinase regulated the molecular size of the product. The dual-enzymatic system achieved greater yield and productivity (67% p/p and 96 g/L/h, respectively) than the enzymatic system with levansucrase alone (3.0% p/p and 0.8 g/L/h, respectively) (Tian et al., 2014).

Other fungi have been reported to be enzyme sources for the production of LFOS. Zambelli et al. (2014) found that two filamentous fungi (*Penicillium sizovae* CK1 and *Cladosporium cladosporioides* CF215) exhibit transfructosylation activity for the production of LFOS. A greater concentration of LFOS was obtained with *C. cladosporioides* due to the 93% conversion of sucrose, indicating the predominance of transglycosylation over hydrolysis. Maximum LFOS yields were 56% for *C. cladosporioides* and 31% for *P. sizovae* (Zambelli et al., 2014).

Another point to consider in the large-scale production of FOS is the use of immobilized enzymes (Bali et al., 2013; Flores-Maltos et al., 2014). For instance, levansucrase Y246S (LsRN-Y246S) from *Bacillus licheniformis* RN-01, which is a mutant enzyme that produces LFOS, was immobilized by covalent bond in chitosan spheres. The stability of immobilized LsRN-Y246S was greater than that of the free enzyme, with the retention of more than 70% of its activity after 13 weeks of storage at 4 °C and 68% after 12 hours of incubation at 40 °C. Moreover, the immobilized

LsRN-Y246S tolerated sucrose concentrations of up to 70% (p/v), retaining more than 85% of its activity (Sangmanee et al., 2016).

The heterologous expression of specific in FOS-producing enzymes has also been described. Avila-Fernandez et al. (2017) described the isolation of a gene from *B. longum* ATCC 15697 that codes a novel β-fructofuranosidase (B.longum_l1). The heterologous expression of this gene in *Escherichia coli* resulted in specificity for LFOS and mixed fructans. B.longum_l1 was able to hydrolyze fructosyl β-(2,1) and β-(2,6) linkages, with its preferential activity for LFOS demonstrated by its specificity and affinity for β-(2,6) linkages (Avila-Fernandez et al., 2017).

LFOS applications depend on their functional properties, which are related to molecular weight and chain length (Mutanda et al., 2014). Kanjanatanin, Pichyangkura and Chunsrivirot (2016) studied the structural and molecular properties of LFOS with chain lengths of 5 ($LFOS_5$), 10 ($LFOS_{10}$) and 15 ($LFOS_{15}$) using replica-exchange molecular dynamics (REMD) as the simulation method in AMBER 14. The authors determined that LFOS tend to form helicoidal structures as the length of the chain increases from five to 15 residues. This information is important for the selection of LFOS with appropriate lengths and properties for pharmaceutical and biological applications (Kanjanatanin, Pichyangkura and Chunsrivirot, 2016).

Production of sucrose analogs and fructosylated compounds

The recent increase in interest for levansucrases is due to the fact that these enzymes have a broad spectrum for fructosyl receptors, which enables generating a variety of different transfructosylation products (Visnapuu, Mardo and Alamae, 2015). For instance, levansucrase from *B. subtilis* promotes the transfructosylation of unconventional acceptors, such as monosaccharides, disaccharides, aliphatic alcohols and aromatic alcohols (Li et al., 2015; Oner, Hernández and Combie, 2016).

Levansucrase from *B. subtilis* is capable of transferring fructosyl from sucrose to monosaccharide acceptors, such as galactose, mannose, xylose and fucose, generating the sucrose analogs Gal-Fru, Man-Fru, Xyl-Fru and Fuc-Fru, respectively (Seibel et al., 2006; Biedendieck et al., 2007; Beine, 2008). Tian et al. (2011) found that levansucrase from *B. amyloliquefaciens* uses raffinose as a donor to produce levan oligosaccharides. Lu et al. (2014) found that levansucrase from *B. licheniformis* exhibits fructosylation activity for L-arabinose, which is an uncommon pentose, but may be a promising calorie-free sweetener (Krog-Mikkelsen et al., 2011). These sucrose analogs represent an interesting source of fructosyl to obtain novel oligosaccharides that could one day be evaluated with regard to their biological role and profile as functional foods (Seibel et al., 2006; Biedendieck et al., 2007; Beine et al., 2008).

The transfructosylation reaction is an important method for the production of lactosucrose, which is a trisaccharide composed of D-glucose, D-galactose and D-fructose (Silvério et al., 2015; Singh et al., 2017). Lactosucrose is a calorie-free sweetener with a prebiotic effect (Mu et al., 2013) that has interesting functional properties, such as the absorption of minerals in the intestine, reduction in the accumulation of body fat and optimization of the metabolism of amino acids, and exhibits an immunoregulating function by inhibiting IgE-mediated allergies (Li et al., 2015; Silvério et al., 2015; Singh et al., 2017).

Due to its beneficial effects, the production of lactosucrose has piqued the interest of researchers (Mu et al., 2013). However, the productivity of this tri-saccharide strongly depends on the microbial strains used (Silvério et al., 2015). Park et al. (2005) studied 11 types of microorganisms with intracellular levansucrase activity. All strains were capable of producing lactosucrose from sucrose and lactose, with the greatest productivity achieved by *B. subtilis* KCCM 32835 (Park et al., 2005).

The production of lactosucrose has recently been studied using levansucrase from *Bacillus methylotrophicus* SK21.002 (Wu et al. 2015) and recombinant levansucrases from *B. subtilis* NCIMB 11871 (Seibel et al., 2006), *Z. mobilis* (Han et al., 2009) and *B. licheniformis* 8-37-0-1 (Li et al., 2015).

Besides the synthesis of levan and oligosaccharides, levansucrase also catalyzes the transference of fructosyl from sucrose to aliphatic and aromatic alcohols (Seibel et al., 2006; Beine et al., 2008; Mena-Arizmendi et al., 2011; Visnapuu et al., 2011). Oseguera et al. (1996) report moderate activity in the transfer of fructosyl unit by levansucrase from *Bacillus circulans* to sorbitol and glycerol. Kim et al. (1998) found that levansucrase from *Rahnella aquatilis* catalyzes the formation of fructoside alkyl from sucrose in the presence of alcohols. In a subsequent study, the authors improved the production of methyl-β-D-fructoside (MF) using methanol as the receptor of fructosyl and sucrose (50 $g.L^{-1}$), achieving a 70% yield of MF (Kim et al., 2000). The enzyme from *B. licheniformis* 8-37-0-1 expressed in *Escherichia coli* exhibited ample transglycosylation capacity for branched alcohol, such as isopropanol and 1-pentanol, forming novel fructoside alkyls (Lu et al., 2014; Li et al., 2015).

Some aromatic alcohols have also been reported to be acceptor molecules in transfructose reactions performed by levansucrase. Kang et al. (2009) report the synthesis of fructoside 4-hydroxyphenyl-β-D-fructofuranoside (HQ-Fru) by levansucrase from *Leuconostoc mesenteroides*, using hydroquinone (HQ) as the acceptor and sucrose as the fructosyl donor. Beginning with 350 mM of HQ and 115 mM of sucrose, the productivity of HQ-Fru was 1.09 $g.L^{-1}$ (Kang et al., 2009). Mena-Arizmendi et al. (2011) report that levansucrase from *B. subtilis* has broad acceptance for aromatic and aliphatic alcohols. According to the authors, aromatic alcohols are more prone to fructosylation than primary alcohols due to the greater reactivity of phenolic groups inherent to their deprotonation capacity (Mena-Arizmendi et al., 2011).

The synthesis of the fructosides Fru-Cat, Fru-Rsr, Fru-4mPh, Fru2-4HB and Fru-Ph from the alcohols catechol (Cat), resorcinol (Rsr), 4-methoxyphenol (4mPh), 4-hydroxybenzylic alcohol (4HB) and phenol (Ph) has also been reported (Mena-Arizmendi et al., 2011). In summary, enzymatic glycosylation by microbial levansucrase is an alternative approach for the production of novel glycosides (Li et al., 2015).

Production of difructose dianhydrides (DFA IV)

In recent years, growing attention has been given to the production of difructose dianhydrides (DFAs) (Wang et al., 2015), which are cyclic disaccharides composed of two fructose residues, with the formation of two reciprocal glycosidic bonds (Saito et al., 2000; Garcia-Moreno et al., 2008). There are three types of DFAs (DFA I, DFA III and DFA V) derived from the conversion of inulin by corresponding enzymes, whereas DFA IV is obtained from the conversion of levan by levan fructosyltransferase (LFTase, EC.4.2.2.16) (Wang et al., 2015; Hang et al., 2017).

DFA IV (β-D-fructofuranose-2',6:2,6'-dianhydride) forms white crystals that are soluble in water, with a fusion point of 177-178 °C and optical rotation of [a] 20-32° (c 1.21, H_2O) (Saito et al., 2000). The DFA IV molecule is very stable due to the presence of the intramolecular dioxane ring, which makes it go undigested in the gastrointestinal tract, where it serves as a functional carbohydrate, improving physiological functions (Hang et al., 2017). For instance, Saito et al. (1999) report the effects of DFA IV on the absorption of calcium and its physiological functions after being metabolized by intestinal microorganisms in rats. The increase in the absorption of calcium was verified both *in vitro* and *in vivo* and was due mainly to the conversion of DFA IV into acetate, butyrate and lactate by microorganisms in the large intestine of the rats (Saito et al., 1999; Saito et al., 2000).

The literature reports the production of DFA IV by LFTase. Tanaka et al. (1981) were the first researchers to report the use of LFTase from *Arthrobacter ureafaciens* to obtain DFA IV from the conversion of levan through an intramolecular transfructosylation reaction (Hang et al., 2017). Saito et al. (1999; 2000) also report the effective production of DFA IV by LFTase from *Arthrobacter nicotinovorans* GS-9 using levan as substrate. Jang et al. (2003) report the isolation of a novel DFA IV-producing LFTase from *Arthrobacter oxydans* J17-21, which is a soil bacterium. The yield of the final conversion of levan into DFA IV was 35%, but the product proved to be very stable in acid pH and at high temperatures (Jang et al., 2003). Kikuchi et al. (2010) developed a novel two-step DFA IV production system. The first step was the conversion of sucrose into levan by *Serratia levanicum* NN and the second step consisted of the conversion of levan into DFA IV by LFTase from *A. nicotinovorans* GS-9. A total of 6.8 kg of DFA IV was obtained, with a conversion yield of 90.7% (p/p) (Kikuchi et al., 2010).

The potential applications of DFA IV remain scarce. Yi et al. (2017) recently investigated the application of DFA IV for *in vitro* fertilization (IVF) in pigs. The total fertilization rate (mono + polysemic oocytes) was greater with the addition of 0.1% DFA IV (94.2%). The oocytes fertilized with this concentration exhibited greater embryonic development and the formation of blastocysts. The authors also found that levels of radical oxygen species during incubation diminished after the addition of DFA IV. Thus, the addition of DFA IV during IVF enhances fertilization and embryonic development.

DFAs are compatible with a broad gamut of reaction conditions due to their stability in the presence of heat and acidic hydrolysis, making these compounds an attractive synthetic intermediary for applications as building blocks in chemical synthesis (García-Moreno et al., 2008). In the food industry, DFA IV is expected to be used as a natural calorie-free sweetener in the years to come due to the fact that it has half the sweetness as sucrose (Hang et al., 2017).

Conclusion and perspectives

Although known for more than a century, levan polysaccharide has regained scientific and industrial interest only within the last few decades. Commercial polysaccharides like starch, chitosan, alginate and cellulose are currently dominating the market however, numerous reports point to the high potential of levan and indeed levan as well as its derivatives are expected to outperform these extensively studied biopolymers on various platforms. Different molecular modification methods can be applied to the biopolymer levan in order to improve its properties, versatility and applications. Chemical modifications of levan with sulfated and acetylated groups have generated pharmacologically active compounds. In addition, derivatives in carboxymethylated and dialdehyde forms have been used in development of controlled release formulations of drugs and enzymatic immobilization.

Physical modification is another promising method and encompasses a large number of innovative techniques. The physical modification of levan has produced FOS of varied sizes and molecular weights through the use of microwaves or radiation.

FOS and low molecular weight levan have also been produced using biological methods through the modification of high molecular weight levan by enzymes, such as levansucrase. These FOS exhibit prebiotic properties and a positive effect on the absorption of minerals, which makes them valuable products in nutraceutical industry.

Further studies on novel modification methods for levan are of considerable importance to broadening its bioactivities and applications in the near future. Trends on combining chemical and physical methods – such as ultrasound or microwaves – could be adopted to obtain more diversified levans. It could result in variations in their structure and composition as well as more promising properties. The use of enzymes is currently limited to certain types of microorganisms. The research and development of other types of enzymes, such as transferase and synthase, can enrich the application of enzymatic technology for the structural modification of levan. The acquisition of further knowledge on the bioactive mechanisms of modified levan could pave the way for new pre-clinical studies and trials and enable its use in the treatment of diseases.

Acknowledgements

The authors acknowledge the financial support given by the Brazilian agencies CNPq and CAPES. E.C.S thanks the concession of the Doctor degree scholarship by CAPES. The authors have been also supported by grants of PFA/UPE 2016.

References

Abdel-Fattah AM, Gamal-Eldeen AM, Helmy WA, Esawy MA. Antitumor and antioxidant activities of levan and its derivative from the isolate *Bacillus subtilis* NRC1aza. Carbohydr Polym [Internet]. 2012;89(2):314–22. http://dx.doi.org/10.1016/j.carbpol.2012.02.041.

Adamberg K, Tomson K, Talve T, Pudova K, Puurand M, Visnapuu T, et al. Levan enhances associated growth of Bacteroides, Escherichia, Streptococcus and Faecalibacterium in fecal microbiota. PLoS One,2015;2:1–18. http://dx.doi.org/10.1371/journal.pone.0144042.

Ahmad NH, Mustafa S, Man YBC. Microbial polysaccharides and their modification approaches: a review. Int J Food Prop, 2015;18(2):332–47. http://dx.doi.org/10.1080/10942912.2012.693561.

Akçay A. Investigation of sulfate-modified levan polysaccharides and their bioactivity: Marmara University; 2014.

Ana Gabriela Bezerra S, Eduardo W, Belian F. Rastreamento de fármacos via espectroscopia de luminescência, 2009. p. 1–4. Available from: http://www.eventosufrpe.com.br/jepex2009/cd/lista_area_13.htm.

Angeli R, Da Paz NVN, Maciel JC, Araújo FFB, Paiva PMG, Calazans GMT, Valente AP, Almeida FCL, Coelho LCBB, Carvalho Jr. LB, Silva MPC, Correia MTS. Ferromagnetic levan composite: an affinity matrix to purify lectin. J Biomed Biotechnol. 2009.doi:10.1155/2009/179106

Arvidson SA, Rinehart BT, Gadala-Maria F. Concentration regimes of solutions of levan polysaccharide from Bacillus sp. Carbohydr Polym. 2006;65(2):144–9. https://doi.org/10.1016/j.carbpol.2005.12.039

Ávila-Fernández, Cuevas-Juárez E, Rodríguez-Alegría ME, Olvera C, López-Munguía A. Functional characterization of a novel β-fructofuranosidase from Bifidobacterium longum subsp. infantis ATCC 15697 on structurally diverse fructans. J Appl Microbiol. 2016;121(1):263–76. https://doi.org/10.1111/jam.13154

Avsar G, Agirbasli D, Agirbasli MA, Gunduz O, Oner ET. Levan based fibrous scaffolds electrospun via co-axial and single-needle techniques for tissue engineering applications. Carbohydr Polym, 2018;193:316–25. https://doi.org/10.1016/j.carbpol.2018.03.075.

Axente E, Sima F, Sima LE, Erginer M, Eroglu MS, Serban N, et al. Combinatorial MAPLE gradient thin fi lm assemblies signalling to human osteoblasts. Biofabrication. 2014;6:1–9. doi: 10.1088/1758-5082/6/3/035010

Bakar B, Kaplan-Türöz B. Structural modelling and structure-function analysis of *Zymomonas mobilis* levansucrase. Journal of Natural and Applied Sciences, 2017;21(1):279-285. https://doi.org/10.19113/sdufbed.81065

Bali V, Panesar PS, Bera MB, Panesar R. Fructo-oligosaccharides: production, purification and potential applications. Crit Rev Food Sci Nutr. 2015;55(11):1475–90. https://doi.org/10.1080/10408398.2012.694084

Barclay T, Ginic-markovic M, Cooper PD, Petrovsky N. The chemistry and sources of fructose and their effect on functionality and health implications. J Excipients Food Chem. 2012;3(June):67–83.

Barone JR, Medynets M. Thermally processed levan polymers. Carbohydr Polym. 2007;69:554–61. https://doi.org/10.1016/j.carbpol.2007.01.017

Bayoumi1 FS, Amir azza ME, S.O2. ED, Zaher HA el, S.A H. Immunomodulatory and chemo preventive activity of Bacillus subtilis sulphated levan. Life Sci J. 2012;9(4):2846–56.

Bayoumi, F.S., El Amir, A.M., El Deeb, S.O., el-Zaher, H.A., Haiam, S.A., 2013. In vitro study of a bacterial polysaccharide as anti-cancer. J. Appl. Sci. Res. 9, 2007–2018.

Beine R, Moraru R, Nimtz M, Na'amnieh S, Pawlowski A, Buchholz K, et al. Synthesis of novel fructooligosaccharides by substrate and enzyme engineering. JBiotechnol.2008;138(1–2):33–41.https://doi.org/10.1016/j.jbiotec.2008.07.1998

Bekers M, Upite D, Kaminska E, Laukevics J, Grube M, Vigants A, et al. Stability of levan produced by Zymomonas mobilis. Process Biochem. 2005;40(5):1535–9. https://doi.org/10.1016/j.procbio.2004.01.052

Berg A, Oner ET, Combie J, Schneider B, Ellinger R, Weisser J, et al. Formation of new, cytocompatible hydrogels based on photochemically crosslinkable levan methacrylates. Int J Biol Macromol, 2018;107:2312–9. http://dx.doi.org/10.1016/j.ijbiomac.2017.10.120

Biedendieck R, Beine R, Gamer M, Jordan E, Buchholz K, Seibel J, et al. Export, purification, and activities of affinity tagged Lactobacillus reuteri levansucrase produced by Bacillus megaterium. Appl Microbiol Biotechnol. 2007;74(5):1062–73. DOI: 10.1007/s00253-006-0756-0

Byun BY, Lee SJ, Mah JH. Antipathogenic activity and preservative effect of levan (β-2,6-fructan), a multifunctional polysaccharide. Int J Food Sci Technol. 2014;49(1):238–45. https://doi.org/10.1111/ijfs.12304

Calazans GMT, Lima RC, De França FP, Lopes CE. Molecular weight and antitumour activity of *Zymomonas mobilis* levans. Int J Biol Macromol. 2000;27(4): 245–7. https://doi.org/10.1016/S0141-8130(00)00125-2

Castillo E, López-Munguía A. Synthesis of levan in water-miscible organic solvents. J Biotechnol. 2004;114:209–17. https://doi.org/10.1016/j.jbiotec.2004.06.003

Chambert R, Gonzy-Treboul G, Dedonder R. Kinetic studies of levanosucrase of *Bacillus subtilis*. Eur J Biochem. 1974;41:285–300. https://doi.org/10.1111/j.1432-1033.1974.tb03269.x

Chambert R, Petit-Glatron M-F. Study of the effect of organic solvents on the synthesis of levan and the hydrolysis of sucrose by *Bacillus subitilis* levan. Carbohydr Res. 1989;191:117–23. https://doi.org/10.1016/0008-6215(89)85051-7

Chatterjee C, Pong F, Sen A. Chemical conversion pathways for carbohydrates. Green Chem. 2015;17(1):40–71. DOI: 10.1039/C4GC01062K

Coimbra CG d O, Lopes CE, Calazans GMT. Three-phase partitioning of hydrolyzed levan. Bioresour Technol. 2010;101(12):4725–8. http://dx.doi.org/10.1016/j.biortech.2010.01.091.

Costa RR, Neto AI, Calgeris I, Correia CR, Pinho ACM, Fonseca J, et al. Adhesive nanostructured multilayer films using a bacterial exopolysaccharide for biomedical applications. J Mater Chem B. 2013;1(18):2367. DOI: 10.1039/C3TB20137F

De Paula VC, Pinheiro IO, Lopes CE, Calazans GMT. Microwave-assisted hydrolysis of *Zymomonas mobilis* levan envisaging oligofructan production. Bioresour Technol. 2008;99(7):2466–70. https://doi.org/10.1016/j.biortech.2007.04.062

Erginer M, Akcay A, Coskunkan B, Morova T, Rende D, Bucak S, et al. Sulfated levan from *Halomonas smyrnensis* as a bioactive, heparin-mimetic glycan for cardiac tissue engineering applications. Carbohydr Polym, 2016;149:289–96. http://dx.doi.org/10.1016/j.carbpol.2016.04.092.

Eroglu MS, Oner ET, Mutlu EC, Bostan MS. Sugar based biopolymers in nanomedicine; new emerging era for cancer imaging and therapy. Curr Top Med Chem. 2017;17:1–14. doi: 10.2174/1568026616666161222101703.

Esawy MA, Abdel-Fattah AM, Ali MM, Helmy WA, Salama BM, Taie HAA, al. Levansucrase optimization using solid state fermentation and levan biological activities studies. Carbohydr Polym, 2013;96(1):332–41. http://dx.doi.org/10.1016/j.carbpol.2013.03.089

Faten S. Bayoumil, Amir AM El, S.O ED, el Zaher HA, Haiam SA. In vitro study of a bacterial polysaccharideas Anti Cancer 1Faten. 2013;9(3):2007–18.

Flores-Maltos DA, Mussatto SI, Contreras-Esquivel JC, Rodríguez-Herrera R, Teixeira JA, Aguilar CN. Biotechnological production and application of fructooligosaccharides. Crit Rev Biotechnol. 2016;36(2):259–67. https://doi.org/10.3109/07388551.2014.953443

Freitas F, Alves VD, Reis MAM. Advances in bacterial exopolysaccharides: from production to biotechnological applications. Trends Biotechnol, 2011;29(8):388–98. http://dx.doi.org/10.1016/j.tibtech.2011.03.008.

García-Moreno MI, Benito JM, Mellet CO, Fernández JMG. Chemical and enzymatic approaches to carbohydrate-derived spiroketals: Di-D-fructose dianhydrides (DFAs). Molecules. 2008;13(8):1640–70. https://doi.org/10.3390/molecules13081640

Goldman D, Lavid N, Schwartz A, Shoham G, Danino D, Shoham Y. Two active forms of *Zymomonas mobilis* levansucrase: an ordered microfibril structure of the enzyme promotes levan polymerization. J Biol Chem. 2008;283(47):32209–17. doi: 10.1074/jbc.M805985200

Gomes TD, Caridade SG, Sousa MP, Azevedo S, Kandur MY, Öner ET, et al.

Adhesive free-standing multilayer films containing sulfated levan for biomedical applications. Acta Biomater, 2018;69:183–95. https://doi.org/10.1016/j.actbio.2018.01.027.

Gonta S, Karsakevida A, Vina I. Periodate oxidation of levan from *Zymomonas mobilis*. Proceeding Latv Acad Sci. 1999;53(5):290–4.

Gonta S, Utinans M, Neilands O, Vina I. Computational analysis of native and modified oligofructosides. J Mol Struct Theochem. 2004;710(1–3):61–4. https://doi.org/10.1016/j.theochem.2004.07.029

González-Garcinuño Á, Tabernero A, Sánchez-Álvarez JM, Galán MA, Martin del Valle EM. Effect of bacteria type and sucrose concentration on levan yield and its molecular weight. Microb Cell Fact. 2017;16(1):1–11. https://doi.org/10.1186/s12934-017-0703-z

González-Garcinuño Á, Tabernero A, Domínguez Á, Galán MA, Martin del Valle EM. Levan and levansucrases: Polymer, enzyme, micro-organisms and biomedical applications. Biocatal Biotransformation.2017;2422(April):1–12. https://doi.org/10.1080/10242422.2017.1314467

Gupta SK, Das P, Singh SK, Akhtar MS, Meena DK, Mandal SC. Microbial levan , an ideal prebiotic and immunonutrient in aquaculture. World Aquac,2011;61–4.

Han YW. Microbial levan. Adv Appl Microbiol. 1990;35:171–94. https://doi.org/10.1016/S0065-2164(08)70244-2

Han W-C. Production of lactosucrose from sucrose and lactose by a levansucrase from *Zymomonas mobilis*. J Microbiol Biotechnol, 2009;19: 1–8. doi: 10.4014/jmb.0901.0045

Hang H. Recent advances on the difructose anhydride IV preparation from levan conversion. Appl Microbiol Biotechnol. 2017;101(20):7477–86. doi: 10.1007/s00253-017-8500-5.

Hebelstrup KH, Sagnelli D, Blennow A. The future of starch bioengineering: GM microorganisms or GM plants? Front Plant Sci. 2015;6:1–6. https://doi.org/10.3389/fpls.2015.00247

Husseiny SM, Sheref FA, Amer H, Elsakhawy TA. Biological activity of chemically modified levan produced by moderately halophilic *Chromohalobacter salexigens* KT989777. 2015;5(3):812–22.

Jalan N, Varshney L, Misra N, Paul J, Mitra D, Rairakhwada DD, et al. Studies on production of fructo- oligosaccharides (FOS) by gamma radiation processing of microbial levan. Carbohydr Polym,2013;96:365–70. https://doi.org/10.1016/

j.carbpol.2013.03.057

Jang KH, Ryu EJ, Park B seek, Song K bang, Kang SA, Kim CH, et al. Levan fructotransferase from *Arthrobacter oxydans* J17-21 catalyzes the formation of the Di-D-fructose dianhydride IV from levan. J Agric Food Chem. 2003;51(9):2632–6. https://doi.org/10.1021/jf026207o

Jensen SL, Diemer MB, Lundmark M, Larsen FH, Blennow A, Mogensen HK, et al. Levanase from *Bacillus subtilis* hydrolyses β-2,6 fructosyl bonds in bacterial levans and in grass fructans. Int J Biol Macromol. 2016;85:514–21. http://dx.doi.org/10.1016/j.ijbiomac.2016.01.008.

Kagimura FY, da Cunha MAA, Barbosa AM, Dekker RFH, Malfatti CRM. Biological activities of derivatized D-glucans: a review. Int J Biol Macromol . 2015;72:588–98. http://dx.doi.org/10.1016/j.ijbiomac.2014.09.008

Kang J, Kim YM, Kim N, Kim DW, Nam SH, Kim D. Synthesis and characterization of hydroquinone fructoside using *Leuconostoc mesenteroides* levansucrase. Appl Microbiol Biotechnol. 2009;83(6):1009–16. doi: 10.1007/s00253-009-1936-5.

Kang SA, Jang K-H, Seo J-W, Kim KH, Kim YH, Rairakhwada D, et al. Levan: applications and perspectives. In: Microbial production of biopolymers and polymer precursors. Norwich: Caister Academic Press.; 2009. p. 145–61.

Kanjanatanin P, Pichyangkura R, Chunsrivirot S. Replica exchange molecular dynamics simulations reveal the structural and molecular properties of levan-type fructo-oligosaccharides of various chain lengths. BMC Bioinformatics.2016;17(1): 4–11. http://dx.doi.org/10.1186/s12859-016-1182-7.

Kasapis S, Morris ER, Gross M, Rudolph K. Solution properties of levan polysaccharide from *Pseudomonas syringae* pv. phaseolicola, and its possible primary role as a blocker of recognition during pathogenesis. Carbohydr Polym. 1994;23(1):55–64. https://doi.org/10.1016/0144-8617(94)90090-6

Katsuichi S, Fusao T. Difructose anhydrides: their mass-production and physiological functions. Biosci Biotechnol Biochem. 2000;64(7):1321–7. https://doi.org/10.1271/bbb.64.1321

Kaur B, Ariffin F, Bhat R, Karim AA. Progress in starch modification in the last decade. Food Hydrocoll. 2012;26(2):398–404. https://doi.org/10.1016/j.foodhyd.2011.02.016

Kekez B, Gojgic-Cvijovic G, Jakovljevic D, Pavlovic V, Beskoski V, Popovic A, et al. Synthesis and characterization of a new type of levan-graft-polystyrene copolymer. Carbohydr Polym.2016;154:20–9. http://dx.doi.org/10.1016/j.carbpol.2016.08.001.

Kim KH, Chung CB, Kim YH, Kim KS, Han CS, Kim CH. Cosmeceutical properties of levan produced by *Zymomonas mobilis*. J Cosmet Sci. 2005;28:231.

Kim S, Bae PK, Chung BH. Self-assembled levan nanoparticles for targeted breast cancer imaging. Chem Commun. 2015;51:107–10. http://dx.doi.org/10.1039/C4CC07679F

Krog-Mikkelsen I, Hels O, Tetens I, Holst JJ, Andersen JR, Bukhave K. The effects of L-arabinose on intestinal sucrase activity: dose-response studies in vitro and in humans. Am J Clin Nutr. 2011;94(2):472–8. https://doi.org/10.3945/ajcn.111.014225

Kurakake M, Masumoto RYO, Maguma K, Kamata A, Saito E, Ukita N, et al. Production of fructooligosaccharides by β-fructofuranosidases from *Aspergillus oryzae* KB. J Agric Food Chem. 2010;58(1):488–92. https://doi.org/10.1002/jctb.967

Kim MG, Seo JW, Song KB, Kim CH, Chung BH, Rhee, SK. Levan and fructosyl derivatives formation by a recombinant levansucrase from *Rahnella aquatilis*. Biotechnology Letters, 1998;20(4):333–336. https://doi.org/10.1023/A:1005310926399

Kim MG, Kim CH, Lee JS, Song KB, Rhee SK. Synthesis of methyl β-D-fructoside catalyzed by levansucrase from *Rahnella aquatilis*. Enzyme Microb Technol. 2000;27(9):646–51. https://doi.org/10.1016/S0141-0229(00)00267-2

Leite Á, Silva JM, Ates O, Eroglu MS, Oner ET, Mano JF. Bioadhesive and bioactive methacrylated nanocomposite levan-based beads for bone tissue applications. In: Oner ET, Flammang P, editors. In: Proceedings of 2nd International Conferance On Biological & Biomimetic Adhesives Istanbul 2014. Istanbul; 2014. p. 85.

Li W, Yu S, Zhang T, Jiang B, Mu W. Recent novel applications of levansucrases. Appl Microbiol Biotechnol. 2015;99:6959–69. https://doi.org/10.1007/s00253-015-6797-5

Li S, Xiong Q, Lai X, Li X, Wan M, Zhang J, et al. Molecular modification of polysaccharides and resulting bioactivities. Compr Rev Food Sci Food Saf. 2016;15(2):237–50. https://doi.org/10.1111/1541-4337.12161

Liu J, Luo J, Ye H, Zeng X. Preparation, antioxidant and antitumor activities in vitro of different derivatives of levan from endophytic bacterium *Paenibacillus polymyxa* EJS-3. Food Chem Toxicol,2012;50(3–4):767–72. http://dx.doi.org/10.1016/j.fct.2011.11.016.

Lu L, Fu F, Zhao R, Jin L, He C, Xu L, et al. A recombinant levansucrase from *Bacillus licheniformis* 8-37-0-1 catalyzes versatile transfructosylation reactions. Process Biochem. 2014;49(9). https://doi.org/10.1016/j.procbio.2014.05.012

Ma L, Chen H, Zhang Y, Zhang N, Fu L. Chemical modification and antioxidant activities of polysaccharide from mushroom Inonotus obliquus. Carbohydr Polym, 2012;89(2):371–8. http://dx.doi.org/10.1016/j.carbpol.2012.03.016.

Maciel JC, Andrad PL, Neri DFM, Carvalho LB, Cardoso CA, Calazans GMT, et al. Preparation and characterization of magnetic levan particles as matrix for trypsin immobilization. J Magn Magn Mater, 2012;324(7):1312–6. http://dx.doi.org/10.1016/j.jmmm.2011.11.017.

Manandhar S, Vidhate S, Souza ND. Water soluble levan polysaccharide biopolymer electrospun fibers. Carbohydr Polym. 2009;78(4):794–8. http://dx.doi.org/10.1016/j.carbpol.2009.06.023.

Mardo K, Visnapuu T, Vija H, Aasamets A, Viigand K, Alamäe T. A highly active endo-levanase BT1760 of a dominant mammalian gut commensal *Bacteroides thetaiotaomicron* cleaves not only various bacterial levans, but also levan of timothy grass. PLoS One. 2017;12(1). https://doi.org/10.1371/journal.pone.0169989

Marx SP, Winkler S, Hartmeier W. Metabolization of β-(2,6)-linked fructose-oligosaccharides by different bifidobacteria. FEMS Microbiol Lett. 2000;182(1):163–9. https://doi.org/10.1111/j.1574-6968.2000.tb08891.x

Mena-Arizmendi A, Alderete J, Águila S, Marty A, Miranda-Molina A, López-Munguía A, et al. Enzymatic fructosylation of aromatic and aliphatic alcohols by *Bacillus subtilis* levansucrase: reactivity of acceptors. J Mol Catal B Enzym. 2011;70(1–2):41–8. https://doi.org/10.1016/j.molcatb.2011.02.002

Méndez-Lorenzo L, Porras-Domínguez JR, Raga-Carbajal E, Olvera C, Rodríguez-Alegría ME, Carrillo-Nava E, et al. Intrinsic levanase activity of *Bacillus subtilis* 168 levansucrase (SacB). PLoS One. 2015;10(11). https://doi.org/10.1371/journal.pone.0143394

Mu W, Chen Q, Wang X, Zhang T, Jiang B. Current studies on physiological functions and biological production of lactosucrose. Appl Microbiol Biotechnol. 2013;97(16):7073–80. DOI https://doi.org/10.1007/s00253-013-5079-3

Muñoz-Gutiérrez I, Rodríguez-Alegría ME, López Munguía A. Kinetic behaviour and specificity of β-fructosidases in the hydrolysis of plant and microbial fructans. Process Biochem. 2009;44(8):891–8. https://doi.org/10.1016/j.procbio.2009.04.013

Mutanda T, Mokoena MP, Olaniran AO, Wilhelmi BS, Whiteley CG. Microbial enzymatic production and applications of short-chain fructooligosaccharides and inulooligosaccharides: recent advances and current perspectives. Vol. 41, Journal of Industrial Microbiology and Biotechnology. 2014. https://doi.org/10.1007/s10295-014-1452-1

Mutlu E. A novel design of levan based tumour targeting nano-scale micelles: Marmara University; 2014.

Nakapong S, Pichyangkura R, Ito K, Iizuka M, Pongsawasdi P. High expression level of levansucrase from Bacillus licheniformis RN-01 and synthesis of levan nanoparticles. Int J Biol Macromol. 2013;54:30–6. http://dx.doi.org/10.1016/j.ijbiomac.2012.11.017.

Oner ET, Hernández L, Combie J. Review of Levan polysaccharide: from a century of past experiences to future prospects. Biotechnol Adv. 2016;34(5):827–44. http://dx.doi.org/10.1016/j.biotechadv.2016.05.002

Ortiz-Soto M, Seibel J. Biotechnological synthesis and transformation of valuable sugars in the food and pharmaceutical industry. Curr Org Chem.2014;18(8):964–86.

Oscarson S, Sehgelmeble FW. Chemical syntheses of inulin and levan structures. J Org Chem. 2002;67(24):8457–62. https://doi.org/10.1021/jo020341q

Osman A, Oner ET, Eroglu MS. Novel levan and pNIPA temperature sensitive hydrogels for 5-ASA controlled release. Carbohydr Polym.2017;165:61–70. http://dx.doi.org/10.1016/j.carbpol.2017.01.097

Ozimek LK, Kralj S, van der Maarel MJEC, Dijhuizen L. The levansucrase and inulosucrase enzymes of *Lactobacillus reuteri* 121 catalyse processive and non-processive transglycosylation reactions. Microbiology. 2006;152(4):1187–96. doi: 10.1099/mic.0.28484-0

Park NH, Choi HJ, Oh DK. Lactosucrose production by various microorganisms harboring levansucrase activity. Biotechnol Lett. 2005;27(7):495–7. https://doi.org/10.1007/s10529-005-2539-6

Perez Oseguera MA, Guereca L, Lopez-Munguia A. Properties of levansucrase from *Bacillus circulans*. Appl Microbiol Biotechnol. 1996;45(4):465–71. https://doi.org/10.1007/BF00578457

Porras-Domínguez JR, Ávila-Fernández Á, Rodríguez-Alegría ME, Miranda-Molina A, Escalante A, González-Cervantes R, et al. Levan-type FOS production using a *Bacillus licheniformis* endolevanase. Process Biochem, 2014;49(5):783–90. http://

dx.doi.org/10.1016/j.procbio.2014.02.005

Porras-Domínguez JR, Ávila-Fernández Á, Miranda-Molina A, Rodríguez-Alegría ME, Munguía AL. *Bacillus subtilis* 168 levansucrase (SacB) activity affects average levan molecular weight. Carbohydr Polym.2015;132:338–44. http://dx.doi.org/10.1016/j.carbpol.2015.06.056.

Raga-Carbajal E, Carrillo-Nava E, Costas M, Porras-Dominguez J, López-Munguía A, Olvera C. Size product modulation by enzyme concentration reveals two distinct levan elongation mechanisms in bacillus subtilis levansucrase. Glycobiology. 2016;26(4). https://doi.org/10.1093/glycob/cwv112

Rahul R, Jha U, Sen G, Mishra S. Carboxymethyl inulin: a novel flocculant for wastewater treatment. Int J Biol , 2014;63:1–7. http://dx.doi.org/10.1016/j.ijbiomac.2013.10.015.

Report. Safety Assessment of Microbial Polysaccharide Gums as Used in Cosmetics. Cosmet Ingred Rev (CIR), 2012;1–52. http:// doi/10.1177/1091581816651606.

Roberts, E.J., Garegg, P.J. Levan derivatives, their preparation, composition and applications including medical and food applications. World patent appl.WO9803184. 1998.

Saito K, Hira T, Suzuki T, Hara H, Yokota A, Tomita F. Effects of DFA IV in rats: calcium absorption and metabolism of DFA IV by intestinal microorganisms. Vol. 63, Bioscience Biotechnology and Biochemistry. 1999. p. 655–61. https://doi.org/10.1271/bbb.63.655

Saito K, Kondo K, Kojima I, Yokota a, Tomita F. Purification and characterization of 2,6-beta-D-fructan 6-levanbiohydrolase from *Streptomyces exfoliatus* F3-2. Appl Environ Microbiol. 2000;66(1):252–6. DOI: 10.1128/AEM.66.1.252-256.2000

Sangmanee S, Nakapong S, Pichyangkura R, Kuttiyawong K. Levan-type fructooligosaccharide production using *Bacillus licheniformis* RN-01 levansucrase Y246S immobilized on chitosan beads. Songklanakarin J Sci Technol. 2016;38(3): 295–303.

Sarilmiser H, Toksoy Oner E. Investigation of anti-cancer activity of linear and aldehyde-activated levan from *Halomonas smyrnensis* AAD6T. Biochem Eng J, 2014;92:28–34. http://dx.doi.org/10.1016/j.bej.2014.06.020.

Sarilmiser H, Ates O, Ozdemir G, Arga KY, Toksoy Oner E. Effective stimulating factors for microbial levan production by *Halomonas smyrnensis* AAD6^{T}. J Biosci Bioeng,2015;119(4):455–63. http://dx.doi.org/10.1016/j.jbiosc.2014.09.019.

Seibel J, Moraru R, Götze S, Buchholz K, Na'amnieh S, Pawlowski A, et al. Synthesis of sucrose analogues and the mechanism of action of *Bacillus subtilis* fructosyltransferase (levansucrase). Carbohydr Res. 2006;341(14):2335–49. https://doi.org/10.1016/j.carres.2006.07.001

Sezer AD, Kazak H, Oner ET, Akbua J. Levan-based nanocarrier system for peptide and protein drug delivery: optimization and influence of experimental parameters on the nanoparticle characteristics. Carbohydr Polym. 2011;84(1):358–63. https://doi.org/10.1016/j.carbpol.2010.11.046

Sezer AD, Kazak Sarılmışer H, Rayaman E, Çevikbaş A, Oner ET, Akbuga J. Development and characterization of vancomycin-loaded levan-based microparticular system for drug delivery. Pharm Dev Technol,2015;7450:1–8. http://doi/full/10.3109/10837450.2015.1116564

Shih I-L, Yu Y-T, Shieh C-J, Hsieh C-Y. Selective production and characterization of levan by *Bacillus subtilis* (Natto) Takahashi. 2005;53(21):8211-8215. https://doi.org/10.1021/jf058084o

Silbir S, Dagbagli S, Yegin S, Baysal T, Goksungur Y. Levan production by *Zymomonas mobilis* in batch and continuous fermentation systems. Carbohydr. Polym. 2014;99:454–461. http://dx.doi.org/10.1016/j.carbpol.2013.08.031.

Silvério SC, Macedo EA, Teixeira JA, Rodrigues LR. Perspectives on the biotechnological production and potential applications of lactosucrose: a review. J Funct Foods,2015;19:74–90. http://dx.doi.org/10.1016/j.jff.2015.09.014.

Sima F, Mutlu EC, Eroglu MS, Sima LE, Serban N, Ristoscu C, et al. Levan nanostructured thin films by MAPLE assembling. Biomacromolecules. 2011;12(6): 2251–6. https://doi.org/10.1021/bm200340b

Sima F, Axente E, Sima LE, Tuyel U, Eroglu MS, Serban N, et al. Combinatorial matrix-assisted pulsed laser evaporation: single-step synthesis of biopolymer compositional gradient thin film assemblies. Appl Phys Lett, 2012;101(233705):1–5. http://dx.doi.org/10.1063/1.4769987

Singh SP, Jadaun JS, Narnoliya LK, Pandey A. Prebiotic oligosaccharides: special focus on fructooligosaccharides, its biosynthesis and bioactivity. Appl Biochem Biotechnol. 2017;183(2):613–35. https://doi.org/10.1007/s12010-017-2605-2

Tabandeh MR, Aminlari M. Synthesis, physicochemical and immunological properties of oxidized inulin-l-asparaginase bioconjugate. J Biotechnol. 2009;141(3–4):189–95. https://doi.org/10.1016/j.jbiotec.2009.03.020

Tabernero A, González-Garcinuno Á, MaSánchez-Álvarez J, Galán MA, del Valle

EMM. Development of a nanoparticle system based on a fructose polymer: stability and drug release studies. Carbohydr Polym. 2017;160:26–33. https://doi.org/10.1016/j.carbpol.2016.12.025

Tanaka T, Yamamoto S, Oi S, Yamamoto T. Structures of heterooligosaccharides synthesized by levansucrase. J Biochem. 1981;90(2):521–6. DOI: 10.1093/oxfordjournals.jbchem.a133500

Tian F, Inthanavong L, Karboune S. Purification and characterization of levansucrases from *Bacillus amyloliquefaciens* in intra- and extracellular forms useful for the synthesis of levan and fructooligosaccharides. Biosci Biotechnol Biochem, 2011;75(10):1929–38. http:// doi/full/10.1271/bbb.110315.

Tian F, Karboune S, Hill A. Synthesis of fructooligosaccharides and oligolevans by the combined use of levansucrase and endo-inulinase in one-step bi-enzymatic system. Innov Food Sci Emerg Technol,2014;22:230–8. http:// doi.org/10.1016/j.ifset.2013.12.004.

Tian F, Khodadadi M, Karboune S. Optimization of levansucrase/endo-inulinase bi-enzymatic system for the production of fructooligosaccharides and oligolevans from sucrose. J Mol Catal B Enzym. 2014;109. https://doi.org/10.1016/j.molcatb.2014.08.005

Tomulescu C, Stoica R, Sevcenco C, Căşărică A, Moscovici M, Vamanu A. Levan - a mini review. Sci Bull Ser F Biotechnol. 2016;XX:309–20.

Ulu A, Ates B. Immobilization of L-Asparaginase on carrier materials: a comprehensive review. Bioconjug Chem. 2017;28(6):1598–610. https://doi.org/10.1021/acs.bioconjchem.7b00217

Vigants A, Upite D, Scherbaka R, Lukjanenko J, Ionina R. An influence of ethanol and temperature on products formation by different preparations of *Zymomonas mobilis* extracellular levansucrase. Folia Microbiol (Praha). 2013;58(1):75–80. https://doi.org/10.1007/s12223-012-0185-7

Vijayendra SVN, Shamala TR. Film forming microbial biopolymers for commercial applications - a review. Crit Rev Biotechnol,2014;34(4):338–57. http:// doi/full/10.3109/07388551.2013.798254

Vina I, Karsakevich A, Bekers M. Stabilization of anti-leukemic enzyme L-asparaginase by immobilization on polysaccharide levan. J Mol Catal - B Enzym. 2001;11(4–6):551–8. https://doi.org/10.1016/S1381-1177(00)00043-6.

Visnapuu T, Mardo K, Alamäe T. Levansucrases of a *Pseudomonas syringae* pathovar as catalysts for the synthesis of potentially prebiotic oligo- and polysaccharides. Vol. 32, New Biotechnology. 2015. p. 597–605. https://doi.org/10.1016/j.nbt.2015.01.009

Wang J, Liu L, Zhang Q, Zhang Z, Qi H, Li P. Synthesized oversulphated, acetylated and benzoylated derivatives of fucoidan extracted from *Laminaria japonica* and their potential antioxidant activity in vitro. Food Chem [Internet]. 2009;114(4):1285–90. http://dx.doi.org/10.1016/j.foodchem.2008.10.082

Wang X, Yu S, Zhang T, Jiang B, Mu W. From fructans to difructose dianhydrides. Appl Microbiol Biotechnol. 2014;99(1):175–88. https://doi.org/10.1007/s00253-014-6238-x

Wang J, Hu S, Nie S, Yu Q, Xie M, Wang J, et al. Reviews on mechanisms of *in vitro* antioxidant activity of polysaccharides. Oxid Med Cell Longev,2016;2016:1–13. http://dx.doi.org/10.1155/2016/5692852

Wang Q, Sheng X, Shi A, Hu H, Yang Y, Liu L, et al. β-Glucans: relationships between modification, conformation and functional activities. Molecules. 2017;22(2):1–13. https://doi.org/10.3390/molecules22020257

Wu C, Zhang T, Mu W, Miao M, Jiang B. Biosynthesis of lactosylfructoside by an intracellular levansucrase from *Bacillus methylotrophicus* SK 21.002. Carbohydr Res, 2015;401:122–6. http://dx.doi.org/10.1016/j.carres.2014.11.001.

Wyrwa R, Berg A, Oner ET, Matthias S. Cross-linkable levan derivatives for biodegradable adhesives. In: Oner ET, Flammang P, editors. In: Proceedings of 2nd International Conferance On Biological & Biomimetic Adhesives Istanbul 2014. Istanbul; 2014. p. 79.

Xue S, Chen X, Lu J, Jin L. Protective effect of sulfated achyranthes bidentata polysaccharides on streptozotocin-induced oxidative stress in rats. Carbohydr Polym, 2009;75(3):415–9. http://dx.doi.org/10.1016/j.carbpol.2008.08.003

Yamamoto Y, Takahashi Y, Kawano M, Iizuka M, Matsumoto T, Saeki S, et al. In vitro digestibility and fermentability of levan and its hypocholesterolemic effects in rats. J Nutr Biochem. 1999;10(1):13–8. https://doi.org/10.1016/S0955-2863(98)00077-1

Yamamoto S, Iizuka M, Tanaka T, Yamamoto T. The Mode of Synthesis of Levan by Bacillus subtilis Levansucrase. Agric Biol Chem. 1985;49(2):343–9. https://doi.org/10.1271/bbb1961.49.343

Yi Y-J, Kamala-Kannan S, Lim J-M, Oh B-T, Lee S-M. Effects of difructose dianhydride (DFA)-IV on in vitro fertilization in pigs. J Biomed Res. 2017;31(5):453–61. doi: 10.7555/JBR.31.20160115

Yoo SH, Yoon EJ, Cha J, Lee HG. Antitumor activity of levan polysaccharides from selected microorganisms. Int J Biol Macromol. 2004;34(1–2):37–41. https://doi.org/10.1016/j.ijbiomac.2004.01.002

Yoon EJ, Yoo SH, Cha J, Lee HG. Effect of levan's branching structure on antitumor activity. Int J Biol Macromol. 2004;34(3):191–4. https://doi.org/10.1016/j.ijbiomac.2004.04.001

Zambelli P, Fernandez-Arrojo L, Romano D, Santos-Moriano P, Gimeno-Perez M, Poveda A, et al. Production of fructooligosaccharides by mycelium-bound transfructosylation activity present in Cladosporium cladosporioides and Penicilium sizovae. Process Biochem. 2014;49(12):2174–80. https://doi.org/10.1016/j.procbio.2014.09.021

Zambelli P, Tamborini L, Cazzamalli S, Pinto A, Arioli S, Balzaretti S, et al. An efficient continuous flow process for the synthesis of a non-conventional mixture of fructooligosaccharides. Food Chem, 2016;190:607–13. http://dx.doi.org/10.1016/j.foodchem.2015.06.002

Zhang Z, Zhang Q, Wang J, Zhang H, Niu X, Li P. Preparation of the different derivatives of the low-molecular-weight porphyran from Porphyra haitanensis and their antioxidant activities in vitro. Int J Biol Macromol. 2009;45(1):22–6. https://doi.org/10.1016/j.ijbiomac.2009.03.009

Cloning and Partial Characterization of an Extracellular Dextransucrase Coding Region (DSR-V) from *Leuconostoc citreum* M-3

11

Reinaldo H. Fraga Vidal[1*], Sandra Pacios Michelena[2], Roberto C. Arísticas Ribalta[1], Lisandra Martínez Valdés[1], Meinardo Lafargue Gámez[1], Amanda Montes Alvarez[1], Magali Remaud-Siméon[3], Pierre Monsan[3,4]

[1] Cuban Research Institute on Sugarcane By-Products (ICIDCA). Havana, Cuba.
[2] Faculty of Biology, University of Havana, Havana, Cuba.
[3] Laboratoire d'Ingénierie des Systèmes Biologiques et des Procédés. Département de Génie Biochimique et Alimentaire INSA 135, Toulouse Cedex 4, France.
[4] Toulouse White Biotechnology (UMS INRA / INSA / CNRS) ,Ramonville Saint-Agne Toulouse Cedex 4, France.
*Corresponding author: reynaldo.fraga@icidca.azcuba.cu

DOI: https://doi.org/10.21775/9781912530267.11

Abstract
The dextransucrase enzymes synthesize dextran, a glucose polymer with broad industrial applications, making the search for new dextransucrases of great interest. The work described aimed at the partial characterizing of a recombinant dextransucrase enzyme from *Leuconostoc citreum* M-3. From the genomic DNA of strain M-3, an amplicon containing a coding region of a dextransucrase called DSR-V was isolated, and deposited in the GenBank™ (Accession number: KF724950). The amino acid sequence alignment of DSR-V with other dextransucrases demonstrated that it shares a 94% identity with the DSR-D of *L. mesenteroides* Lcc4 and the DSR-S of *L. mesenteroides* NRRL B-512F. The DSR-V was cloned and expressed in *Escherichia coli* JM109 facilitating the formation and detection of DSR-

V specific dextran. The SDS-PAGE soluble fraction zymography of *E. coli* DSR-V and the ^{13}C-NMR spectra of dextran polymers synthesized by this clone confirm that *L. citreum* M-3 *dsrV* gene codes for a different dextransucrase synthesizing linear dextrans with mainly α(1-6) linkages.

Introduction

Lactic acid bacteria (LAB) have great potential to produce exopolysaccharides (EPS) and oligosaccharides, which have been the subject of extensive research efforts due to their physicochemical and bioactive properties (Pu et al., 2001). EPS represent a group of very diverse polymers, whose individual functional properties are determined by their chemical composition, molecular weights, types of glycosidic linkages as well as degree and arrangement of branches (Zannini et al., 2016). As a result of the unmatched variety of possible osidic bonds between sugar monomers, the structural diversity of these polysaccharides offers an extensive range of functionalities of interest for food, feed, pharmaceuticals, cosmetics and chemicals industries (Xu and Zhang, 2016).

Among the first microbial polymers produced at industrial scale are the dextrans (Naessens et al., 2005). These polysaccharides are produced by microorganisms of the genera *Lactobacillus*, *Streptococcus*, *Weissella*, *Leuconostoc, Pediococcus and Oenococcus* (Daudé et al., 2014). The most common dextran of commercial use is synthesized by *Leuconostoc mesenteroides* strain NRRL B-512F. The biopolymer contains a(1-6) linked glucosyl residues in the main chain with only 5% of a(1-3) linked branches (Naessens et al., 2005). Dextran fractions of controlled molecular weight and their numerous derivatives are mainly used in medicine, pharmaceuticals and fine chemistry (Zdolsek et al., 2011). The polymer production is attributed to the extracellular glucansucrase (dextransucrase) DSR-S, a 6-a-D-glucosyltransferase (EC 2.4.1.5) (Monchois et al., 1997). This enzyme belongs to glycoside hydrolase family 70 (GH70) according to the CAZy classification (http://www.cazy.org/) (Lombard et al., 2014). The GH70 family consists of a large and diverse group of polymerases and branching enzymes, some of them being mainly active on sucrose and others on starch substrate (Passerini et al., 2015, Gangoiti et al., 2017). In the catalytic domain of GH70 enzymes can be identified three regions which contain the amino acids involved in the catalysis (D551, E589, D662, DSR-S numbering) forming the catalytic triad and the domain presents the typical $(b/a)_8$ barrel of glucansucrases (Macgregor et al., 1996, Vujicić-Zagar and Dijkstra 2006 , Pijning et al., 2008).

Biochemical and molecular studies of these enzymes have allowed identifying some of determinants of enzyme specificity and are promising to increase the prospects of their practical applications (Daudé et al., 2014). The study of GH70 enzymes with different specificity from new LAB strains could provide new insights in structure-function relationships of glucansucrases as well as enlarge the dextransucrase toolbox available for industrial application. In this sense, the Cuban Research Institute on Sugarcane By-products (ICIDCA) has a collection of LAB strains. Some of them have already been characterized (Fraga et al., 2011a, Fraga et al., 2011b). They have been isolated from sugarcane and sugarcane derivatives which includes the *Leuconostoc citreum* strain M-3. In the present study, we describe the cloning, sequencing, and heterologous expression in *Escherichia coli* of a novel extracellular dextransucrase coding region (*dsrV*) from *L. citreum* M-3, and the partial characterization the polymer produced by the recombinant dextransucrase. The alignment of the deduced amino acid sequence with other dextransucrase sequences is presented as well as a 3D structural model of DSR-V.

Materials and methods

Bacterial strains and culture media

The *Leuconostoc citreum* strain M-3 was obtained from the Cuban Research Institute on Sugarcane By-products (ICIDCA) culture collection. It was isolated from sugar cane juice. Stock cultures were maintained at 80°C in 15% (v/v) glycerol. The cultures were grown in a rotary shaker at 30°C, 175 cycles min^{-1} in standard medium M1 (20 $g.L^{-1}$ sucrose, 20 $g.L^{-1}$ potassium hydrogen phosphate, 20 $g.L^{-1}$ yeast extract, 0.2 $g.L^{-1}$ $MgSO_4$ $7H_2O$, 0.01 $g.L^{-1}$ $MnSO_4$ H_2O, 0.01 $g.L^{-1}$ NaCl, 0.02 $g.L^{-1}$ $CaCl_2$, 0.01 $g.L^{-1}$ $FeSO_4$ $7H_2O$) with an adjusted pH of 6.9. Glucose instead of sucrose was added in the standard medium for chromosomal DNA purification purposes.

Dextransucrase enzyme activity assays

The enzyme activity was determined by measuring the release of reducing sugars with the dinitrosalicylic acid method (Sumner and Howell, 1935). One unit of glucansucrase enzyme (U) is defined as the amount of enzyme that catalyzes the formation of 1 $mmol.min^{-1}$ fructose at 30°C in 20 $mmol.L^{-1}$ (pH 5.4) sodium acetate buffer, 2 mmol.L^{-1} $CaCl_2$, with 100 g.L^{-1} sucrose and 1 mL of enzymatic preparation in a total reaction volume of 5 mL.

SDS-PAGE and zymogram analysis

Protein electrophoresis under denaturing conditions (SDS-PAGE) was performed with the Mini-Cell SureLock™ XCell system with polyacrylamide gels (8%, v/v) - SDS (Novex Tris-Acetate 1.5 mm thick, Invitrogen Corp.) NuPAGE antioxidant, and NuPAGE running buffer Tris-acetate - SDS (for Tris-Acetate gels) from Invitrogen Corp. The NuPAGE reducing agent (DTT 1 mol.L^{-1}) 3mL and 7.5 mL of loading buffer NuPAGEâLDS (4X) were mixed with 20 mL of the sample and heated to 70°C for 10 min prior to loading on the gel. Samples containing resuspended cells were centrifuged (Eppendorf centrifuge 5804 R) 10,000 x *g* for 5 minutes before being applied. About 2 mU of enzyme was applied to the gels and electrophoresis was performed for approximately one hour at 150 volts. Gels were treated for in situ detection of glucansucrase-dextransucrase activity (zymogram) using sucrose as the substrate (Miller and Robyt, 1986) in combination with the technique for reversibly negative staining protein (Fernandez-Patrón et al., 1995). High Molecular Weight Marker (Amersham Biosciences) was included in the electrophoresis runs.

DNA extraction and purification

Leuconostoc citreum M-3 DNA was purified using the "Blood and Cell Culture DNA Maxi kit" (Qiagen, Germany), and plasmid DNA purification was done using the "QIAprep Spin Plasmid kit" (Qiagen, Germany). LA-PCR (Long and Accurate PCR) amplification products and gel extraction were performed using "QIAquick gel extraction kit" (Qiagen, Germany).

PCR amplification and cloning of the dextransucrase encoding region

Primers were designed based on the sequence of the dextransucrase gene *dsrS* deposited in GenBank™ (accession number: I09598). The recognized sequence of the restriction endonuclease *Cfr*9I (*Sma*I) was inserted in both primers to flank the amplicon. DNA fragments were generated by LA PCR (Long and Accurate PCR) using the Expand High Fidelity PCR System (Roche Applied Science) with a Perkin Elmer Gene Amp PCR System 2400 thermocycler and 50 ng of genomic DNA, and 10 µmol.L^{-1} of forward and reverse primers:

dsrS-dir-PS (5'-CCCGGG*ATG*CCATTTACAGAAAAAGTAATGCGGA-3')
dsrS-inv-PS (5'- CCCGGG*TTA*TGCTGACACAGCATTTCCATTATT-3')

(*ATG* and *TTA* in italic type represent, respectively, the start and stop codons. Nucleotides underlined represent mismatches with the sequence and the cleavage site of restriction enzyme of *Cfr*9I.

The thermal cycling was 1 cycle at 94°C for 2 min, ten cycles of 94°C for 15 s, 50°C for 30 s, 68°C for 10 min, 20 cycles of 94°C for 15 s, 50°C for 30 s, 68°C for 10 min plus 5 s for each cycle, and 1 cycle 68°C for 15 min.

The amplicon was digested with the restriction endonuclease *Cfr*9I (MBI Fermentas) and cloned in the vector pGEM-3Zf(+) (Promega). The transformation of *Escherichia coli* JM109 with the construction was achieved by the procedure of Hanahan, (1983). The screening of *E. coli* colonies producing functional dextransucrase enzymes was done basically as described by Hernández et al. (2002). LBT medium (10 g tryptone, 5 g yeast extract, and 10 g NaCl, 100 mmol.L^{-1} Tris-HCl, pH 6,4) with ampicillin 100 µg.mL^{-1} was used instead of Luria-Bertani (LB) medium (Monchois et al., 1997). The method relies on the addition of the pH indicator bromothymol blue to LBT media, supplemented with 5% sucrose and 1% glycerol as extra carbon sources 2.4 mM isopropyl-1-thio-b-D-galactopyranoside (IPTG). *E. coli* was grown and maintained on LB medium supplemented when needed with ampicillin (100 µg.mL^{-1}). All *E. coli* strains were stored at -80°C in 15% glycerol.

Amplicon DNA sequence analysis

Amplicons were sequenced in both directions by Genome Express, France. Nucleotide sequence analysis was performed with Vector NTI 8.0 software (Invitrogen- InforMax, USA). The amino acid sequence encoded by the amplicon was subjected to a BLAST 2.2.7 search (Altschul et al., 1997). COBALT – constraint-based multiple alignment tool (Papadopoulos and Agarwala 2007) was run on BLAST search result and was used to build the phylogenetic tree using the neighbor joining method with a maximum sequence difference (0.85) and the distance estimated according the model of Kimura (1980). Final edition of the phylogenetic tree was conducted using MEGA version X (Kumar et al., 2018). The MultAlin (http:// multalin.toulouse.inra.fr/multalin/) internet software of Corpet (1988) was used to perform the sequence alignments of the dextransucrases and the result was visualized using the EsPript 3.0 software (http://espript.ibcp.fr/ESPript/cgi-bin/ESPript.cgi) (Robert and Gouet 2014). The cleavage site of the peptidase signal was predicted by the SignalP 4.1 software (Petersen et al., 2011).

Inducible production of dextransucrase in Escherichia coli JM109

For the expression of recombinant dextransucrase, a single colony of a clone was inoculated into 10 ml of 2xYT/T medium (16 g tryptone, 10 g yeast extract, 5 g NaCl, complete to 1L with 100 mmol.L^{-1} Tris/HCl, pH 6.4 buffer) supplemented with 100 µg.mL^{-1} of ampicillin and grown at 30°C, 175 cycles.min^{-1} overnight. The cultures were then diluted 1:100 with 2xYT/T medium supplemented with 200 µg.ml^{-1} ampicillin, and grown at 30°C and 175 cycles.min^{-1} until the OD_{600} reached 0.5. Then IPTG was added to a final concentration of 2.4 mmol.L^{-1} and the cultures were incubated 12-14 h at 20°C at 175 cycles.min^{-1}. The cells were centrifuged at 10,000 x *g* 15 min, 4°C and resuspended in a rupture solution [50 mmol.L^{-1} NaAc, pH 5.4, Tritón X-100 0.1 % (v/v), $CaCl_2$ 0.05 g.L^{-1}, inhibitors cocktail of proteases (Roche)] until the OD_{600} reached 80. Finally the cells were disrupted by sonication (MSP, England). The crude extract was centrifuged at 21,390 x *g* for 40 min. The soluble and insoluble fractions of every crude extract were separated, recovered, and kept at -20°C until they were analyzed.

Isolation of the exopolysaccharide produced by the recombinant dextransucrase in Escherichia coli JM109

A positive colony was spread on solid LBT medium and grown at 30°C for one week. The cells and exopolysaccharides were recovered, resuspended in 1 mL of deionized (Mili Q) water and mixed by vortexing. The cellular and the supernatant fractions were separated by centrifugation (Eppendorf 5804 R) at 10,000 x *g* at 4°C for 15 min. The soluble exopolysaccharide of the supernatant fraction was precipitated in 60 % (v/v) ethanol at room temperature. The precipitate was recovered by centrifugation (Eppendorf 5804 R) at 7,000 x *g* 10 min at 20°C and was resuspended in deionized (Mili Q) water, and this step was repeated two times. The polymer was finally resuspended in deuterium oxide (D_2O).

^{13}C-NMR analyses of dextran polymers

All NMR spectra were obtained from solutions of 20 mg.mL^{-1} in deuterium oxide (D_2O). The NMR 1D spectra were obtained using inverse sequences taken from the Bruker pulse sequences library and using 90° pulse, 25,000 Hz sweep width, 2.5 s of relaxation delay and 0.63 s acquisition time. A total of 30,000 scans were recorded. Signal assignments were made by the comparison of spectra from commercially available dextran, inulin and levan, (Sigma-Aldrich). Some signals of dextran were assigned according to Maina et al. (2008).

Results and discussions

Amplification, cloning and expression to a coding DNA fragment for DSR-V dextransucrase

Primers were designed based on the nucleotide sequence of the N-terminal and C-terminal regions of the dextransucrase gene *dsrS* from *L. mesenteroides* NRRL B-512F. This method was previously used in the isolation and cloning of the gene *dsrF* and its truncated variant *dsr*F-ΔSP-ΔGBD from *L. citreum* B/110-1-2 (Fraga et al., 2011a). Briefly, a DNA fragment of 4602 bp containing an open reading frame (ORF) of 4587 bp putatively encoding a glucansucrase was amplified by PCR and cloned into pGEM3Zf (+) (Figure 1A, lane 2).

E. coli JM109 clones expressing glucansucrase activity (DSR-V) were selected using a simple phenotypic screening procedure. First, the positive clones turned the medium color from the initial green (pH 6.5–7.0) to yellow (pH 6.0) due to medium acidification resulting from the consumption of the glucose and fructose released from sucrose (Hernández et al., 2002). In addition, the positive clones showed the typical dextran polymer formation phenotype and negative clones appeared as flat colonies without dextran (Figure 1B). The dextransucrase activity in the soluble fraction of sonicated cell extracts was low (0.032 U.mL^{-1}) as compared with other production levels of recombinant dextransucrase in *E. coli* (Moulis et al., 2006, Yalin et al., 2008). A possible cause of these levels may be the presence of 53 rare codons in the *dsrV* sequence including 2 Arginine-AGG (2), AGA (4); Isoleucine-AUA (9); Leucine-CUA (12); Proline-CCC (2); Glycine-GGA (24) and two in tandem (AGAAUA). This tandem codon is especially unfavorable because it may cause an early end of the translation process (Rogulin et al., 2004). In other study, an *E. coli* clone expressing the dextransucrase DSR-F was supplemented with the pRARE vector, which carries rare RNAt genes, resulting in an increase in enzymatic activity of 66%, indicating that rare codons have a negative effect on the production of dextransucrase enzymes (Fraga 2015).

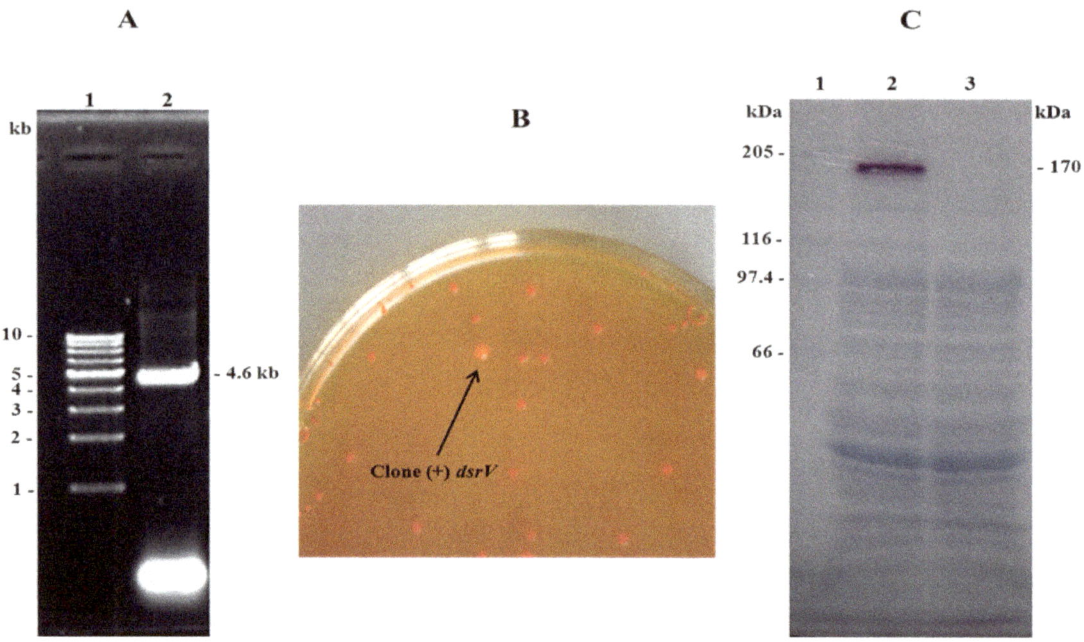

Figure 1. Cloning and detection of glucansucrases in *E. coli* JM109. (A): Lane 1: Protein marker 1 Kb DNA Step Ladder (Promega). Lane 2: DNA electrophoresis of 4.6 kb amplicons encoding for glucansucrase enzyme amplified by LA PCR. (B) Histochemical detection of dextransucrase enzyme activity in *E. coli.* (C) Zymogram for detection of dextransucrase activity in the clone pGEM-3Zf(+) *dsrV*. Lane 1: Protein marker HMW Maker. Lane 2: *E. coli* JM109 pGEM-3Zf(+) *dsrV*. Lane 3: *E. coli* JM109 pGEM-3Zf(+).

Amino acid sequence analysis of DSR-V dextransucrase

The *dsrV* gene, whose sequence was deposited in GenBank^{TM} data base (accession number: KF724950), encoding the protein DSR-V (GenBank^{TM} accession number: AHC31982) of 1530 amino acids, has a theoretical molecular weight of 170466.27 Da (~ 170 kDa) and an isoelectric point of 4.25 according to information provided by the Vector NTI 8.0 software. The deduced amino acid sequence of DSR-V showed a high level of identity of 94% with several characterized dextransucrases of *L. mesenteroides* strains like: DSR-D, a dextransucrase of 1527 aa of *L. mesenteroides* Lcc4 (Neubauer et al., 2003), with the DSR-S of *L. mesenteroides* NRRL B-512F (Moulis et al., 2006), with DexYG, a dextransucrase of 1527 aa of *L. mesenteroides* 0323 (Zhang et al., 2008) (Figure 2). The highest level of identity of DSR-V (99%) was with a recently reported uncharacterized hydrolase of 1525 aa from an annotated genome of *L. mesenteroides* subsp. *suionicum* deposited in GenBank™: API72687.1. Sequence alignment showed that DSR-V mainly differed from DSR-S in 89 amino acids all over the sequence. Concerning *L. citreum*, the highest identity was 70% with an uncharacterized dextransucrase of 1381 aa deposited in GenBank™: CDX65712. The characterized dextransucrase of 1477 aa from *L. citreum* strain HJ-P4, showed an identity of 67% (Yi et al., 2009).

Furthermore, the three amino acids (D551, E589, D662, DSR-S and DSR-V numbering) forming the catalytic triad are highly conserved in GH70 family. It has been proposed that the first conserved residue D551 (Asp) operates in the nucleophilic attack, upon forming the intermediate enzyme substrate and activity transferase (Macgregor et al., 1996). The second amino acid E589 (Glu) is involved in the acid/general base catalysis (Macgregor et al., 1996, Vujicić-Zagar and Dijkstra 2006, Pijning et al., 2008). The conserved residue D662 (Asp) is the catalytic residue stabilizing the transition state intermediate (Macgregor et al., 1996, Vujicić-Zagar and Dijkstra 2006, Yi et al., 2009). Located downstream of the catalytic triad amino acid D662 is the dextransucrase consensus sequence SEV (Ser663-Glu664-Val665) which seems to be the responsible for a(1-6) linkage formation (Moulis et al., 2006). All the characteristic motifs responsible for a(1-6) linkages specificity, were actually observed in the predicted amino acid sequence of DSR-V (Figure 3). The highly conserved regions (II, III, IV) (Vujicić-Zagar and Dijkstra, 2006) of the catalytic domain of DSRV are identical to DSR-S dextransucrase ones (Figure 3).

Figure 2. Phylogenetic tree of DSR-V (arrow) with members of the GH-70 family. *Enterococcus* (Ent.), *Lactobacillus* (Lc.), *Leuconostoc mesenteroides* (Leu. mesent.), *Oenococcus* (O.), *Weissella* (W.).

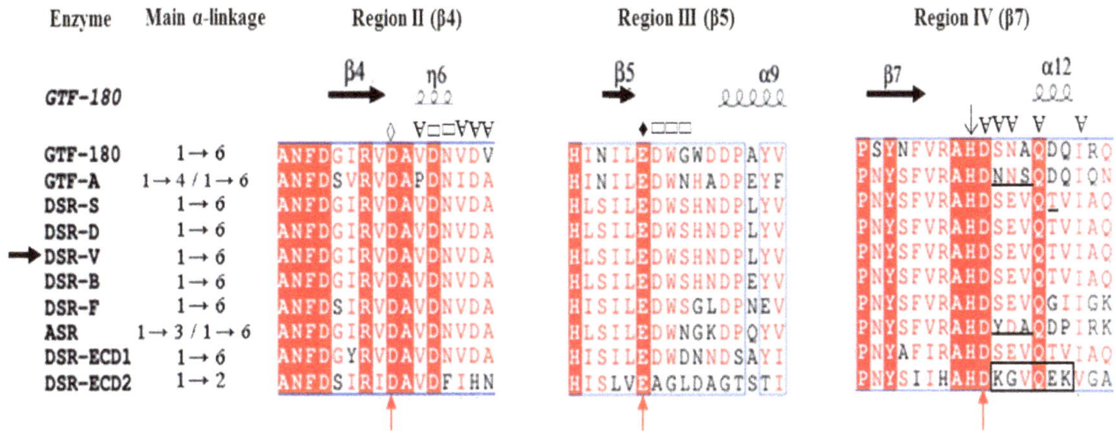

Figure 3. Amino acid sequence alignment of highly conserved regions (II, III, IV) in the catalytic domains of dextran-, mutan-, alternan-, and reuteransucrase enzymes of lactic acid bacteria (including DSR-V) and residues with functional roles identified by site directed mutagenesis. Key: GTF-180 *Lb. reuteri* 180, GTF-A *Lb. reuteri* 121, Dsr-S, *L. mesenteroides* NRRL B-512F; Dsr-D *L. mesenteroides* Lcc4; DSR-V *L. citreum* M-3; Dsr-B 742, *L. citreum* NRRL B-742; Dsr-F, *L. citreum* B/110-1-2; ASR, *L. citreum* NRRL B-1355; Dsr-E CD1 y CD-2, *L. citreum* NRRL B-1299. Symbols → dextransucrase from *L. citreum* strain M-3, ↑ catalytic triad amino acid, ◊ putative catalytic nucleophile, ♦ putative acid/base catalyst, ↓ putative residue stabilizing the transition state intermediate (Macgregor et al., 1996, Ito et al., 2011), ∀ putative sugar-binding/glycosyl transfer sites (Devulapalle et al., 1997), ∀ residues involved in linkage specificity and glucan solubility. Amino acids involved in enzyme linkage specificity (Corpet 1988, Kralj et al., 2005) are in bold underlined type.

Primary structure analyses revealed that DSR-V displays the same organization as that of all the dextransucrases from family GH-70 (EC 2.4.1.5) (Figure 4).

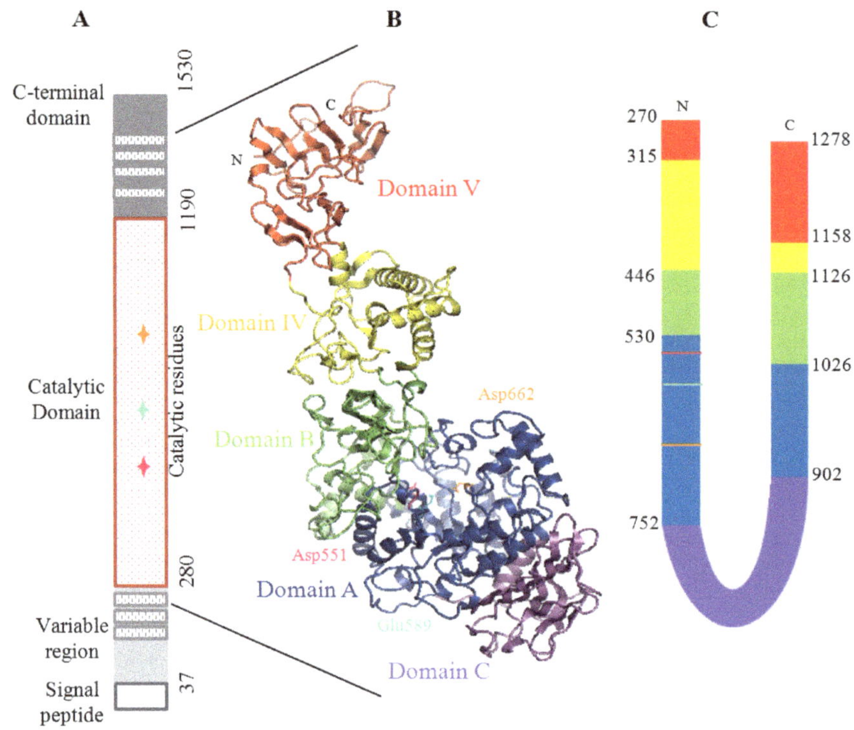

Figure 4. Representation of DSR-V from *L. citreum* M-3. (A): Schematic representation of the primary structure organization of DSR-V from *L. citreum* M-3. Repeat regions CW are shown. (B): Model of the 3D structure to DSR-V based in the 3D structure of GTF-180 from *Lactobacillus reuteri* 180 (3HZ3) obtained by SWISS-MODEL, https://swissmodel.expasy.org. (C): "U" conformation based on the 3D structure.

In the N-terminal region, a Gram-positive signal peptide with a cleavage site followed by the amino acid 37 was identified. These are typical regions of Gram positive bacteria dextransucrases. Previous deletion studies of this region in DSR-B of *L. citreum* NRRL B-1299 have shown that its absence does not affect the enzyme activity (Monchois et al., 1998). Following the signal peptide it was identified the variable region consisting of 243 aa with three conserved regions corresponded to sequences binding to the cell wall (CW) consisting of repeating units YG-consensus. These regions have homology with other proteins binding to the cell wall; available three-dimensional structures of such proteins show a b-solenoid folding where the repeating units for binding to the cell wall are binding sites of choline like the choline binding domain of LytA autolysin from *Streptococcus pneumoniae* (Fernandez-

Tornero et al., 2001) or a trisaccharide (C-terminal domain of *Clostridium difficile* toxin A [Greco et al., 2006]) present in the cell wall components.

Adjacent to the variable region is the catalytic domain that has 910 aa and is the region of greatest homology with other known dextransucrases as the DSR-S of *L. mesenteroides* NRRL B-512F and DSR-D of *L. mesenteroides* Lcc4 (Figure 3). Amino acid alignment sequence analysis of DSR-V compared to other dextransucrases of the GH70 family was performed to detect the correspondence between the predicted secondary structure elements of DSR-V and those determined from the 3D structure of GTF-180 (PDB: 3HZ3). From sequence alignment, DSR-V can be predicted to adopt the typical U-shape fold adopted by GH70 glucansucrases which comprises five domains: the catalytic A domain, domain B and C that are related to GH13 family enzymes and the domains IV and V specific of GH70 family (Figure 4).

Finally, the C-terminus, also named glucan binding domain (GBD), is 340 residues in size. Once again four putative CW units can be localized at the N-terminal region of this domain. Truncation and deletion studies showed that GBD is involved in binding to the polymerization products, constituting functional domains that can bind to the glucans themselves, but they are not directly involved in catalysis (Suwannarangsee et al., 2007, Brison et al., 2012). The GBD contributes to increasing the concentration of the carbohydrate chain to the vicinity of the active site and plays a role in the presentation of these chains to the catalytic site for elongation and branching (Brison et al., 2016). Truncations in the GBD of DSR-S, synthesizes dextrans with lower molar mass than complete enzymes, having an effect on the elongation of the polymer without affecting the catalysis reaction (Moulis et al., 2006).

^{13}C-NMR analyses of dextran polymers synthetized by recombinant DSR-V

The polymer formed by the recombinant DSR-V was analyzed and compared to the dextran T70 synthesized by dextransucrases DSR-S from *L. mesenteroides* NRRL B-512F (ATCC 10830), as shown in Figure 5.

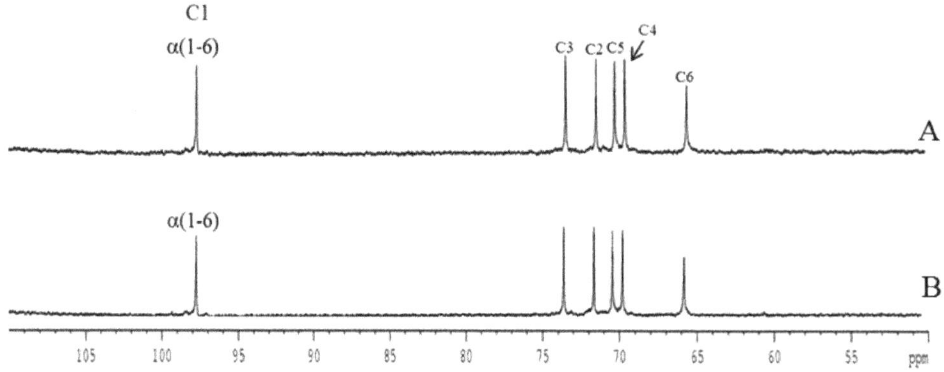

Figure 5. ^{13}C-NMR spectra of dextran polymers. (A) Dextran polymers synthetized by recombinant DSR-V in *E. coli* JM109. (B) Dextran T-70 produced from *L. mesenteroides* NRRL B-512F DSR-S dextransucrase.

Dextrans have their signals ^{13}C NMR anomeric with chemical shifts after 90 ppm, while at C-2, C-3, C-4 and C-5 the chemical shift appears in the range of 70-85 ppm, which is the resonance region known for branched links, and the C-6 is normally before 60 ppm (Gorin, 1981, Ahmed et al., 2012).

The ^{13}C-NMR spectrum of the dextran produced by recombinant dextransucrase DSR-V in the presence of sucrose, has 6 prominent resonances, characteristic of linear dextrans (Maina et al., 2008), showing a very similar maximum spectrum to dextran T-70. In addition to these six peaks, the spectrum doesn't contain peaks indicative of lower branches. The maximum resonance 98.4 ppm indicates the presence of α(1−6) links. This has also been described for DSR-B (Monchois et al., 1998) and DSR-M (Passerini et al 2015), both of the *L. citreum* NRRL B-1299. Overall, these results strongly support the proposal that *L. citreum* M-3 it is probably the first *L. citreum* strain that possesses a DSR-S like protein (DSR-V) with specificity like DSR-B and DSR-M.

Conclusions

On the basis of its amino acid sequence and specificity, it is concluded that DSR-V, is a novel dextransucrase enzyme (EC 2.4.1.5), which must be classified in family GH70. The data obtained in this study enhances the understanding of this dextransucrase and enriches the great biodiversity of this group of enzymes. Further studies of new chimeric fusion enzymes obtained by the substitution/addition of different Carbohydrate Binding Modules (CBM) will permit the design of rationally/semi-rationally engineered dextransucrases producing a wide range of glucose polymers with novel physicochemical properties.

Conflict of interest

No conflict of interest exist since the authors declare no competing financial interest, and manuscript is approved by all authors for publication

Acknowledgements

Reinaldo Fraga and Roberto C. Aríticas were supported by a Development Research Project financed by the Development Cooperation Program (2017-2021) of ARES (Académie de Recherche et d'Enseigment Supérieur) from Belgium. Reinaldo Fraga thanks William David Rau (MS Rau Antiques, New Orleans, USA) for his kind support. We express our gratitude to Sandrine Morel for her technical support and advice for the ^{13}C-NMR analyses of dextran polymers. We thank to Aidín Martínez for providing the strain *Leuconostoc citreum* M-3. We also thank Dr. Joan Combie for the style correction of the manuscript.

References

Ahmed, R.Z., Siddiqui, K., Arman, M., Ahmed, N. (2012). Characterization of high molecular weight dextran produced by *Weissella cibaria* CMGDEX3. Carbohydr. Res., 343, 1237-1250. https://doi.org/10.1016/j.carbpol.2012.05.063

Altschul, S.F., Madden, T.L., Schäffer, A.A., Zhang, J., Zhang, Z., Miller, W., Lipman, D.J. (1997). Gapped BLAST and PSI-BLAST: a new generation of protein database search programs. Nucl. Acids Res., 25, 3389-3402.

Brison, Y., Malbert, Y., Czaplicki, G., Mourey, L., Remaud-Simeon, M., Tranier, S. (2016). Structural insights into the carbohydrate binding ability of an α-(1→2) branching sucrose from glycoside hydrolase family 70. J. Biol. Chem., 291, 7527-7540. doi: 10.1074/jbc.M115.688796

Brison, Y., Pijning, T., Malbert, Y., Fabre, E., Mourey, L., Morel, S., Potocki-Veronese, G., Monsan, P., Tranier, S., Remaud-Siméon, M., Dijkstra, B.W. (2012). Functional and structural characterization of α-(1→2) branching sucrase derived from DSR-E glucansucrase. J. Biol. Chem., 287, 7915-7924. doi: 10.1074/jbc.M111.305078

Corpet, F. (1988). Multiple sequence alignment with hierarchical clustering. Nucl. Acids Res., 16(22), 10881-10890.

Daudé, D., André, I., Monsan, P., Remaud-Siméon, M. (2014). Successes in engineering glucansucrases to enhance glycodiversification. In Carbohydrate Chemistry: Volume 40, A. P. Rauter, T. Lindhorst, Y. Queneau, eds. (Royal Society of Chemistry) pp. 624-645. http://dx.doi.org/10.1039/9781849739986-00624

Devulapalle, K.S., Goodman, S.D., Gao, Q., Hemsley, A., Mooser, G. (1997). Knowledge-based model of a glucosyltransferase from the oral bacterial group of mutans streptococci. Protein Sci., 6, 2489-2493. https://doi.org/10.1002/pro.5560061201

Fernandez-Patrón, C., Hardy, E., Seoane, J., Castellanos, L. (1995). Double staining of coomassie blue-stained polyacrylamide gels by imidazole-sodium dodecyl sulfate-zinc reverse staining: Sensitive detection of coomassie blue-undetected proteins. Anal. Biochem., 224(1), 263-269. https://doi.org/10.1006/abio.1995.1039

Fraga, R. (2015). Aislamiento, caracterización y expresión heteróloga del gen de una nueva dextranasacarasa de la cepa industrial de *Leuconostoc* sp. B/110-1-2. Ph.D. Thesis Dissertation, University of Havana (2015).

Fraga, R., Moulis, C., Escalier, P., Remaud-Simeón, M., Monsan, P. (2011). Isolation of a gene from *Leuconostoc citreum* B/110-1-2 encoding a novel dextransucrase enzyme. Curr. Microbiol., 62, 1260-1266. https://doi.org/10.1007/s00284-010-9851-7

Fraga, R., Martínez, A., Moulis, C., Escalier, P., Morel, S., Remaud-Simeón, M., Monsan, P. (2011). A novel dextransucrase is produced by *Leuconostoc citreum* strain B/110-1-2: an isolate used for the industrial production of dextran and dextran-derivatives. J. Ind. Microbiol. Biotechnol., 38, 1499-1503. https://doi.org/10.1007/s10295-010-0936-x

Fernandez-Tornero, C., Lopez, R., Garcıa, E., Gimenez-Gallego, G., Romero, A. (2001). A novel solenoid fold in the cell wall anchoring domain of the pneumococcal virulence factor LytA. Nat. Struct. Biol., 8, 1020-1024. https://doi.org/10.1038/nsb724

Gangoiti, J., van Leeuwen, S., Gerwig, G., Duboux, S., Vafiadi, C., Pijning, T., Dijkhuizen, L. (2017). 4,3-α-Glucanotransferase, a novel reaction specificity in glycoside hydrolase family 70 and clan GH-H. Sci. Rep., 7, 39761. doi: 10.1038/srep39761.

Gorin, P.A.J. (1981). Carbon-13 nuclear magnetic resonance spectroscopy of polysaccharides. Adv. Carbohydr. Chem. Biochem., 38, 13-104. https://doi.org/10.1016/S0065-2318(08)60309-1

Greco, A., Ho, J.G., Lin, S.J., Palcic, M.M., Rupnik, M., Ng, K.K. (2006). Carbohydrate recognition by *Clostridium difficile* toxin A. Nat. Struct. Mol. Biol., 13, 460-461. https://doi.org/10.1038/nsmb1084

Hanahan, D. (1983). Studies on transformation of *Escherichia coli* with plasmids. J. Mol. Biol., 166(4), 557-580. https://doi.org/10.1016/S0022-2836(83)80284-8

Hernández, L., Menéndez, C., Soto, M., Trujillo, L.E., Arrieta, J.G. (2002). A simple method for direct screening of *Escherichia coli* and *Pichia pastoris* expressing recombinant saccharolytic enzymes. Biotecnología Habana 2002. Agro-Biotech in the New Millennium. ISBN 959-235-021-3 Elfos Scientiae.

Ito, K., Ito, S., Shimamura, T., Weyand, S., Kawarasaki, Y., Misaka, T., Abe, K., Kobayashi, T., Cameron, A.D., Iwata, S. (2011). Crystal structure of glucansucrase from the dental caries pathogen *Streptococcus mutans*. J. Mol. Biol., 408(2), 177-86. https://doi.org/10.1016/j.jmb.2011.02.028

Kimura, M. (1980). A simple method for estimating evolutionary rate of base substitutions through comparative studies of nucleotide sequences. Journal of Molecular Evolution, 16, 111-120.

Kralj, S., van Geel-Schutten, G.H., Faber, E.J., van der Maarel, M.J., Dijkhuizen, L. (2005). Rational transformation of a reuteransucrase into a dextransucrase: analysis of mutant enzymes and their glucan and oligosaccharides products. Biochemistry, 44, 9206-9216. **DOI:** 10.1021/bi050447q

Kumar, S., Stecher, G., Li, M., Knyaz, C., Tamura, K. (2018). MEGA X: Molecular evolutionary genetics analysis across computing platforms. Molecular Biology and Evolution, 35, 1547-1549. https://doi.org/10.1093/molbev/msy096

Lombard, V., Golaconda Ramulu, H., Drula, E., Coutinho, P.M., Henrissat, B. (2014). The Carbohydrate-active enzymes database (CAZy) in 2013. Nucl. Acids Res., 42, D490-D495. https://doi.org/10.1093/nar/gkt1178

Macgregor, E.A., Jespersen, H.M., Svensson, B. (1996). A circularly permuted alpha-amylase-type alpha/beta-barrel structure in glucan-synthesizing glucosyltransferases. FEBS Lett., 378, 263-266. https://doi.org/10.1016/0014-5793(95)01428-4

Maina, N.H., Tenkanen, M., Maaheimo, H., Juvonen, R., Virkki, L. (2008). NMR spectroscopic analysis of exopolysaccharides produced by *Leuconostoc citreum* and *Weissella confuse*. Carbohydr. Res., 343, 1446-1455. https://doi.org/10.1016/j.carres.2008.04.012

Miller, A.W., Robyt, J.F. (1986). Detection of dextransucrase and levansucrase on polyacrylamide gels by the periodic acid-Schiff stain: staining artifacts and their prevention. Anal. Biochem., 156, 357-363. https://doi.org/10.1016/0003-2697(86)90266-6

Monchois, V. Remaud-Simeon, M., Monsan, P., Willemot, R.M. (1998). Cloning and sequencing of a gene coding for an extracellular dextransucrase (DSR-B) from *Leuconostoc mesenteroides* NRRL B1299-synthesizing only a(1–6) glucan. FEMS Microbiol. Lett., 159, 307-315. https://doi.org/10.1111/j.1574-6968.1998.tb12876.x

Monchois, V., Remaud-Simeon, M., Russel, R.R., Monsan, P., Willemot, R.M. (1997). Characterization of *Leuconostoc mesenteroides* NRRL B-512F dextransucrase (DSRS) and identification of amino-acid residues playing a key role in enzyme activity. Appl. Microbiol. Biotechnol., 48(4), 465-72. DOI:10.1007/s002530051081

Moulis, C., Joucla, G., Harrison, D., Fabre, E., Potocki-Veronese, G., Monsan, P., Remaud-Simeon, M. (2006). Understanding the polymerization mechanism of glycoside-hydrolase family 70 glucansucrases. J. Biol. Chem., 281, 31254-31267. doi: 10.1074/jbc.M604850200

Naessens, M., Cerdobbek, A., Soetaert, W., Vandamme, E.J. (2005). *Leuconostoc* dextransucrase and dextran: production, properties and applications. J. Chem. Technol. Biotechnol., 80, 845-860. https://doi.org/10.1002/jctb.1322

Neubauer, H., Bauche, A., Mollet, B. (2003). Molecular characterization and expression analysis of the dextransucrase DsrD of *Leuconostoc mesenteroides* Lcc4 in homologous and heterologous *Lactococcus lactis* cultures. Microbiology, 149, 973-982. doi: 10.1099/mic.0.26029-0

Papadopoulos, J.S., Agarwala, R. (2007). COBALT: constraint-based alignment tool for multiple protein sequences. Bioinformatics, 23(9), 1073-1079. https://doi.org/10.1093/bioinformatics/btm076

Passerini, D., Vuillemin, M.I., Ufarté, L., Morel, S., Loux, V., Fontagné-Faucher, C., Monsan, P., Remaud-Siméon, M., Moulis, C. (2015). Inventory of the GH70 enzymes encoded by *Leuconostoc citreum* NRRL B-1299–identification of three novel-transglucosylases. FEBS Journal, 282(11), 2115-2130. https://doi.org/10.1111/febs.13261

Petersen, T.N., Brunak, S., von Heijne, G., Nielsen, H. (2011). SignalP 4.0: discriminating signal peptides from transmembrane regions. Nature Methods, 8, 785-786. https://doi.org/10.1038/nmeth.1701

Pijning, T., Vujicic-Zagar, A., Kralj, S., Eeuwema, W., Dijkhuizen, L., Dijkstra, B.W. (2008). Biochemical and crystallographic characterization of a glucansucrase from *Lactobacillus reuteri* 180. Biocat. Biotrans., 26(1-2), 12-17. https://doi.org/10.1080/10242420701789163

Pu, Y., Zou, Q., Hou, D., Zhang, Y., Chen, S. (2017). Molecular weight kinetics and chain scission models for dextran polymers during ultrasonic degradation. Carbohydr. Polym., 156, 71-76. https://doi.org/10.1016/j.carbpol.2016.09.017

Robert, X., Gouet, P. (2014). Deciphering key features in protein structures with the new ENDscript server, Nucl. Acids Res., 42(W1), W320-W324. https://doi.org/10.1093/nar/gku316

Rogulin, E., Perevyazova, T., Zheleznaya, L., Matvienko, N. (2004). Plasmid pRARE as a vector for cloning to construct a superproducer of the site-specific Nickase N.BspD6I. Biochemistry (Moscow), 69, 1123-1127.

Sumner, J., Howell, S. (1935). A method for determination of saccharase activity. J. Biol. Chem., 108, 51-54.

Suwannarangsee, S., Moulis, C., Potocki-Veronese, G., Monsan, P., Remaud-Simeon, M., Chulalaksananukul, W. (2007). Search for a dextransucrase minimal motif involved in dextran binding. FEBS Letters, 581, 4675-4680. https://doi.org/10.1016/j.febslet.2007.08.062

Vujicić-Zagar, A., Dijkstra, B.W. (2006). Monoclinic crystal form of *Aspergillus niger* alpha-amylase in complex with maltose at 1.8 angstroms resolution. Acta Crystallogr. Sect. F Struct. Biol. Cryst. Commun., 62(Pt 8), 716-21. https://doi.org/10.1107/S1744309106024729

Xu, L., Zhang, J. (2016). Bacterial glucans: production, properties, and applications. Appl. Microbiol. Biotechnol., 100(21), 9023-9036. https://doi.org/10.1007/s00253-016-7836-6

Yalin, Y., Jin, L., Jianhua, W., Da, T., Zigang, T. (2008). Expression and characterization of dextransucrase gene *dsr*X from *Leuconostoc mesenteroides* in *Escherichia coli*. J. Biotechnol., 133, 505-512. https://doi.org/10.1016/j.jbiotec.2007.12.002

Yi, A.R., Lee, S.R., Jang, M.U., Park, J.M., Eom, H.J., Han, N.S., Kim, T.J. (2009). Cloning of dextransucrase gene from *Leuconostoc citreum* HJ-P4 and its high-level expression in *E. coli* by low temperature induction. J. Microbiol. Biotechnol., 19(8), 829-35.

Zannini, E., Waters, D.M., Arendt, E.K. (2016). Production, properties, and industrial food application of lactic acid bacteria-derived exopolysaccharides, Appl. Microbiol. Biotechnol., 100(3), 1121-1135. https://doi.org/10.1007/s00253-015-7172-2

Zdolsek, H.J., Vegfors, M., Lindahl, T.L., Tornquist, T., Bortnik, P., Hahn, R.G. (2011). Hydroxyethyl starches and dextran during hip replacement surgery: effects on blood volume and coagulation. Acta Anaesthesiologica Scandinavica, 55(6), 677-685. https://doi.org/10.1111/j.1399-6576.2011.02434.x

Zhang, H., Hu, Y., Zhu, C., Zhu, B., Wang, Y. (2008). Cloning, sequencing and expression of a dextransucrase gene (dexYG) from *Leuconostoc mesenteroides*. Biotechnol. Lett., 30, 1441-1446. https://doi.org/10.1007/s10529-008-9711-8

www.ingramcontent.com/pod-product-compliance
Ingram Content Group UK Ltd.
Pitfield, Milton Keynes, MK11 3LW, UK
UKHW062003041026
15591UKWH00016B/1151

9 781912 530267